Lightweight Design

Andreas Öchsner

Lightweight Design

An Introduction Based on One-Dimensional Structures

Springer Vieweg

Andreas Öchsner
Fakultät Maschinen und Systeme
Hochschule Esslingen
Esslingen am Neckar, Germany

ISBN 978-3-658-48161-2   ISBN 978-3-658-48162-9   (eBook)
https://doi.org/10.1007/978-3-658-48162-9

This Springer Vieweg imprint is published by the registered company Springer Fachmedien Wiesbaden GmbH, part of Springer Nature.
The registered company address is: Abraham-Lincoln-Str. 46, 65189 Wiesbaden, Germany

Preface

Lightweight design concepts can be understood as the application of classic engineering concepts and disciplines to reduce structural weight. Here, in particular, basic knowledge of applied mechanics, materials science, manufacturing technology and design theory is used. In addition to classic applications in the aerospace industry or in automotive engineering, these now also extend to quite different areas, such as sports equipment or medical prostheses. Due to the reduced weight, a reduction in fuel consumption and thus a reduction in pollutants can be achieved in the transport sector. This results in economic as well as ecological advantages.

In engineering practice, however, there are usually quite complex systems, also optimized using commercial program packages. However, to introduce the different lightweight design concepts as part of a university engineering course, one can also first use simple structural elements that are presented in the context of applied mechanics. The simplest elements here are bars and beams, which, along with springs, are assigned to the one-dimensional structural elements. Based on these elements, questions about the selection of materials and the geometric design and optimization of load-bearing structures, can be discussed quite clearly. This textbook therefore offers simple and comprehensive 'instructions' for the application of the various lightweight design concepts, with the focus on material and form lightweight construction, or the combination of these two concepts. It should be noted here that a compact presentation of the same topic is also available as *essentials*, see [1].

If the classic one-dimensional structural elements are used in a lightweight design lecture, the basics of applied mechanics are further consolidated and a contribution is made to the vertical integration of engineering knowledge.

The pedagogical methodology proposed in the first edition of the German textbook, i.e. the focus on classical one-dimensional structural elements was well received by the students in the lectures on structural optimization (Bachelor's program) and lightweight design (Master's program), see [2]. The second edition was expanded by about 90 pages. Many additional examples were included to practice the various lightweight design concepts computationally. In addition, an introduction to classical optimization problems, i.e. the formulation of an objective function (e.g. the weight of a structure) and corresponding restrictions, was included. However, the consideration is limited to one- or two-dimensional design spaces, i.e. with a maximum of two design variables. For such simpler cases, the minimum of the objective function can often be determined using analytical or graphical methods. Last but not least, the entire content was critically reviewed. The current German edition (see [3]) has been supplemented with additional exercises and forms the basis for this English textbook.

Andreas Öchsner

References

[1] Öchsner, A.: Leichtbaukonzepte: Eine Einführung anhand einfacher Strukturelemente für Studierende. Springer Vieweg, Wiesbaden (2018)
[2] Öchsner, A.: Leichtbaukonzepte anhand einfacher Strukturelemente: Neuer didaktischer Ansatz mit zahlreichen Übungsaufgaben. Springer Vieweg, Wiesbaden (2019)
[3] Öchsner, A.: Stoff- und Formleichtbau: Leichter Einstieg mit eindimensionalen Strukturen. Springer Vieweg, Wiesbaden (2022)

Contents

Symbols and Abbreviations[1]

Latin Symbols (Capital Letters)

A	Area, Cross-sectional area
A_s	Shear area
AG	Shear stiffness
B_1	Factor
B_1'	Factor
C	Constant
E	Young's modulus
$\overline{EA}$	Average tensile stiffness
$\overline{EI}_y$	Average bending stiffness
F	Force, Yield condition
F_g	Dead weight
F_{cr}	Critical force (buckling force)
G	Shear modulus
I	Axial second moment of area
I_p	Polar second moment of area
K	Buckling coefficient
L	Length
M	Lightweight index, Moment
N	Normal force
Q	Shear force
$R_{p0.2}$	Initial yield stress
SEA	Specific energy absorption
V	Volume

Latin Symbols (Small Letters)

a	Geometric dimension
b	Geometric dimension
c	Damper constant, Constant of integration

---

[1]The following lists explain the most important symbols and abbreviations that are used in the course of this book.

d	Diameter
f	Function
g	Gravitational acceleration, Function
h	Geometric dimension
h_{c}	Average sandwich thickness
Δh	Layer thickness (sandwich)
Δh^{F}	Face sheet thickness (sandwich)
$\Delta h^{\mathrm{F,n}}$	Length specific face sheet thickness (sandwich)
Δh^{C}	Core thickness (Sandwich)
$\Delta h^{\mathrm{C,n}}$	Length specific core thickness (sandwich)
k_{s}	Spring constant, Geometric and material parameter, Shear yield stress, Shear correction factor
k_{t}	Tensile yield stress
$k_{\mathrm{t}}^{\mathrm{init}}$	Initial tensile yield stress
m	Mass, Length specific moment
m^{n}	Length specific mass
p	Distributed load in x-direction
q	Distributed load in z-direction
r	Radius
s	Geometric dimension
t	Geometric dimension
u	Displacement
w	Geometric dimension
x	Cartesian coordinate
y	Cartesian coordinate
z	Cartesian coordinate

Greek Symbols (Capital Letters)

Θ	Argument ($\Theta = \frac{2\pi \, \Delta h^{\mathrm{C}}}{\lambda}$)
Π	Strain energy

Greek Symbols (Small Letters)

α	Parameter, Angle
γ	Shear strain
γ_{aB}	Shear strain at failure
ε	Normal strain
ε_{A}	Strain at failure
$\varepsilon_{\mathrm{A_t}}$	Total strain at failure (including elastic part)

$\varepsilon_{\text{p0.2}_t}$	0.2-% strain limit (corresponding to $R_{\text{p0.2}}$)
κ	Curvature
λ	Parameter, Wave length
ν	Poisson's ratio
ϱ	Density
π	Volume specific strain energy
π^{el}	Elastic part of volume specific strain energy
π^{pl}	Plastic part of volume specific strain energy
π^{s}	Deviatoric part of the volume specific strain energy
π°	Hydrostatic part of the volume specific strain energy
σ	Stress, normal stress
σ_{cr}	Critical stress
σ_{eff}	Equivalent stress (effective stress)
σ_i	Principal stress ($i = 1, 2, 3$)
τ	Shear stress
τ_{aB}	Shear strength
τ_{p}	Shear yield stress
ϕ	Rotation angle, rotation
ϕ_{F}	Fiber volume fraction
φ	Rotation angle, rotation, Angle of torsion

Mathematical Symbols

$\times$	Multiplication sign

Indices, Superscripted

$\ldots^{\text{C}}$	Core
$\ldots^{\text{c}}$	Compression
$\ldots^{\text{E}}$	Euler
$\ldots^{\text{el}}$	Elastic
$\ldots^{\text{F}}$	Face sheet
$\ldots^{k}$	Layer index
$\ldots^{\text{pl}}$	Plastic
$\ldots^{\text{t}}$	Tension

Indices, Subscripted

$\ldots_{\text{o}}$	Outer
$\ldots_{\text{b}}$	Bending
$\ldots_{\text{F}}$	Face sheet
$\ldots_{\text{i}}$	Internal
$\ldots_{\text{C}}$	Core
$\ldots_{\text{m}}$	Average value

$\ldots_{max}$ Maximum value
$\ldots_{ref}$ Reference
$\ldots_{s}$ Shear
$\ldots_{t}$ Torsion, Tension
$\ldots_{c}$ Center

Abbreviations

1D   One-dimensional
CFRP   Carbon fiber-reinforced plastic
UD   Unidirectional

About the Author

Professor Professor h.c. Dr.-Ing. Dr. h.c. Andreas Öchsner, D.Sc. (UoN)
Lightweight Design / Structural Simulation
Esslingen University of Applied Sciences
Faculty of Mechanical and Systems Engineering
Kanalstr. 33
73728 Esslingen
Germany
E-Mail: andreas.oechsner@gmail.com
url: https://scholar.google.com/citations?user=-jQHnjUAAAAJ&hl=en

Introduction and Motivation

Abstract

In this chapter, various motivations are given for dealing with the topic of lightweight design. Furthermore, the corresponding literature on the topic is briefly presented and the current work is classified.

Lightweight design plays a central role in transportation (e.g. in the aerospace industry or in automotive engineering), since a reduction in weight is directly reflected in a reduction in fuel costs. As a rough estimate of the influence of weight on the fuel consumption of aircrafts, a reduction of 1% in weight can result in fuel savings—depending on the engine type—from 0.75 to 1% (see [8]). If one takes the total fuel consumption of the Lufthansa fleet in 2015 of 8,947.766 tons as an example (see [6]), depending on the kerosene price (see [3]), there is a savings potential of several million euros per year.

Figure 1.1 shows a much simpler example of a steel plate. The figure on the left (a) is a plate made of solid material, which has a mass of around 7.7 kg for the dimensions given. If the plate is designed as a hollow sphere structure (see [7]) with the same external dimensions, the result is a significantly reduced mass of around 0.446 kg or a reduction of 94%. From this it can be concluded that not only the material itself, but also other factors, such as the shape or the mesostructure, can have an impact on the 'lightweight potential' of a structure.

There is also specialized literature, for example, with a focus on the automotive industry, see [2, 4, 10]. It should also be noted that lightweight design includes different disciplines, such as strength of materials (see [1, 5]), materials science (see [11]) and design (see [9]).

A typical design principle in the context of lightweight structures is to compose different materials, many times in the form of layers, to so-called sandwich structures, see Fig. 1.2 for some typical examples. The idea is to combine the different advantages of the single constituents and to achieve overall properties, which are better than the ones of the single components.

A. Öchsner, *Lightweight Design*,
https://doi.org/10.1007/978-3-658-48162-9_1

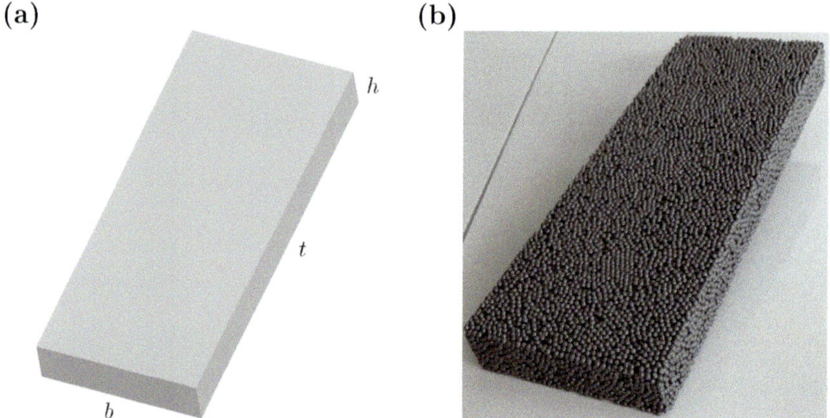

Fig. 1.1 (a) Steel plate with external dimensions $b = 11$ cm, $t = 30$ cm and $h = 3$ cm. Mass: $m \approx$ 7.7 kg; (b) Hollow sphere structure made of steel with the same dimensions. Mass: $m \approx 0.446$ kg

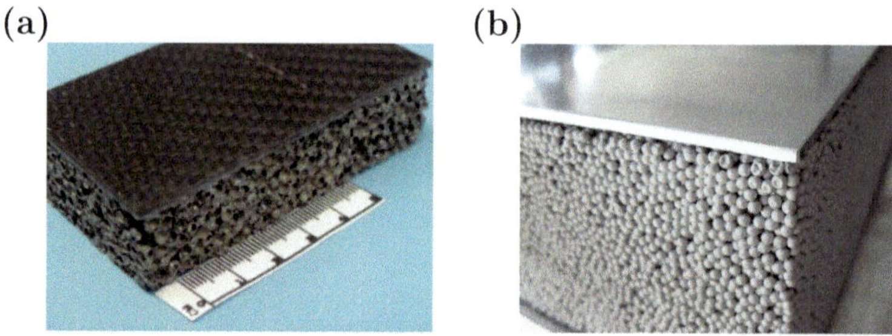

Fig. 1.2 Examples of sandwich plates with metallic hollow sphere core: (a) carbon-fiber reinforced face sheets; (b) aluminium face sheets. Adapted from [7], reprinted with permission from Springer Nature publishers

This textbook focuses purely on one-dimensional structural elements, i.e. bars and beams, and thus offers a new didactic approach to conveying the basic ideas of lightweight design. The restriction to one-dimensional elements allows a relatively simple representation using equations that is easy for students to understand. This means that the focus is on lightweight design concepts and the application of the principles of applied mechanics and not on complicated mathematical derivations or algorithms. Those who have mastered these basics can also relatively easily familiarize themselves with more complicated topics in lightweight construction, such as plane two-dimensional structures.

References

1. Altenbach, H.: Holzmann/Meyer/Schumpich Technische Mechanik Festigkeitslehre. Springer Vieweg, Wiesbaden (2018)
2. Friedrich, H.E.: Leichtbau in der Fahrzeugtechnik. Springer Vieweg, Wiesbaden (2017)
3. IATA: Jet fuel price monitor. http://www.iata.org/publications/economics/fuel-monitor/Pages/index.aspx. Accessed 15 April 2017
4. Kurek, R.: Karosserie-Leichtbau in der Automobilindustrie. Vogel, Würzburg (2011)
5. Linke, M., Eckart Nast, E.: Festigkeitslehre für den Leichtbau: Ein Lehrbuch zur Technischen Mechanik. Springer Vieweg, Wiesbaden (2015)
6. Lufthansa Group: Balance Issue 2016. https://www.lufthansagroup.com/fileadmin/downloads/en/LH-sustainability-report-2016.pdf. Accessed 25 April 2017
7. Öchsner, A., Augustin, C. (eds.): Multifunctional Metallic Hollow Sphere Structures: Manufacturing, Properties and Application. Springer, Berlin (2009)
8. Ohrn, K.E.: Aircraft energy use. In: Capehart, B.L. (eds.) Encyclopedia of Energy Engineering and Technology Vol. 1. pp. 24–30. CRC Press, Boca Raton (2007)
9. Pahl, G., Beitz, W.: Konstruktionslehre: Methoden und Anwendung. Springer, Berlin (1997)
10. Siebenpfeiffer, W.: Leichtbau–Technologien im Automobilbau. Springer Vieweg, Wiesbaden (2014)
11. Weißbach, W.: Werkstoffkunde: Strukturen, Eigenschaften, Prüfung. Springer Vieweg, Wiesbaden (2012)

Basics of Mechanics of Materials

2

Abstract

This chapter deals with the continuum mechanical principles of bars and various beams. For bars, a distinction is made between tensile, compressive and torsional loading. The beam theories according to Euler-Bernoulli, Timoshenko and Levinson are then discussed. The chapter closes with a brief description of the equivalent stress hypotheses according to von Mises and Tresca.

2.1 Tensile, Compressive and Torsional Loads on the Bar

Let us consider first, a pure tension and compression bar. This is a prismatic body that can only be loaded and deformed (displacement $u_x(x)$) along its bar axis (here: x), see Fig. 2.1. Individual forces F_0 and continuously distributed loads $p_x(x)$ are considered as external loads. In the simplest case, the geometry is described by the length L and the constant cross-sectional area A. The material behavior is described by one-dimensional Hooke's law with the constant modulus of elasticity (Young's modulus) E as the material parameter.

If the bar is cut free at an arbitrary position x, the internal reactions become visible as normal forces $N_x(x)$, see Fig. 2.2. These internal reactions always occur in pairs, but are oppositely directed and point in the same direction as the outward-pointing normal vectors.

Fig. 2.1 General configuration of a tension and compression member: Example of geometric boundary conditions and external loads

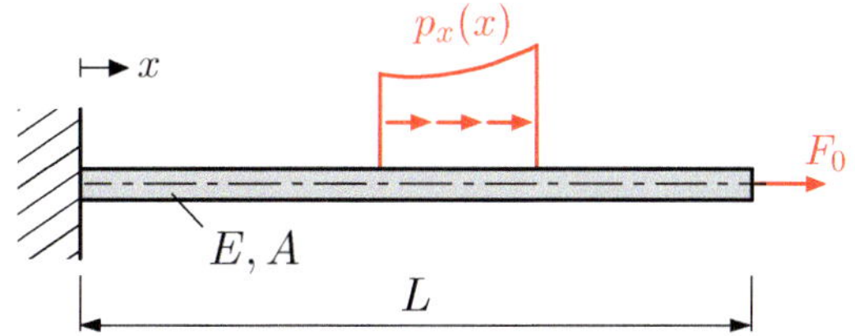

A. Öchsner, *Lightweight Design*,
https://doi.org/10.1007/978-3-658-48162-9_2

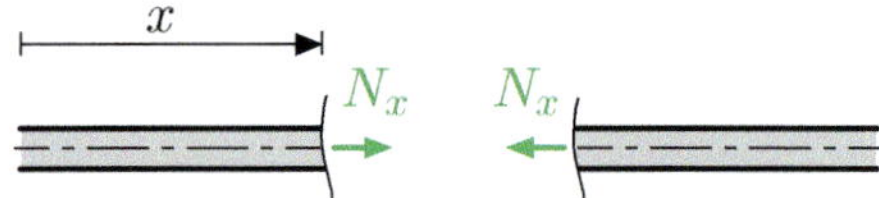

Fig. 2.2 Internal reactions for tensile and compression bar

The describing partial differential equation results from the combination of the basic equations of continuum mechanics, i.e. the kinematics equation, the constitutive law and equilibrium equation (see [8,9]), to:

$$\frac{\mathrm{d}}{\mathrm{d}x}\left(E(x)A(x)\frac{\mathrm{d}u_x}{\mathrm{d}x}\right) + p_x(x) = 0\,,\tag{2.1}$$

from where the general normal force distribution is obtained by one-time integration for constant tensile stiffness EA and constant distributed load p_0:

$$N_x(x) = EA\frac{\mathrm{d}u_x(x)}{\mathrm{d}x} = -p_0 x + c_1\,.\tag{2.2}$$

A further integration gives the general distribution of the displacement field as

$$u_x(x) = \frac{1}{EA}\left(-\frac{1}{2}p_0 x^2 + c_1 x + c_2\right),\tag{2.3}$$

where the constants of integration c_1 and c_2 must be adjusted under consideration of the boundary conditions. The stress distribution is obtained from the normal force distribution according to Eq. (2.2) or from Hooke's law (see also Fig. 2.3):

$$\sigma_x(x) = \frac{N_x(x)}{A} = \varepsilon_x(x) \times E = \frac{\mathrm{d}u_x(x)}{\mathrm{d}x} \times E\,.\tag{2.4}$$

Now let us look on a torsion bar. This is a prismatic body, which can only be loaded and deforms along its principal axis (angle of torsion $\varphi_x(x)$), see Fig. 2.4. External loads are applied as single torsional moments M_t and distributed torsional moments $m_x(x)$. The geometry is described, in the simplest case, by the length L

Fig. 2.3 Axially loaded bar: **(a)** strain and **(b)** stress distribution

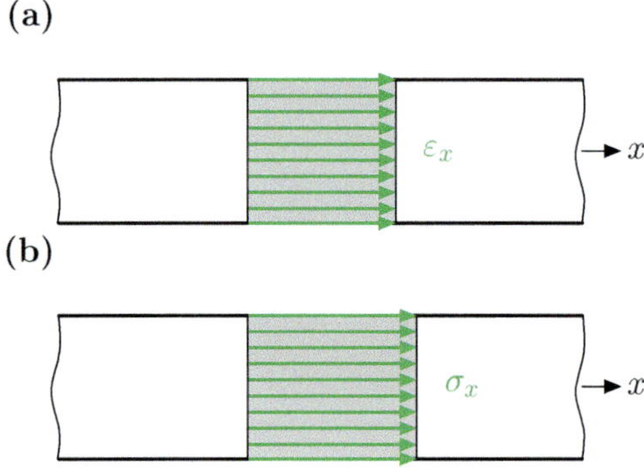

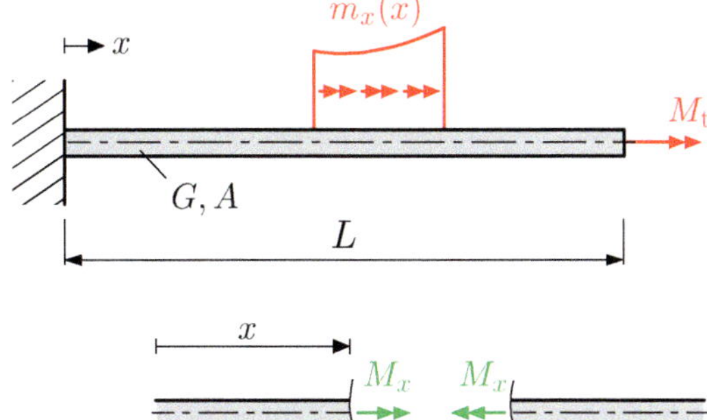

Fig. 2.4 Schematic representation of a torsion bar: examples of boundary conditions and external loads

Fig. 2.5 Internal reactions for torsion bar

and the constant polar second moment of area I_p. The material behavior is described by the one-dimensional Hooke's law with the constant shear modulus G.

If the bar is cut free at an arbitrary position x, the internal reactions become visible as torsional moments $M_x(x)$, see Fig. 2.5.

The describing partial differential equation is obtained by combining the basic equations of continuum mechanics, i.e. the kinematics equation, the constitutive law and the equilibrium equation (see [1,6]), as:

$$\frac{\mathrm{d}}{\mathrm{d}x}\left(G(x)I_\mathrm{p}(x)\frac{\mathrm{d}\varphi_x}{\mathrm{d}x}\right) + m_x(x) = 0\,, \tag{2.5}$$

from where the general torsional moment distribution is obtained by one-time integration for constant torsional stiffness GI_p and constant distributed moment m_0:

$$M_x(x) = GI_\mathrm{p}\frac{\mathrm{d}\varphi_x(x)}{\mathrm{d}x} = -m_0 x + c_1\,. \tag{2.6}$$

A further integration, under the assumption of constant torsional stiffness and constant distributed moment, gives the general distribution of the angle of torsion as:

$$\varphi_x(x) = \frac{1}{GI_\mathrm{p}}\left(-\frac{1}{2}m_0 x^2 + c_1 x + c_2\right)\,, \tag{2.7}$$

where the constants of integration c_1 and c_2 must be adjusted under consideration of the boundary conditions. The shear stress distribution results from the distribution of the torsional moment according to Eq. (2.6) or Hooke's law (see Fig. 2.6) as:

$$\tau(x,r) = \frac{M_x(x)}{I_\mathrm{p}} \times r = G\frac{\mathrm{d}\varphi_x(x)}{\mathrm{d}x} \times r\,. \tag{2.8}$$

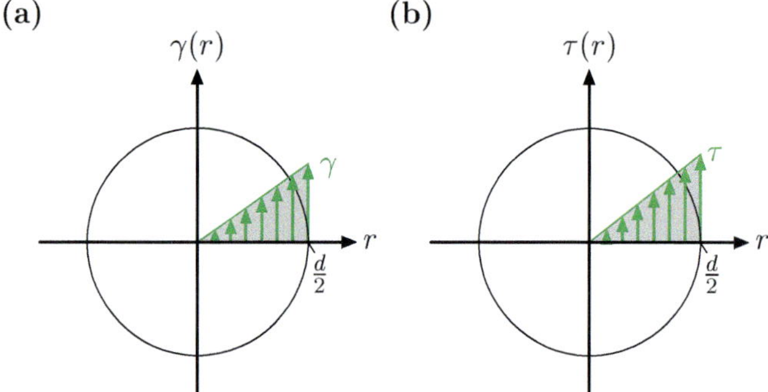

Fig. 2.6 Torsion bar: (**a**) shear strain and (**b**) shear stress distribution. Both distributions are proportional to the radius

2.2   Beams

2.2.1   Euler-Bernoulli Beam Theory

A beam in bending is a prismatic body, which is loaded and deforms perpendicular to its principal axis (deflection $u_z(x)$), see Fig. 2.7. External loads are applied as single forces F_z and moments M_y as well as distributed forces $q_z(x)$ and moments $m_y(x)$. The geometry is described, in the simplest case, by the length L and the constant second moment of area I_y. The material behavior is described by the one-dimensional Hooke's law with the constant Young's modulus E.

The Euler-Bernoulli beam theory assumes that the shear stress (or the shear force) has no influence on the deformation. This is generally true for slender and homogeneous beams with $L \gg h$ (factor: ≈ 10).

If we cut the member at an arbitrary position x into two parts, we can see the internal reactions as shear forces $Q_z(x)$ and bending moments $M_y(x)$, see Fig. 2.8.

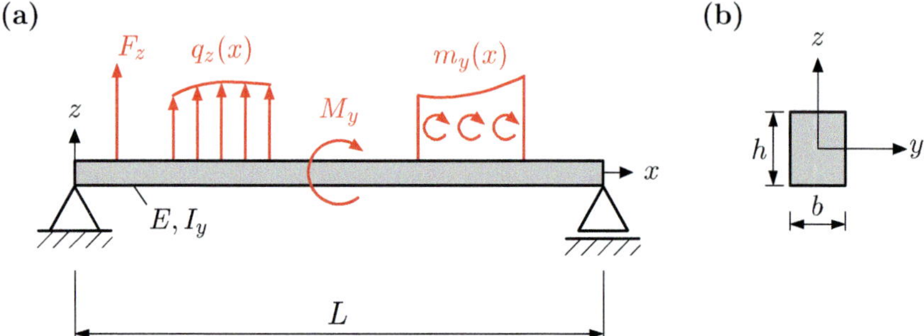

Fig. 2.7 General configuration for an Euler-Bernoulli beam: (**a**) example of boundary conditions and external loads; (**b**) cross-sectional area

Fig. 2.8 Internal reactions for an Euler-Bernoulli beam

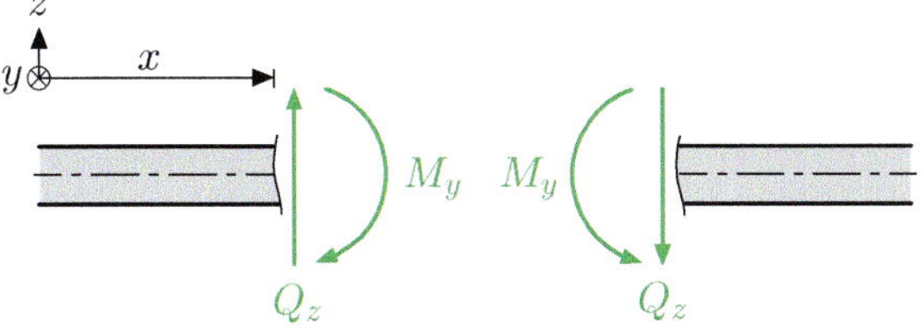

The describing partial differential equation is obtained by combining the basic equations of continuum mechanics, i.e. the kinematics equation, the constitutive law and the equilibrium equation (see [1,9]), in its different formulations as:

$$\frac{d^2}{dx^2}\left(EI_y\frac{d^2u_z(x)}{dx^2}\right) = q_z(x)\,, \tag{2.9}$$

$$\frac{d}{dx}\left(EI_y\frac{d^2u_z(x)}{dx^2}\right) = -Q_z(x)\,, \tag{2.10}$$

$$EI_y\frac{d^2u_z(x)}{dx^2} = -M_y(x)\,, \tag{2.11}$$

whereas gradual integration, under the assumption of constant bending stiffness EI and constant distributed load ($q_z = $ const.), gives the general distribution of the shear force, bending moment and angle of rotation:

$$Q_z(x) = -q_zx - c_1\,, \tag{2.12}$$

$$M_y(x) = -\frac{q_zx^2}{2} - c_1x - c_2\,, \tag{2.13}$$

$$\varphi_y(x) = -\frac{1}{EI_y}\left(\frac{q_zx^3}{6} + \frac{c_1x^2}{2} + c_2x + c_3\right)\,. \tag{2.14}$$

The last integration gives, under the assumption of constant bending stiffness and constant distributed load, the general distribution of the deflection as

$$u_z(x) = \frac{1}{EI_y}\left(\frac{q_zx^4}{24} + \frac{c_1x^3}{6} + \frac{c_2x^2}{2} + c_3x + c_4\right)\,, \tag{2.15}$$

where the constants of integration $c_1, \ldots, c_4$ must be adjusted under consideration of the boundary conditions.

For the example of a cantilever with end load F_0 at $x = L$, we obtain the deflection of the load application point as:

Fig. 2.9 Different stress distributions of an Euler-Bernoulli beam with rectangular cross section and linear-elastic material behavior: (**a**) normal stress and (**b**) shear stress (bending occurs in the x-z plane)

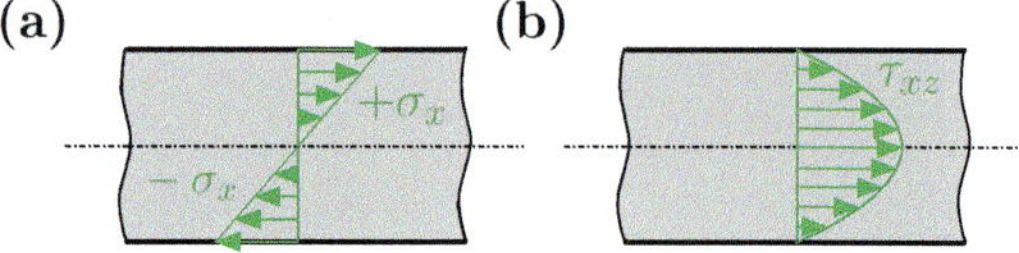

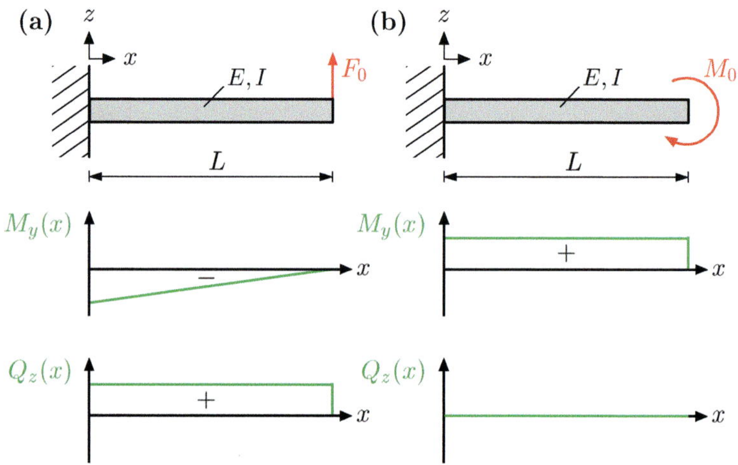

Fig. 2.10 Bending moment and shear force distribution for a cantilever beam: (**a**) loading by a single force F_0; (**b**) loading by a single moment M_0

$$u_z(x = L) = \frac{F_0 L^3}{3 E I_y}. \tag{2.16}$$

The stress distributions are obtained from the bending moment and shear force distribution according to Eqs. (2.12) and (2.13) (see Fig. 2.9):

$$\sigma_x(x, z) = \frac{M_y(x)}{I_y} \times z(x), \quad \tau_{xz}(x, z) = \frac{Q_z(x)}{2 I_y} \times \left[\left(\frac{h}{2} \right)^2 - z^2 \right]. \tag{2.17}$$

Figure 2.10 shows the internal reactions, i.e., bending moment and shear force distributions, for a cantilever beam. The derivation of such distributions is essential for the investigation of the lightweight potential of simple structures.

Finally, some typical properties of different materials are summarized in Table 2.1. The following calculations are based on the mean values derived from the corresponding intervals. However, some parameters are subject to quite large ranges, so that a different choice of the reference value could produce significantly different results.

Table 2.1 Properties of different materials: E: Young's modulus; ν: Poisson's ratio; $R_{\text{p0.2}}$: initial yield stress; ε_{A}: strain at failure. The intervals are taken from [2]

	$E,\ \frac{\text{N}}{\text{mm}^2}$	$\nu,\ -$	$\varrho,\ 10^{-6}\ \frac{\text{kg}}{\text{mm}^3}$	$R_{\text{p0.2}},\ \frac{\text{N}}{\text{mm}^2}$	$\varepsilon_{\text{A}},\ -$
Stainless steel (austenitic)	195,000 190,000 ... 200,000	0.30	7.8 7.5 ... 8.1	393 286 ... 500	0.55 0.45 ... 0.65
Al alloys	74,000 69,000 ... 79,000	0.33	2.75 2.6 ... 2.9	364 100 ... 627	0.175 0.05 ... 0.3
Ti alloys	105,000 80,000 ... 130,000	0.35	4.7 4.3 ... 5.1	750 180 ... 1320	0.18 0.06 ... 0.3

2.2.2 Timoshenko Beam Theory

The Timoshenko beam theory is an extension of the theory for thin beams (see [12, 13]), which considers the influence of the shear force on the deformation. The general configuration is shown in Fig. 2.11. New is that two further geometric properties (cross sectional area A and shear correction factor k_{s}) and a further material parameters (shear modulus $G = \frac{E}{2(1+\nu)}$ for isotropic materials) are considered.

However, it is assumed for simplicity that an equivalent *constant* shear stress and strain is acting in the cross section, see Fig. 2.12. This constant shear stress is obtained by dividing the shear force by an equivalent cross section, the so-called shear area A_{s}:

$$\tau_{xz} = \frac{Q_z}{A_{\text{s}}} = \frac{Q_z}{k_{\text{s}}A}, \tag{2.18}$$

where the relation between the shear area A_{s} and the real cross sectional area A is called the shear correction factor k_{s} ($\frac{5}{6}$ for rectangular cross sections), see [8].

If we cut the member at an arbitrary position x into two parts, we can see the internal reactions as shear forces $Q_z(x)$ and bending moments $M_y(x)$, see Fig. 2.13.

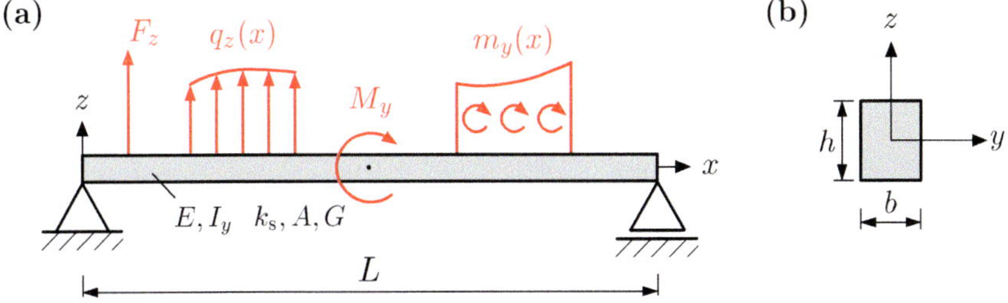

Fig. 2.11 General configuration for a Timoshenko beam: **(a)** example of boundary conditions and external loads; **(b)** cross-sectional area

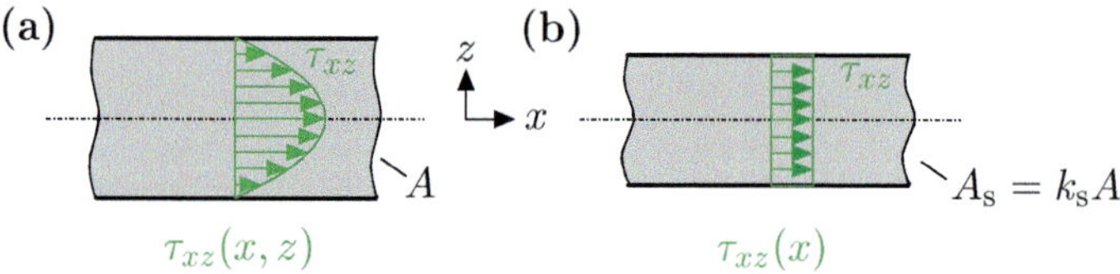

Fig. 2.12 Shear stress distribution for rectangular cross section: **(a)** real distribution (parabolic); **(b)** Approximation according to Timoshenko (constant)

Fig. 2.13 Internal reactions for a Timoshenko beam

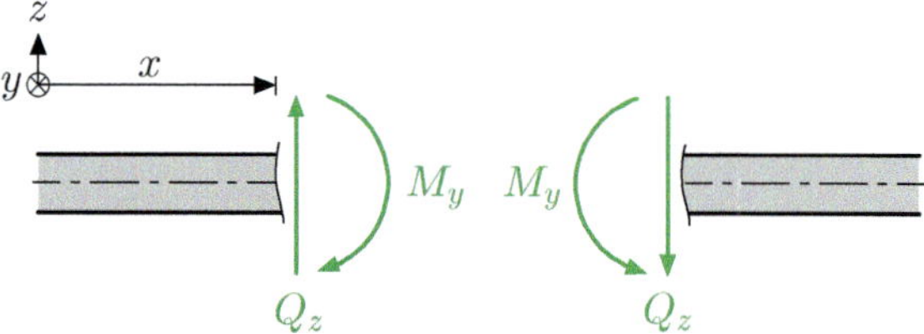

The describing differential equations (coupled, 2nd order) are obtained by combining the basic equations of continuum mechanics, i.e. the kinematics equation, the constitutive law and the equilibrium equation (see [3,9]), for the special case of $EI_y = \text{const.}$, $k_{\mathrm{s}}AG = \text{const.}$ and $m_y = 0$ as:

$$EI_y \frac{\mathrm{d}^2\phi_y}{\mathrm{d}x^2} - k_{\mathrm{s}}GA \left(\frac{\mathrm{d}u_z}{\mathrm{d}x} + \phi_y \right) = 0 \tag{2.19}$$

$$- k_{\mathrm{s}}GA \left(\frac{\mathrm{d}^2u_z}{\mathrm{d}x^2} + \frac{\mathrm{d}\phi_y}{\mathrm{d}x} \right) - q_z = 0 , \tag{2.20}$$

or combined to a single equation:

$$EI_y \frac{\mathrm{d}^4u_z(x)}{\mathrm{d}x^4} = q_z(x) - \frac{EI_y}{k_{\mathrm{s}}AG} \frac{\mathrm{d}^2q_z(x)}{\mathrm{d}x^2} . \tag{2.21}$$

Application of a computer algebra system (for example Maple®, Matlab® or Maxima, see Fig. 2.14) gives the general solution for constant distributed load and constant material and geometry properties as:

$$u_z(x) = \frac{1}{EI_y} \left(\frac{q_z x^4}{24} + c_1 \frac{x^3}{6} + c_2 \frac{x^2}{2} + c_3 x + c_4 \right) , \tag{2.22}$$

$$\phi_y(x) = -\frac{1}{EI_y} \left(\frac{q_z x^3}{6} + c_1 \frac{x^2}{2} + c_2 x + c_3 \right) - \frac{q_z x}{k_{\mathrm{s}}AG} - \frac{c_1}{k_{\mathrm{s}}AG} , \tag{2.23}$$

where the constants of integration $c_1, \ldots, c_4$ must be adjusted under consideration of the boundary conditions.

```
>    sys_T := -E*I*((D@@2) (p)(x)) + G*A*K*((D@@1) (u)(x) + (p)(x)) = 0,
     -G*A*K*((D@@2) (u)(x) + (D@@1) (p)(x)) = q;
```

$$sys\_T = -EI(\mathrm{D}^{(2)})(p)(x) + GAK(\mathrm{D}(u)(x) + (p)(x)) = 0,$$
$$-GAK((\mathrm{D}^{(2)})(u)(x) + \mathrm{D}(p)(x)) = q$$

```
>    dsolve( {sys_T}, {p(x), u(x)} );
```

$$\{\mathrm{p}(x) = -\frac{\_C1x^2}{2} - \frac{qx^3}{6EI} - \_C2x - \_C3 - \frac{EI\_C1}{GAK} - \frac{qx}{GAK},$$
$$\mathrm{u}(x) = \frac{qx^4}{24EI} + \frac{\_C1x^3}{6} + \frac{\_C2x^2}{2} + \_C3x + \_C4\}$$

Fig. 2.14 Solution of the differential equations for the Timoshenko beam based on the symbolic and numeric computing environment Maple®

A formally slightly different solution approach is obtained using the computer algebra system Maxima, see Fig. 2.15. However, both approaches can be converted into each other after a short calculation.

Based on Eqs. (2.22)–(2.23), we can state the distribution of the bending moment and the shear force as:

$$M_y(x) = -\left(\frac{q_z x^2}{2} + c_1 x + c_2\right) - \frac{q_z EI_y}{k_s AG}, \tag{2.24}$$

$$Q_z(x) = -(q_z x + c_1) . \tag{2.25}$$

In the case of a cantilever beam with a single force F_0 at the free end, we obtain the maximum deflection at the load application point as:

$$u_z(x = L) = \frac{F_0 L^3}{3EI_y} + \frac{F_0 L}{k_s AG}. \tag{2.26}$$

The stress distributions are obtained from the bending moment and shear force distributions according to $M_y(x) = +EI_y \frac{\mathrm{d}\phi_y(x)}{\mathrm{d}x}$ or $Q_z(x) = +EI_y \frac{\mathrm{d}^2\phi_y(x)}{\mathrm{d}x^2}$ (see Fig. 2.16):

$$\sigma_x(x, z) = \frac{M_y(x)}{I_y} \times z(x) , \; \tau_{xz} = \frac{Q_z(x)}{A_s} = \frac{Q_z(x)}{k_s A}. \tag{2.27}$$

2.2.3 Levinson Beam Theory

The Levinson beam theory is a higher order theory and an extension of the theory for thick beams which considers a more realistic shear stress distribution than the

Solution of the coupled DEs for a Timoshenko beam with kAG, EI, q, m = const

(%i1)　　eqn_1: -kAG*'diff(u(x),x,2)=q+kAG*'diff(phi(x),x);

(eqn_1)

$$-kAG\left(\frac{\mathrm{d}^2}{\mathrm{d}x^2}\mathrm{u}(x)\right) = kAG\left(\frac{\mathrm{d}}{\mathrm{d}x}\phi(x)\right) + q$$

(%i2)　　eqn_2: -kAG*'diff(u(x),x)=-m-EI*'diff(phi(x),x,2)+kAG*phi(x);

(eqn_2)

$$-kAG\left(\frac{\mathrm{d}}{\mathrm{d}x}\mathrm{u}(x)\right) = -EI\left(\frac{\mathrm{d}^2}{\mathrm{d}x^2}\phi(x)\right) + kAG\,\phi(x) - m$$

(%i3)　　desolve([eqn_1, eqn_2], [u(x), phi(x)]);

(%o3)

$$[\mathrm{u}(x) = -\frac{x^3\left(kAG\left(\frac{\mathrm{d}}{\mathrm{d}x}\mathrm{u}(x)\Big|_{x=0}\right) - m + \phi(0)kAG\right)}{24EI} - \frac{kAGx^3\left(\frac{\mathrm{d}}{\mathrm{d}x}\mathrm{u}(x)\Big|_{x=0}\right)}{8EI}$$

$$+ x\left(\frac{\mathrm{d}}{\mathrm{d}x}\mathrm{u}(x)\Big|_{x=0}\right) - \frac{x^2\left(2kAG\left(\frac{\mathrm{d}}{\mathrm{d}x}\phi(x)\Big|_{x=0}\right) + 2q\right)}{24kAG} - \frac{x^2\left(kAG\left(\frac{\mathrm{d}}{\mathrm{d}x}\phi(x)\Big|_{x=0}\right) + q\right)}{6kAG}$$

$$- \frac{x^2\left(\frac{\mathrm{d}}{\mathrm{d}x}\phi(x)\Big|_{x=0}\right)}{4} + \frac{q\,x^4}{24EI} + \frac{mx^3}{8EI} - \frac{\phi(0)kAGx^3}{8EI} - \frac{qx^2}{4kAG} + \mathrm{u}(0),$$

$$\phi(x) = \frac{x^2\left(kAG\left(\frac{\mathrm{d}}{\mathrm{d}x}\mathrm{u}(x)\Big|_{x=0}\right) - m + \phi(0)kAG\right)}{6EI} + \frac{kAGx^2\left(\frac{\mathrm{d}}{\mathrm{d}x}\mathrm{u}(x)\Big|_{x=0}\right)}{3EI}$$

$$+ x\left(\frac{\mathrm{d}}{\mathrm{d}x}\phi(x)\Big|_{x=0}\right) - \frac{qx^3}{6EI} - \frac{mx^2}{3EI} + \frac{\phi(0)kAGx^2}{3EI} + \phi(0)]$$

Fig. 2.15 Solution of the differential equations for the Timoshenko beam based on the symbolic and numeric computing environment Maxima

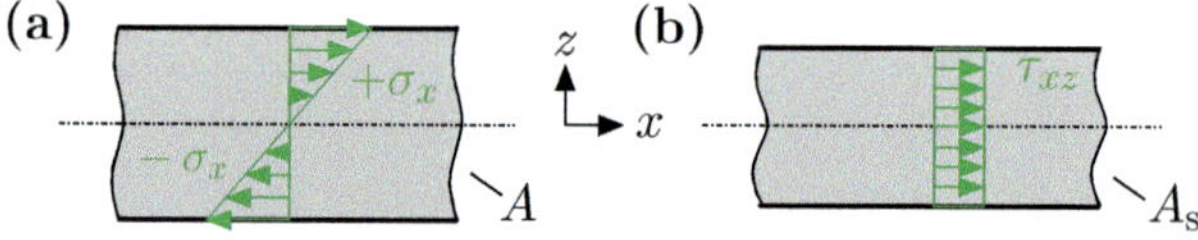

Fig. 2.16 Different stress distributions of a Timoshenko beam with rectangular cross section and linear-elastic material behavior: **(a)** normal stress and **(b)** shear stress (bending occurs in the x-z plane)

theory for thick beams according to Timoshenko (see Sect. 2.2.2), [4, 10, 14]. The general configuration is shown in Fig. 2.17. New is that there is no need for a shear correction factor and only two factors need to be considered (cross sectional area A and shear modulus $G = \frac{E}{2(1+\nu)}$).

If we cut the member at an arbitrary position x into two parts, we can see the internal reactions as shear forces $Q_z(x)$ and bending moments $M_y(x)$, see Fig. 2.18.

The coupled system of differential equations is obtained by combining the basic equations of continuum mechanics, i.e. the kinematics equation, the constitutive law and the equilibrium equation (see [10]), as:

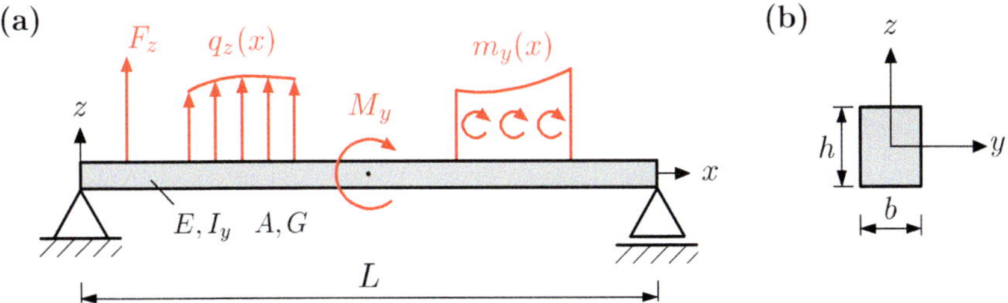

Fig. 2.17 General configuration for a Levinson beam: **(a)** example of boundary conditions and external loads; **(b)** cross-sectional-area

Fig. 2.18 Internal reactions for a Levinson beam

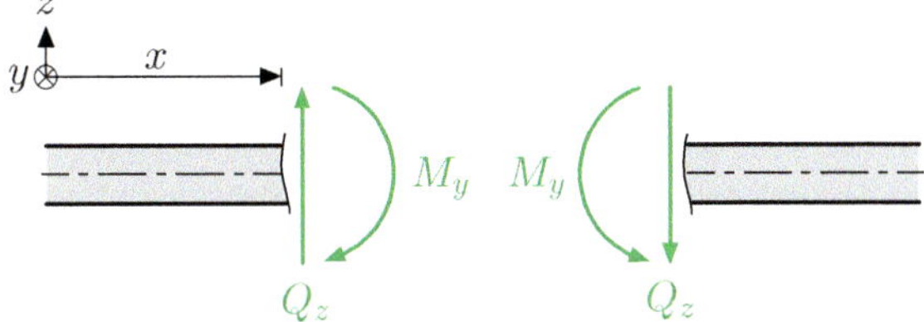

$$\frac{1}{5}\frac{\partial}{\partial x}\left[EI_y\left(4\frac{\partial\phi_y}{\partial x}-\frac{\partial^2 u_z}{\partial x^2}\right)\right]-\frac{2}{3}GA\left(\frac{\partial u_z}{\partial x}+\phi_z\right)=0\,, \tag{2.28}$$

$$-\frac{2}{3}\frac{\partial}{\partial x}\left[AG\left(\frac{\partial u_z}{\partial x}+\phi_y\right)\right]-q_z(x)=0\,. \tag{2.29}$$

Under the assumption of constant material (E, G) and geometry parameters (I_y, A), one can combine, as outlined in Sect. 2.2.2, this system to a single equation:

$$EI_y\frac{\partial^4 u_z(x)}{\partial x^4}=q_z(x)-\frac{6EI_y}{5GA}\frac{\partial^2 q_z(x)}{\partial x^2}\,. \tag{2.30}$$

Application of a computer algebra system (for example Maple®, Matlab® or Maxima) gives for constant material and geometry parameters EI_y, AG and constant distributed load q_z the solution as:

$$u_z(x)=\frac{1}{EI_y}\left(\frac{q_z x^4}{24}+c_1\frac{x^3}{6}+c_2\frac{x^2}{2}+c_3 x+c_4\right), \tag{2.31}$$

$$\phi_y(x)=-\frac{1}{EI_y}\left(\frac{q_z x^3}{6}+c_1\frac{x^2}{2}+c_2 x+c_3\right)-\frac{q_z x}{\frac{2}{3}AG}-\frac{c_1}{\frac{2}{3}AG}\,. \tag{2.32}$$

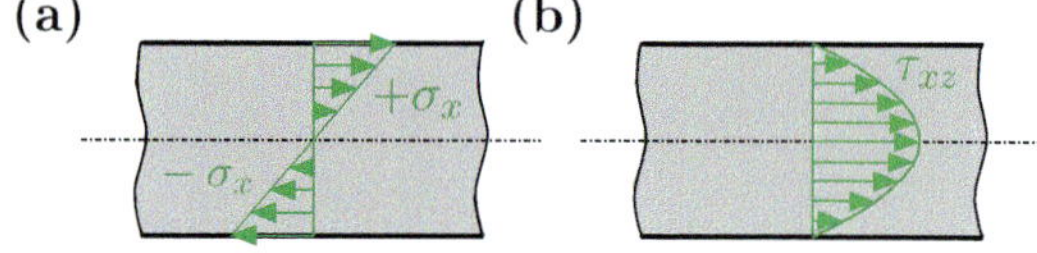

Fig. 2.19 Different stress distributions of a Levinson beam with rectangular cross section and linear-elastic material behavior: **(a)** normal stress and **(b)** shear stress

Under consideration of the relation between internal reactions and stresses (see [10]), we obtain the distribution of the bending moment and the shear force as:

$$M_y(x) = -\left(\frac{q_z x^2}{2} + c_1 x + c_2\right) - \frac{6 q_z E I_y}{5 A G}, \tag{2.33}$$

$$Q_z(x) = -(q_z x + c_1) . \tag{2.34}$$

For the example of a cantilever beam (rectangular cross section) with a single force F_0 at the free end, one obtains the maximum deflection at the load application point as:

$$u_z(x = L) = \frac{F_0 L^3}{3 E I_y} + \frac{F_0 L^3 (1 + \nu)}{4 E I_y}\left(\frac{h^2}{L^2}\right) = \frac{F_0 L^3}{3 E I_y} + \frac{3 F_0 L}{2 A G}. \tag{2.35}$$

The stress distributions (see Fig. 2.19) are obtained from the bending moment distribution $M_y(x) = \frac{E I_y}{5}\left(4\frac{\partial \phi_y(x)}{\partial x} - \frac{\partial^2 u_z(x)}{\partial x^2}\right)$ and the shear force distribution $Q_z(x) = \frac{2}{3} G A \left(\phi_y(x) + \frac{\partial u_z(x)}{\partial x}\right)$ for a rectangular cross section (see [10]):

$$\sigma_x(x, z) = \frac{M_y(x)}{I_y} \times z(x) , \; \tau_{xz}(x, z) = \frac{3}{2} \times \frac{Q_z(x)}{A} \times \frac{h^2 - 4z(x)^2}{h^2} . \tag{2.36}$$

Let us illustrate at the end of this section the influence of the different beam theories in the case of a cantilever beam with a single end force (see Fig. 2.10a). In this case, the boundary conditions are $u_z(0) = 0$, $\varphi_y(o) = 0$, $M_y(L) = 0$ and $Q_z(L) = F_0$.

Thus, we can determine the constants of integration in the general solutions according to Eqs. (2.15), (2.22) and (2.31), see Table 2.2. Using the relations $G = \frac{E}{2(1+\nu)}$ and $I_y = \frac{bh^3}{12} = \frac{A h^2}{12}$ (rectangular cross section with width b and height h) and normalizing the maximum deflections with the value of the Euler-Bernoulli theory, one obtains the normalized deflections at load application point as a function of the slenderness ratio, see Table 2.2.

One can extract from Fig. 2.20 that all theories give the same result for slender beams (i.e. $0 < \frac{h}{L} \leq 0.1$). However, compact beams (i.e. $\frac{h}{L} > 0.1$) may require the application of the theories according to Timoshenko or Levinson.

The influence of the different beam theories on the stress is examined in more detail in the following section.

Table 2.2 Constants of integration and normalized deflections according to the different beam theories for a cantilever beam with point load

Theory	c_1	c_2	c_3	c_4	$\dfrac{u_z(L)}{\frac{F_0 L^3}{3EI_y}}$
Euler-Bernoulli	$-F_0$	$F_0 L$	0	0	1
Timoshenko	$-F_0$	$F_0 L$	$\dfrac{EI_y F_0}{k_s AG}$	0	$1 + \dfrac{3(1+v)}{5}\left(\dfrac{h}{L}\right)^2$
Levinson	$-F_0$	$F_0 L$	$\dfrac{3EI_y F_0}{2AG}$	0	$1 + \dfrac{3(1+v)}{4}\left(\dfrac{h}{L}\right)^2$

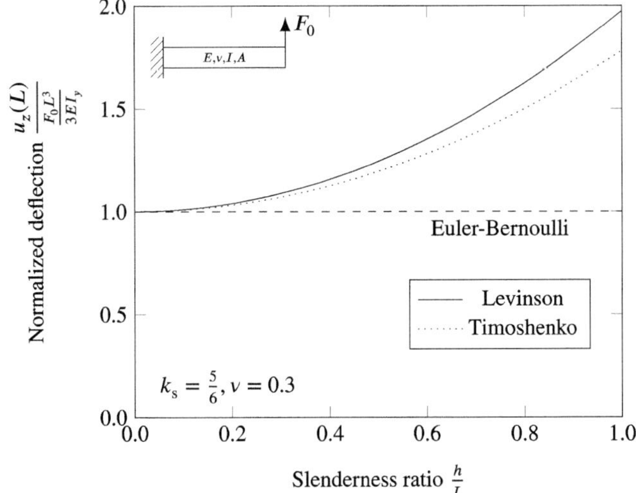

Fig. 2.20 Comparison of the end displacements of a cantilever beam with point load according to the theories of Euler-Bernoulli, Timoshenko and Levinson

2.3   Yield Conditions

The consideration of a multi-axial stress state (see Eq. (2.17), (2.27) or (2.36)) can be done by so-called equivalent stress hypotheses. In the case of ductile materials, the von Mises hypothesis (see [8]) can be stated as (see Fig. 2.21)

$$\underbrace{\sqrt{\sigma^2 + 3\tau^2}}_{\sigma_{\text{eff}}} = k_t \,, \tag{2.37}$$

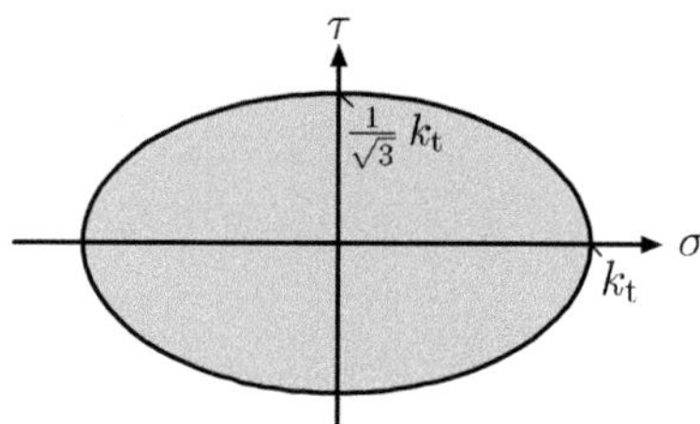

Fig. 2.21 Graphical representation of the yield condition according to von Mises in the σ-τ plane

whereas the special case of a single normal and a single shear stress component is considered. k_t denotes the tensile yield stress (experimental value) and σ_eff the equivalent stress or effective stress. Alternatively, we may formulate Eq. (2.37) in units of a volume-specific energy (see [7, 11]), i.e.

$$\underbrace{\frac{1}{6G} \times \left(\sigma^2 + 3\tau^2\right)}_{\pi^\mathrm{s}} = \frac{k_\mathrm{t}^2}{6G}, \tag{2.38}$$

where π^s is the deviatoric part of the volume specific strain energy[1].

One should note that we can extract from Fig. 2.21 the relationship between the shear and tensile yield stress: $k_\mathrm{s} = k_\mathrm{t}/\sqrt{3}$.

For the general three-dimensional case, we can express the theory according to von Mises as[2]

$$F(\sigma_{ij}) = \underbrace{\sqrt{\frac{1}{2}\left((\sigma_x - \sigma_y)^2 + (\sigma_y - \sigma_z)^2 + (\sigma_z - \sigma_x)^2\right) + 3\left(\sigma_{xy}^2 + \sigma_{yz}^2 + \sigma_{xz}^2\right)}}_{\sigma_\mathrm{eff}} - k_\mathrm{t} = 0,$$

$$\tag{2.39}$$

whereas the graphical representation in the principal stress space is given in Fig. 2.22.

Table 2.3 illustrates that we must consider an equivalent stress in the case of a multi-axial stress state. A sole consideration of single stress components is not sufficient. Consider in the case of Table 2.3 the general definition of the stress tensor:

$$\sigma_{ij} = \begin{bmatrix} \sigma_x & \sigma_{xy} & \sigma_{xz} \\ \sigma_{xy} & \sigma_y & \sigma_{yz} \\ \sigma_{xz} & \sigma_{yz} & \sigma_z \end{bmatrix}. \tag{2.40}$$

One may use as an alternative the hypothesis according to Tresca. Yielding starts as soon as the largest shear stress component reaches a critical value (see [1,5]). For the

---

[1] The volume specific strain energy π or specific work of the internal reactions can be split into a volume specific hydrostatic part π° and a volume specific deviatoric part π^s: $\pi = \pi^\circ + \pi^\mathrm{s}$.

[2] Note the notation $\sigma_{xy} = \tau_{xy}$ and the symmetry $\sigma_{xy} = \sigma_{yx}$.

Fig. 2.22 Graphical representation of the yield condition according to von Mises in the principal stress space

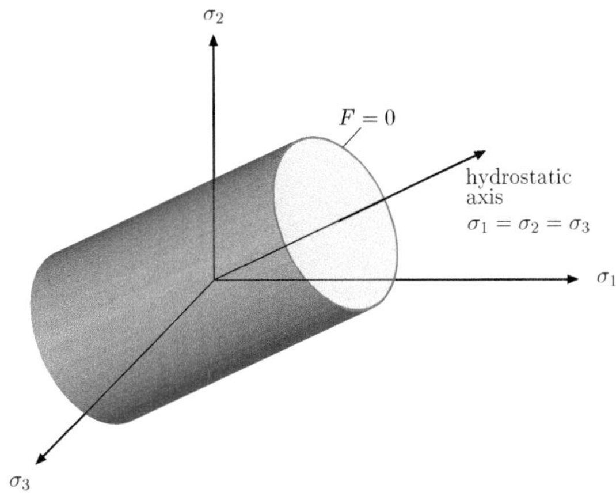

Table 2.3 Equivalent stress according to von Mises for different stress states

Stress Tensor σ_{ij}	von Mises Stress Eq. (2.39)	Range ($k_{\mathrm{t}}^{\mathrm{init}} = 150$)
$\begin{bmatrix} 100 & 0 & 0 \\ 0 & 100 & 0 \\ 0 & 0 & 0 \end{bmatrix}$	100	Elastic
$\begin{bmatrix} 100 & 0 & 0 \\ 0 & -100 & 0 \\ 0 & 0 & 0 \end{bmatrix}$	173.2	Plastic
$\begin{bmatrix} 200 & 0 & 20 \\ 0 & 80 & 20 \\ 20 & 20 & 90 \end{bmatrix}$	125.3	Elastic
$\begin{bmatrix} 200 & 0 & 20 \\ 0 & 80 & 20 \\ 20 & 20 & 200 \end{bmatrix}$	129.3	Elastic
$\begin{bmatrix} 100 & 0 & 20 \\ 0 & 80 & 20 \\ 20 & 20 & -80 \end{bmatrix}$	177.8	Plastic

special case of a single normal and a single shear stress component, one can write (see Fig. 2.23):

$$\underbrace{\sqrt{\sigma^2 + 4\tau^2}}_{\sigma_{\mathrm{eff}}} = k_{\mathrm{t}} . \tag{2.41}$$

One should note that we can extract from Fig. 2.23 the relationship between the shear and tensile yield stress: $k_{\mathrm{s}} = k_{\mathrm{t}}/2$.

For the general three-dimensional case, we can express the theory according to Tresca as

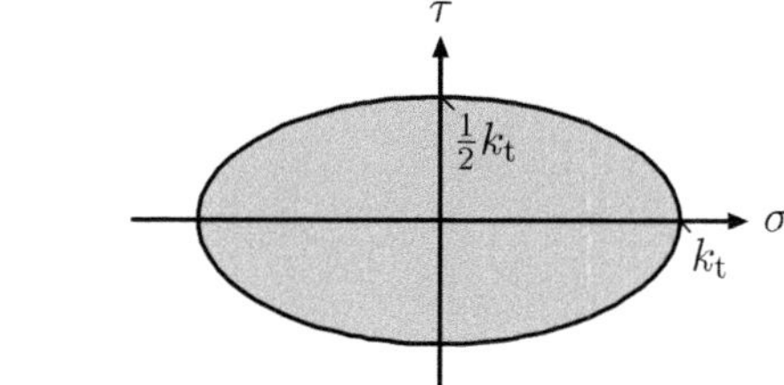

Fig. 2.23 Graphical representation of the yield condition according to Tresca in the σ-τ plane

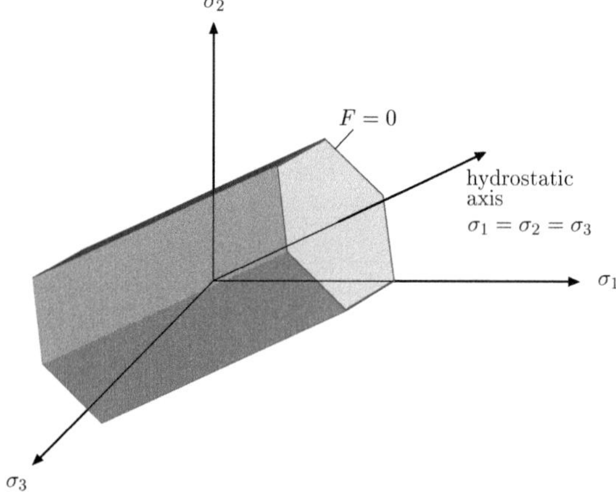

Fig. 2.24 Graphical representation of the yield condition according to Tresca in the principal stress space

$$F(\sigma_i) = \max\left(\frac{1}{2}|\sigma_1 - \sigma_2|, \frac{1}{2}|\sigma_2 - \sigma_3|, \frac{1}{2}|\sigma_3 - \sigma_1|\right) - k_\text{s} = 0, \qquad (2.42)$$

whereas the graphical representation in the principal stress space is given in Fig. 2.24.

2.4   Exercises

2.4.1   Knowledge Questions

- How many material parameters are required for the one-dimensional Hooke's law?
- State the one-dimensional Hooke's law for a uniaxial tensile test.
- Hooke's law can be written as $\sigma(x) = E\varepsilon(x)$ for a special case. State two common assumptions for this formulation.
- Explain the assumptions for (a) an 'isotropic' and (b) a 'homogeneous' material.
- State the major characteristic of an *elastic* material.
- State a common value for the Young's modulus of (a) steel, (b) aluminum, and (c) titanium.
- State a common value for the mass density of (a) steel, (b) aluminum, and (c) titanium.

- The following Fig. 2.25 shows two structural members of the same length L. State for this problem the condition between the spring constant k and properties E, A and L to obtain the same mechanical response.
- Consider again the two structural members shown in Fig. 2.25. Assume linear-elastic material behavior and sketch an ideal force-displacement diagram for the spring and a stress-strain diagram for the bar. Indicate the material constants in the diagrams. What is represented by the areas under the graphs?
- State the three (3) basic equations of continuum mechanics which are required to derive the partial differential equation of a static problem.
- State the internal reaction to describe a tension member.
- Explain the meaning of the (a) kinematics, (b) constitutive and (c) equilibrium equation.
- Name the primary unknown in the partial differential equation of a bar member.
- Give in words the definition of a bar.
- Name the different types of external loads that can be applied to a tension bar.
- Name possible failure modes of a thin bar under compressive load.
- Characterize in words the stress and strain distribution in an elastic bar.
- Sketch (a) the normal strain and (b) the normal stress distribution in the square cross section of a bar under tensile load. Explain in words how these quantities are connected.
- State the major assumptions for the Euler-Bernoulli beam theory.
- State the major assumptions for the Timoshenko beam theory.
- Name the primary unknown in the partial differential equation of an Euler-Bernoulli beam.
- Name the internal reactions to describe an Euler-Bernoulli beam.
- Sketch (a) the normal and (b) the shear stress distribution in a square cross section of an Euler-Bernoulli beam under a general bending load.
- Sketch (a) the normal and (b) the shear stress distribution in a square cross section of an Euler-Bernoulli beam under a *pure* bending load (M_0).
- Name the different types of external loads that can be applied to an Euler-Bernoulli beam.
- Describe in words the meaning of the axial second moment of area (I)
- Why is the shear stress usually not taken into account in a strength analysis of an Euler-Bernoulli beam?
- Sketch (a) the normal and (b) the shear stress distribution of a Timoshenko beam under bending load.

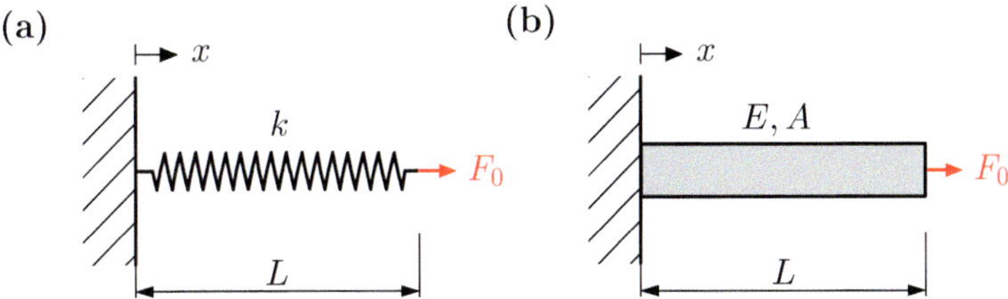

Fig. 2.25 Simple members: (a) Spring and (b) bar

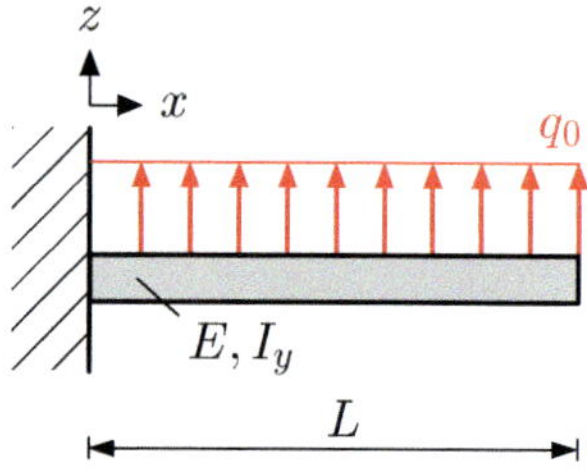

Fig. 2.26 Cantilever beam loaded by a constant distributed load

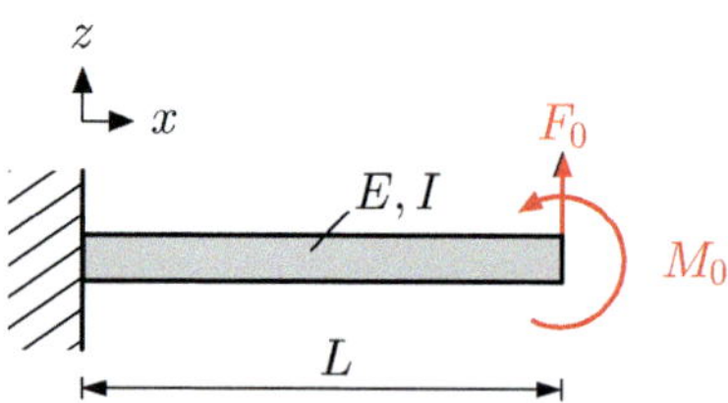

Fig. 2.27 Cantilever beam loaded by point loads

- Consider a beam bending problem. To which internal reactions do the (a) normal stress and (b) shear stress refer?
- The following Fig. 2.26 shows a cantilever beam which is loaded by a constant distributed load q_0.
  Sketch schematically (without calculation) the distribution of the internal shear force $Q_z(x)$ and bending moment $M_y(x)$ (based on the theory of continuum mechanics).
- The following Fig. 2.27 shows a cantilever beam which is simultaneously loaded by point loads F_0 and M_0.
  Sketch schematically (without calculation) the distribution of the internal shear force $Q_z(x)$ and bending moment $M_y(x)$ (based on the theory of continuum mechanics).
- The following Fig. 2.28 shows a generalized cantilever beam which is simultaneously loaded at its right-hand end by a horizontal force F_0 and a moment M_0.
  Sketch schematically (without calculation) the distribution of the internal shear force $Q_z(x)$, bending moment $M_y(x)$, and normal force $N_x(x)$ (based on the theory of continuum mechanics).
- The following Fig. 2.29 shows different cantilever beams which are loaded at their right-hand ends by a horizontal force F_0.
  Sketch schematically (without calculation) the distribution of the internal shear force $Q_z(x)$ and bending moment $M_y(x)$ for all three configurations (based on the theory of continuum mechanics).
- Name the primary unknowns in the partial differential equations of a Timoshenko beam.
- Name the internal reactions to describe a Timoshenko beam.
- State the one-dimensional Hooke's law for a pure shear state in common variables. Which material parameter is involved?
- State for isotropic materials the relationship between the shear modulus G, Young's modulus E, and Poisson's ratio ν.

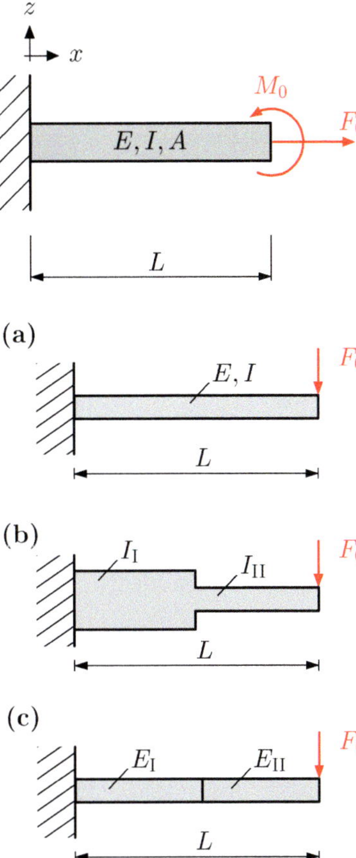

Fig. 2.28 Generalized cantilever beam loaded by point loads

Fig. 2.29 Different cantilever beam problems with point loads: (**a**) homogeneous; (**b**) different cross-sections; (**c**) different materials

- Give practical limits of the Poisson's ratio ν. Give an example of a material that is representative of the lower or upper limit.
- Give typical values for the Poisson's ratio ν of steel or aluminum.
- For a metallic material, the elastic modulus (E) is known to be 210000 MPa. Estimate the value for the shear modulus (G).
- Explain in words the main difference between the Euler-Bernoulli and Timoshenko beam theories.
- Consider a beam bending problem which is described based on the Euler-Bernoulli and Timoshenko beam theories. Which theory gives the larger deflection and why?
- State the rule of thumb in terms of geometric properties that allows to distinguish between the application of the Euler-Bernoulli and the Timoshenko beam theory.
- State the main assumptions for the Levinson beam theory.
- Name the primary unknowns in the partial differential equations of a Levinson beam.
- Sketch (a) the normal and (b) the shear stress distribution of a Levinson beam under a general bending load.
- Name the internal reactions to describe a Levinson beam.

- Sketch (a) the shear strain and (b) the shear stress distribution in the circular cross-section of a bar under torsional loading. Explain in words how these quantities are related.
- Name the primary unknown in the partial differential equation of a torsion bar.
- For a ductile material, the initial yield stress under pure tensile loading is known to be 600 MPa. Use the maximum shear stress theory (Tresca) to estimate the value of the initial yield stress under pure torsional loading.
- For a ductile material, the initial yield stress under pure tensile loading is known to be 519.62 MPa. Use the maximum distortion energy theory (von Mises) to estimate the value of the initial yield stress under pure torsional loading.
- Which of the two equivalent stress hypotheses for ductile materials, i.e. von Mises or Tresca, leads to a conservative estimate (i.e. higher values of the equivalent stress)?
- What is the maximum difference between the equivalent stress according to Tresca and von Mises?
- Describe in words the geometric shape of the yield surface according to von Mises in the three-dimensional principal stress space $(\sigma_1, \sigma_2, \sigma_3)$.
- Describe in words the geometric shape of the yield surface according to Tresca in the three-dimensional principal stress space $(\sigma_1, \sigma_2, \sigma_3)$.

2.4.2   Calculation Problems

2.4.1 Simplified model of a tower under dead weight

Given is a simplified model of a tower which is deforming under the influence of its dead weight, cf. Fig. 2.30. The tower is of original length L, cross-sectional area A, Young's modulus E, and mass density ϱ. The standard gravity is given by g. Calculate:

a) The stress distribution $\sigma_x(x)$ in the member.
b) The reduced length $L' = L - u_x(L)$ due to the acting dead weight.
c) The maximum length $L_{\max}$ if a given stress limit $\sigma_{\max}$ at the foundation ($x = 0$) cannot be exceeded.

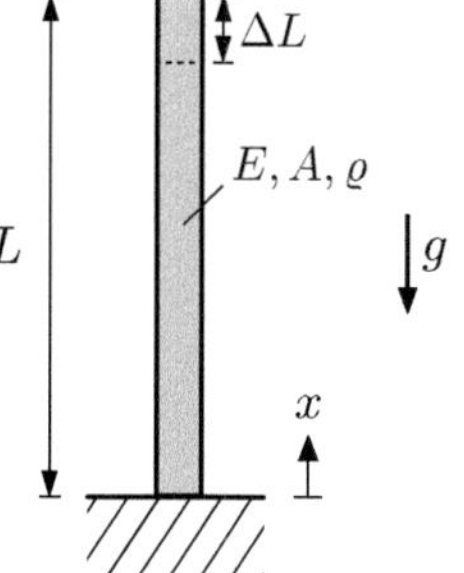

Fig. 2.30 Simplified model of a tower under its dead weight

2.4.2 Course of the bending line and maximum stress for a cantilever beam with different loads

For the cantilever beam in Fig. 2.31, calculate and sketch the course of the bending line and the maximum stress (σ_x) along the longitudinal axis for both load cases according to the Euler-Bernoulli theory. The geometric dimensions (I, L) and the material parameter (E) are assumed to be given and constant. For simplification, a square cross-section (side length: a) can also be assumed.

2.4.3 Course of bending line, stress and strain for a bi-material cantilever beam

For the bi-material cantilever beam in Fig. 2.32, calculate the course of the bending line, the stress and strain on the top surface along the longitudinal axis according to the Euler-Bernoulli theory. The geometric dimensions (I, L) and the two material parameters (E_I and E_II) are assumed to be given and constant. For simplification, a square cross-section (side length: a) can also be assumed. To graphically represent the three courses, $E_\mathrm{I} = 2E_\mathrm{II}$ can be set.

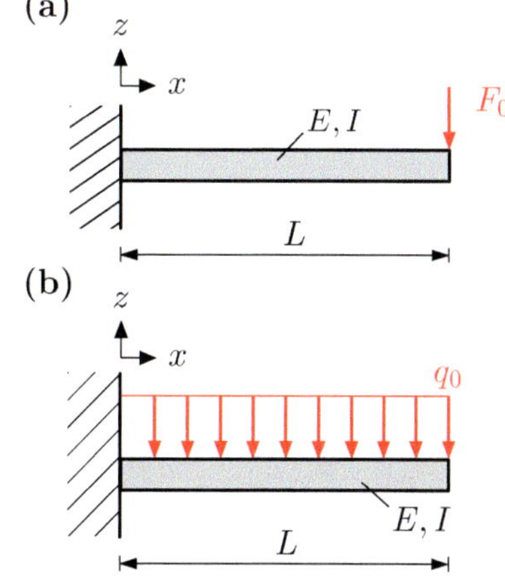

Fig. 2.31 Cantilever beam loaded by (**a**) a single force F_0 and (**b**) a constant distributed load q_0

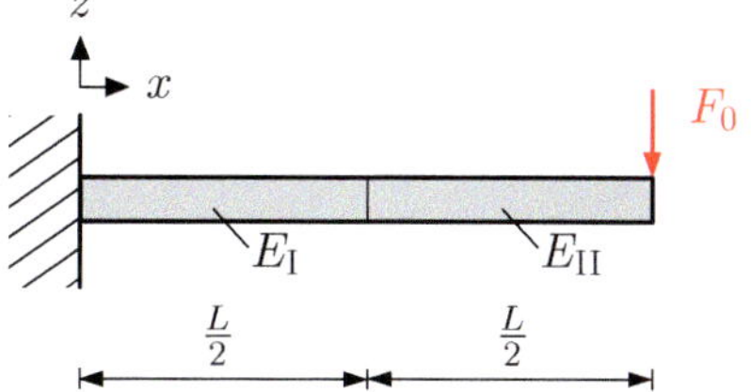

Fig. 2.32 Bi-material cantilever beam loaded by a single force F_0

2.4.4 Comparison of the maximum deflections of a cantilever beam with constant line load according to different beam theories

For a cantilever beam with constant line load q_0 (rectangular cross-section with width b and height h, beam length L, q_0 in negative z-direction), calculate the normalized maximum deflection according to the theories of (a) Euler-Bernoulli, (b) Timoshenko ($k_\mathrm{s} = 5/6$) and (c) Levinson as a function of the slenderness ratio h/L. The maximum deflection according to Euler-Bernoulli should be used as the normalization factor.

2.4.5 Second moment of area for a circular ring

Calculate the second moment of area I for a circular ring, first for a thick wall thickness s (see Fig. 2.33a). Then simplify the result for a thin circular ring, i.e. $s \ll r_\mathrm{m}$ (see Fig. 2.33b). For which ratio of $\frac{r_\mathrm{m}}{s}$ is the relative error of the simplification smaller than 1%?

Finally, corresponding equations for the cross-sectional area A should be given.

2.4.6 Second moment of area for a square box profile

Calculate the second moment of area I for a square box profile, first for a thick wall thickness s (see Fig. 2.34a). Then simplify the result for a thin box profile, i.e. $s \ll a_\mathrm{m}$ (see Fig. 2.34b). For which ratio of $\frac{a_\mathrm{m}}{s}$ is the relative error of the simplification smaller than 1%?

Finally, corresponding equations for the cross-sectional area A should be given.

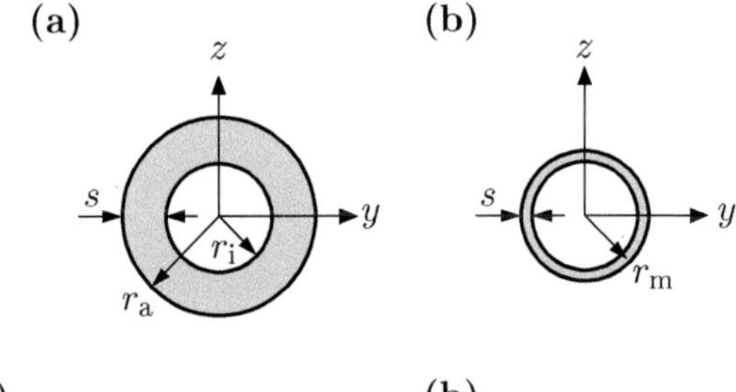

Fig. 2.33 Second moment of area for a circular ring: **(a)** thick and **(b)** thin wall thickness

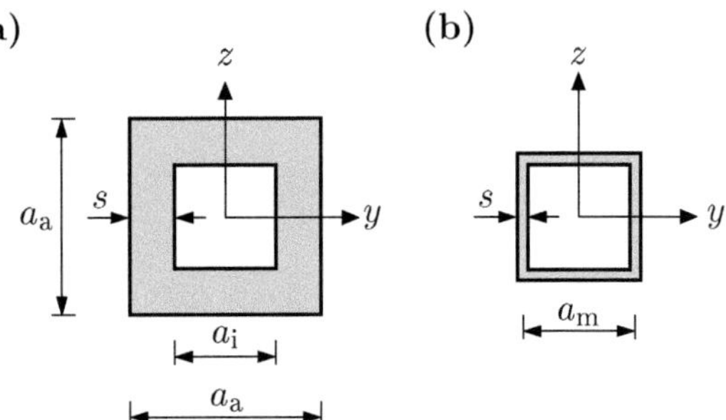

Fig. 2.34 Second moment of area for a squared box section: **(a)** thick and **(b)** thin wall thickness

2.4.7 General solution of the differential equations of the Timoshenko beam using a computer algebra systems

The coupled differential equations of the Timoshenko beam according to Eq. (2.19)–(2.20) can be solved quite easily using computer algebra systems. By using Maple$^{\circledR}$ one obtains the following representation:

$$
\mathrm{u}(x) = \frac{qx^4}{24EI} + \frac{C1x^3}{6} + \frac{C2x^2}{2} + C3x + C4\,,
\tag{2.43}
$$

$$
\phi(x) = -\frac{qx^3}{6EI} - \frac{C1x^2}{2} - C2x - C3 - \frac{qx}{GAK} - \frac{EIC1}{GAK}\,.
\tag{2.44}
$$

However, a calculation using Maxima yields the following representation ($m = 0$):

$$
\mathrm{u}(x) = -\frac{x^3\left(kAG\left(\frac{\mathrm{d}}{\mathrm{d}x}\mathrm{u}(x)\big|_{x=0}\right) + \phi(0)kAG\right)}{24EI} - \frac{kAGx^3\left(\frac{\mathrm{d}}{\mathrm{d}x}\mathrm{u}(x)\big|_{x=0}\right)}{8EI}
$$

$$
+ x\left(\frac{\mathrm{d}}{\mathrm{d}x}\mathrm{u}(x)\bigg|_{x=0}\right) - \frac{x^2\left(2kAG\left(\frac{\mathrm{d}}{\mathrm{d}x}\phi(x)\big|_{x=0}\right) + 2q\right)}{24kAG}
$$

$$
- \frac{x^2\left(kAG\left(\frac{\mathrm{d}}{\mathrm{d}x}\phi(x)\big|_{x=0}\right) + q\right)}{6kAG} - \frac{x^2\left(\frac{\mathrm{d}}{\mathrm{d}x}\phi(x)\big|_{x=0}\right)}{4}
$$

$$
+ \frac{q\,x^4}{24EI} - \frac{\phi(0)kAGx^3}{8EI} - \frac{qx^2}{4kAG} + \mathrm{u}(0),
\tag{2.45}
$$

$$
\phi(x) = \frac{x^2\left(kAG\left(\frac{\mathrm{d}}{\mathrm{d}x}\mathrm{u}(x)\big|_{x=0}\right) + \phi(0)kAG\right)}{6EI} + \frac{kAGx^2\left(\frac{\mathrm{d}}{\mathrm{d}x}\mathrm{u}(x)\big|_{x=0}\right)}{3EI}
$$

$$
+ x\left(\frac{\mathrm{d}}{\mathrm{d}x}\phi(x)\bigg|_{x=0}\right) - \frac{qx^3}{6EI} + \frac{\phi(0)kAGx^2}{3EI} + \phi(0)\,.
\tag{2.46}
$$

Show that both representations can be transformed into each other.

2.4.8 Maximum difference between von Mises and Tresca

Calculate the maximum absolute error between the equivalent stress hypotheses according to von Mises and Tresca in the σ-τ stress plane as a percentage of k_t. Furthermore, state the maximum relative error.

2.4.9 Bar under tensile and torsional loading

For the bar shown in Fig. 2.35 (length L, cross-sectional area A, elastic modulus E, and shear modulus G), calculate the maximum equivalent stress according to von Mises. The bar is loaded by a single force F_0 and by a torsional moment M_t.

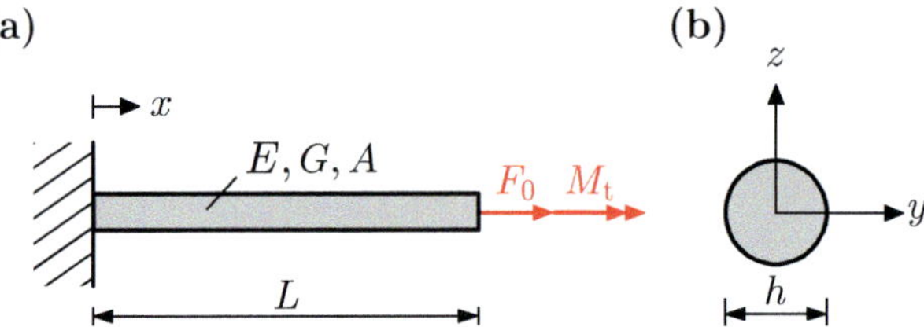

Fig. 2.35 Bar under tensile and torsional loading: **(a)** general configuration and **(b)** cross-sectional profile

References

1. Altenbach, H.: Holzmann/Meyer/Schumpich Technische Mechanik Festigkeitslehre. Springer Vieweg, Wiesbaden (2016)
2. Ashby, M.F., Jones, D.R.H.: Engineering Materials 1: An Introduction to Properties, Applications and Design. Elsevier, Amsterdam (2005)
3. Gross, D., Hauger, W., Schröder, J., Wall, W.A.: Technische Mechanik 2: Elastostatik. Springer Vieweg, Berlin (2014)
4. Levinson, M.: A new rectangular beam theory. J. Sound Vib. 74(1), 81–87 (1981). https://doi.org/10.1016/0022-460X(81)90493-4
5. Mang, H.A., Hofstetter, G.: Festigkeitslehre. Springer Vieweg, Berlin (2013)
6. Merkel, M., Öchsner, A.: Eindimensionale Finite Elemente: Ein Einstieg in die Methode. Springer Vieweg, Berlin (2014)
7. Nash, W.A.: Schaum's Outline of Theory and Problems of Strength of Materials. McGraw-Hill, New York (1998)
8. Öchsner, A.: Elasto-Plasticity of Frame Structure Elements: Modeling and Simulation of Rods and Beams. Springer, Berlin (2014)
9. Öchsner, A.: Computational Statics and Dynamics: An Introduction Based on the Finite Element Method. Springer, Singapore (2016)
10. Öchsner, A.: Theorie der Balkenbiegung: Einführung und Modellierung der statischen Verformung und Beanspruchung. Springer Vieweg, Wiesbaden (2016)
11. Öchsner, A.: Continuum Damage and Fracture Mechanics. Springer, Singapore (2016)
12. Timoshenko, S.P.: On the correction for shear of the differential equation for transverse vibrations of prismatic bars. Philos. Mag. 41(245), 744–746 (1921). https://doi.org/10.1080/14786442108636264
13. Timoshenko, S.P.: On the transverse vibrations of bars of uniform cross-section. Philos. Mag. 43(253), 125–131 (1922). https://doi.org/10.1080/14786442208633855
14. Wang, C.M., Reddy, J.N., Lee, K.H.: Shear Deformable Beams and Plates: Relationships with Classical Solution. Elsevier, Oxford (2000)

Lightweight Design Concepts: Material Selection

3

Abstract

This chapter takes a closer look at the lightweight design in regard to the material selection. The lightweight potential is mainly determined by the choice of the material. The geometry of a structural member therefore initially remains unchanged. Using the lightweight index and the specific energy absorption, different configurations are assessed with regard to their lightweight potential.

3.1 Technical Question

The different lightweight design concepts are explained in the following using the example of a cantilever member with different external loads, see Fig. 3.1. These are the basic load cases of tension, torsion and bending.

The following quantities can be considered as descriptive variables, whereby a possible dependence on the longitudinal axis was deliberately included (see [5]):

- Material i with elastic modulus $E_i(x)$, shear modulus $G_i(x)$ and density $\varrho_i(x)$,
- cross-section with area $A(x)$ and axial second moment of area $I_y(x)$ for rectangular cross-section with width $b(x)$ and height $h(x)$ (tension/compression and bending), polar second moment of area for circular cross-section with diameter $d(x)$ (torsion),
- length L,
- boundary conditions and loads.

The basic idea here is to use the simple one-dimensional structural elements of bars and beams, which are known from the theory of strength of materials, and to introduce the basic concepts and ideas of lightweight design. In this case, we can avoid the use of complex software packages or numerical optimization algorithms.

A. Öchsner, *Lightweight Design*,
https://doi.org/10.1007/978-3-658-48162-9_3

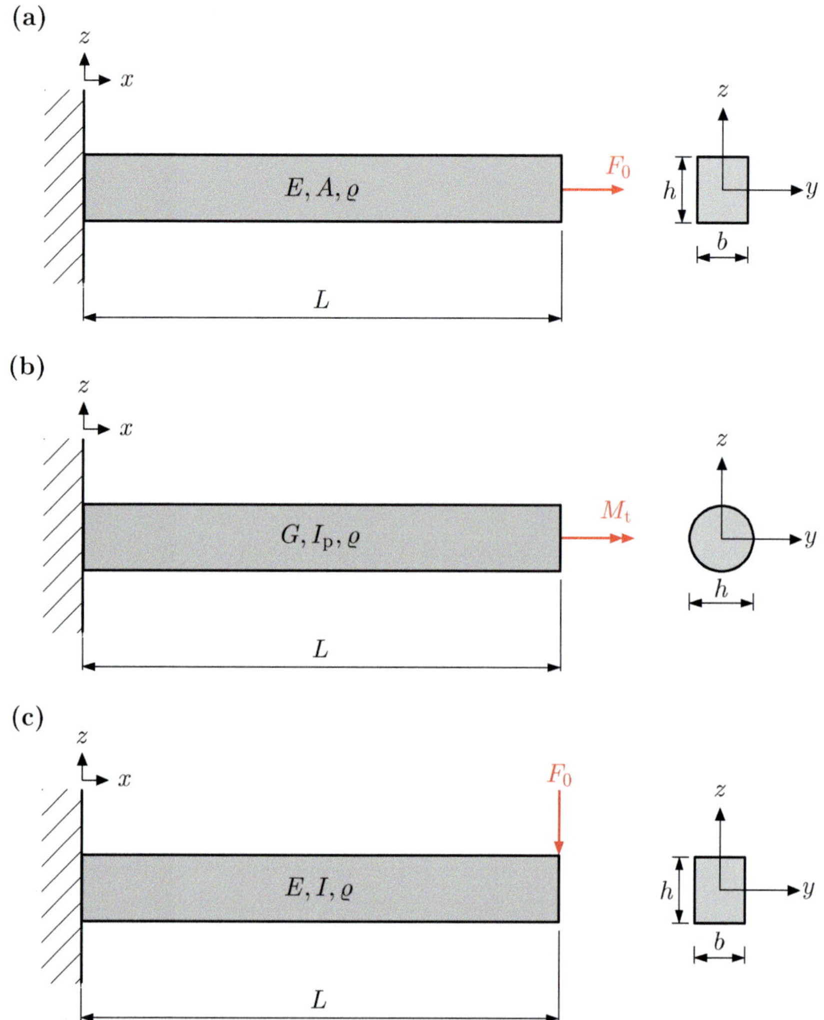

Fig. 3.1 General configuration of a cantilever with different external loads: (**a**) tensile force; (**b**) torsional moment; (**b**) shear force (bending)

In lightweight design in regard to the material the lightweight potential is mainly determined by the choice of the material. Thus, the geometry of a component initially remains unchanged.

3.2   Evaluation Using the Lightweight Index

In order to characterize the performance of a lightweight construction, the so-called lightweight index M according to Klein [3] is used first:

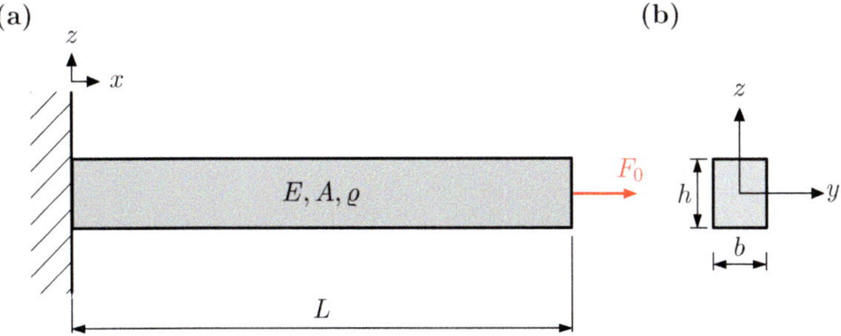

Fig. 3.2 Tensile bar: **(a)** general configuration with support and external point load F_0; **(b)** cross section

$$M = \frac{F_0}{F_G},\qquad(3.1)$$

where the following comments should be done:

- external force F_0 (multiplied with safety factor SF),
- dead weight $F_G = mg = V\varrho g$, for mass $m = \text{const.}$,
- dimensionless performance index; the larger the value, the more efficient a lightweight construction.
- Alternative formulation according to Ashby [1]: mass m and not F_G.

The concept of the lightweight index and the associated estimation of lightweight potential can also be applied to other problem areas. For *stability problems*, i.e. for the buckling of slender bars or thin-walled tubes or plates, the external force F_0 in Eq. (3.1) can be replaced by the so-called critical force F_{cr} (buckling force): $M = \frac{F_{cr}}{F_G}$.

For the example of the tensile bar shown in Fig. 3.2a, a constant normal force distribution results and thus a constant normal stress along the entire tension bar.

Thus, the *limit of the load-carrying capacity* can be defined by the state where the tensile stress reaches a material characteristic (here the 0.2-% strain limit or initial yield stress $R_{p0.2}$, see Table 2.1:

$$\sigma_x = \frac{N_x}{A} = \frac{F_0}{A} \overset{!}{=} R_{p0.2}\,.\qquad(3.2)$$

Therefore, in the definition of the lightweight index according to Eq. (3.1), the external force F_0 can be replaced by material and geometry parameters:

$$M = \frac{F_0}{F_G} = \frac{A R_{p0.2}}{A L \varrho g} = \frac{R_{p0.2}}{\varrho g L} \sim \frac{R_{p0.2}}{\varrho}\,.\qquad(3.3)$$

Based on the definition according to Eq. (3.3) and the characteristic values according to Table 2.1, the lightweight potential of stainless steel (St), aluminum (Al) and

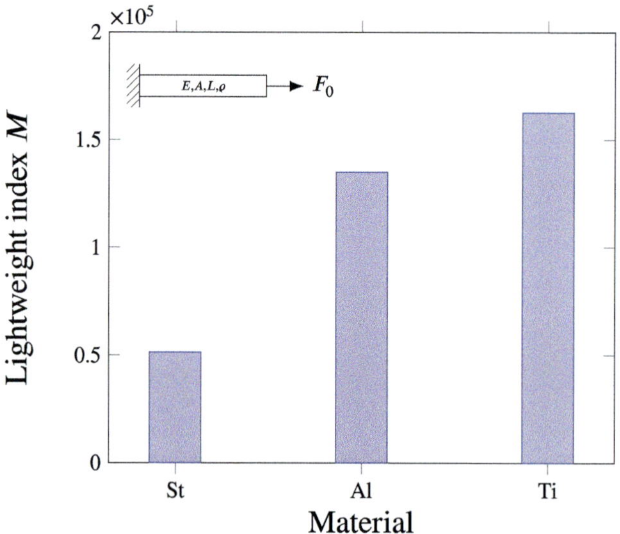

Fig. 3.3 Lightweight index for a tension bar as a function of the material at constant geometry ($L = 100$ mm, $b = h = 10$ mm) and *stress criterion*

Table 3.1 External loads and dead weights for the calculation of the lightweight index according to Fig. 3.3

	F_0, N	F_G, N	M, –
Stainless steel	39,300	0.76518	51360,5
Al alloy	36,400	0.26978	134927.3
Ti alloy	75,000	0.46107	162665.1

titanium (Ti) alloys is shown comparatively in Fig. 3.3. It can be seen that the highest lightweight potential is not achieved by the lightest material (Al), but by the Ti alloy.

It is important that the results in Fig. 3.3 do not compare the same configurations. External loads and dead weights are different (but with the same bar geometry), since the maximum possible lightweight index was evaluated, see Table 3.1.

As an alternative limit criterion, it may be required that the *maximum elongation* does not exceed a predefined value:

$$u_x(x = L) = \frac{F_0 L}{E A} \overset{!}{=} u_{\max} .$$

(3.4)

Thus, in the definition of the lightweight index according to Eq. (3.1), the external force F_0 can be replaced by material and geometry parameters or by the maximum displacement:

$$M = \frac{F_0}{F_G} = \frac{E A u_{\max}}{L(AL\varrho g)} = \frac{E u_{\max}}{\varrho g L^2} \sim \frac{E}{\varrho} .$$

(3.5)

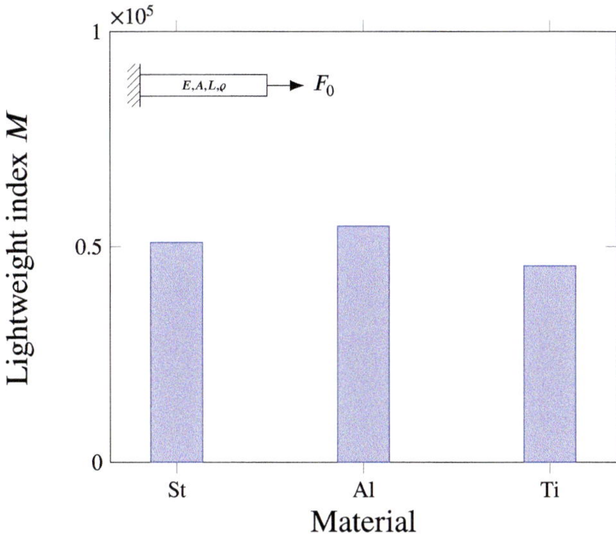

Fig. 3.4 Lightweight index for a tension bar as a function of the material at constant geometry ($L = 100$ mm, $b = h = 10$ mm) and *deformation criterion* ($u_{\max} = 0.2$ mm and $\sigma_{\max} < R_{\mathrm{p}\,0.2}$)

Based on this definition as a deformation criterion ($u_{\max}$), the lightweight indices of stainless steel (St), Al and Ti alloys are shown comparatively in Fig. 3.4. For this case, the lightest material (Al) has now the largest lightweight potential.

As a further alternative limit criterion, it may be required that the *external load (force)* does not exceed a predefined value:

$$F_0 \overset{!}{=} F_{\max} \,. \tag{3.6}$$

With this condition, the definition of the lightweight index is given (see also Fig. 3.5):

$$M = \frac{F_0}{F_{\mathrm{G}}} = \frac{F_{\max}}{\varrho A L g} \quad \sim \frac{1}{\varrho}. \tag{3.7}$$

Based on this assumption, the lightest material ($\rightarrow$ density) has the maximum lightweight potential. Finally, it should be noted here that when a distributed load (see Fig. 2.4, $p_x(x)$) occurs, the application of the lightweight index according to Eq. (3.1) can be difficult, since a distributed load with the unit force per unit length is not provided in the definition. An integration over the distributed load could be used here as an approximation.

For the example of the torsion bar according to Fig. 3.1b an application of the lightweight index according to Eq (3.1) cannot be done, because the external torsional moment cannot be introduced in the numerator ($\rightarrow F_0$). At the end of this subsection, an alternative approach is presented to overcome this limitation.

For the example of the bending beam shown in Fig. 3.6, the maximum bending moment and thus the maximum bending stress results at the fixed support, that is,

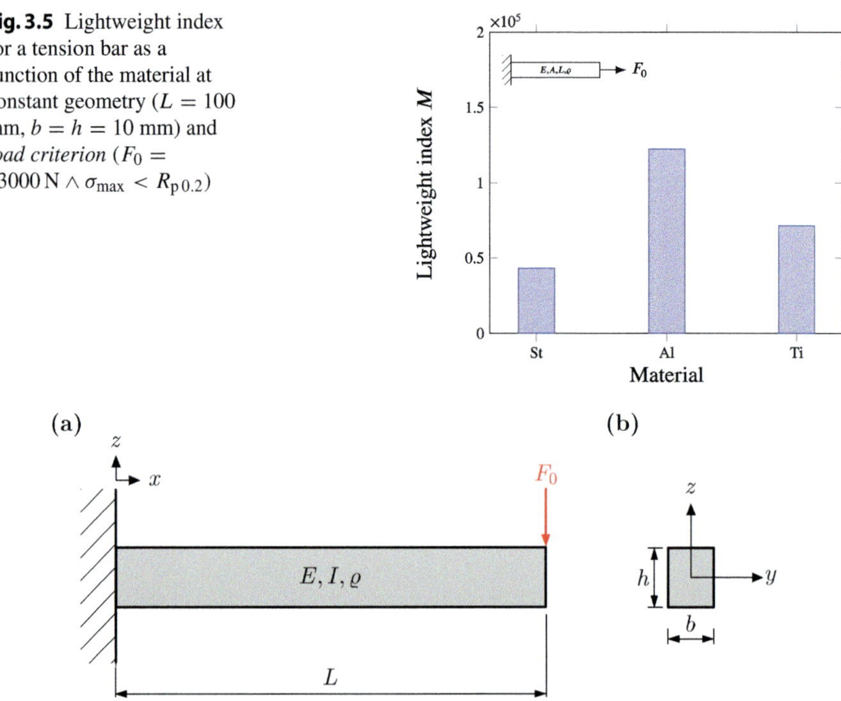

Fig. 3.5 Lightweight index for a tension bar as a function of the material at constant geometry ($L = 100$ mm, $b = h = 10$ mm) and *load criterion* ($F_0 = 33000\,\mathrm{N} \wedge \sigma_{\max} < R_{\mathrm{p}0.2}$)

Fig. 3.6 (**a**) General configuration of a thin cantilever beam with shear force and (**b**) cross section

at $x = 0$. Thus, the limit of the load-carrying capacity can be defined by the state where the tensile stress reaches a material characteristic (here the 0.2-% strain limit or initial yield stress $R_{\mathrm{p}0.2}$, see Table 2.1):

$$\sigma_{x,\max} = \frac{M_y(x = 0)}{I_y} \times \frac{h}{2} = \frac{F_0 L}{I_y} \times \frac{h}{2} = \frac{6 F_0 L}{b h^2} \overset{!}{=} R_{\mathrm{p}0.2}\,. \tag{3.8}$$

Therefore, in the definition of the lightweight index according to Eq. (3.1), the external force F_0 can be replaced by material and geometry parameters:

$$M = \frac{F_0}{F_G} = \frac{R_{\mathrm{p}0.2}}{6 \varrho g \frac{L^2}{h}} \quad \sim \frac{R_{\mathrm{p}0.2}}{\varrho}\,. \tag{3.9}$$

Alternatively, according to Ashby [1], the elimination of h in Eq. (3.9) yields the following relationship: $M \sim \dfrac{(R_{\mathrm{p}0.2})^{2/3}}{\varrho}$. Based on the definition according to Eq. (3.9) and the characteristic values according to Table 2.1, the lightweight potential of stainless steel (St), Al and Ti alloys is shown comparatively in Fig. 3.7. It can be seen that the greatest lightweight potential is not achieved by the lightest material (Al), but by the Ti alloy.

The maximum deflections and the corresponding external loads based on the stress criterion according to Eq. (3.9) are shown comparatively in Fig. 3.8.

Fig. 3.7 Lightweight index for cantilever beams (Euler-Bernoulli) as a function of different materials at constant geometry ($L = 100$ mm, $b = h = 10$ mm) and *stress criterion*

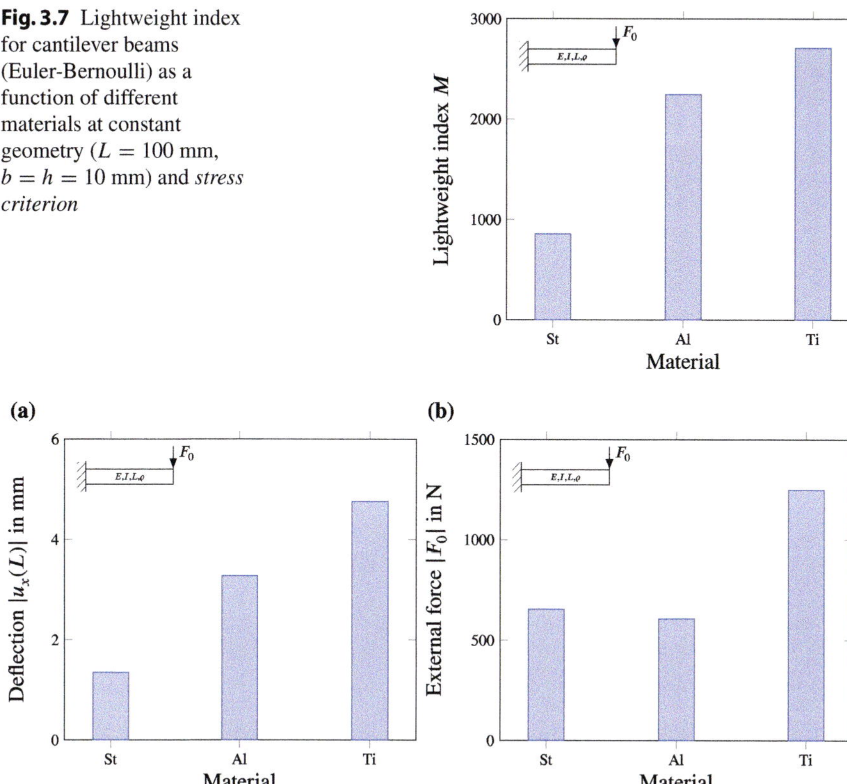

Fig. 3.8 (**a**) Maximum deflection and (**b**) external force for cantilever beams (Euler-Bernoulli) as a function of the material at constant geometry ($L = 100$ mm, $b = h = 10$ mm) and *stress criterion* ($\sigma_{\max} = R_{\mathrm{p}\,0.2}$)

Finally, it should be noted that the results in Figs. 3.7 and 3.8 do not compare the same configurations. External loads and dead weights are different (but with the same beam geometry), as the *maximum possible lightweight index* has been evaluated, see Table 3.2.

Table 3.2 External loads and dead weights for the calculation of the lightweight index according to Fig. 3.7

	F_0, N	F_{G}, N	M, –
Stainless steel	655	0.76518	856.0
Al alloy	606.7	0.26978	2248.8
Ti alloy	1250	0.46107	2711.1

Fig. 3.9 Lightweight index for cantilever beams (Euler-Bernoulli) as a function of the material at constant geometry ($L = 100$ mm, $b = h = 10$ mm) and *deformation criterion* ($u_{\max} = -1$ mm and $\sigma_{\max} < R_{p\,0.2}$)

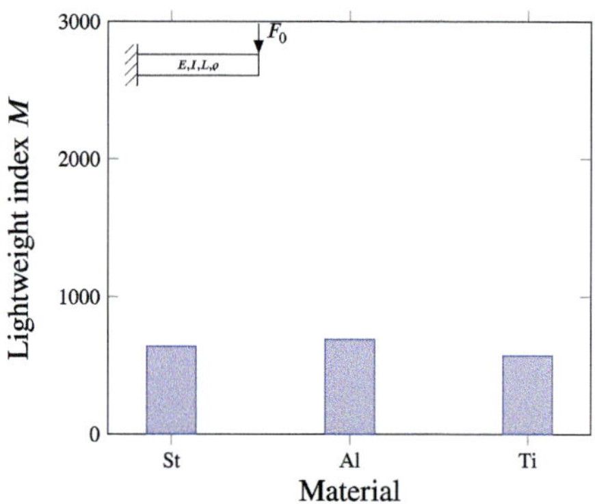

As an alternative limit criterion, it may be required that the maximum deflection does not exceed a predefined value ($u_{\max}$):

$$u_z(x = L) = \frac{|F_0|L^3}{3EI_y} = \frac{4|F_0|L^3}{Ebh^3} \overset{!}{=} u_{\max}\,. \tag{3.10}$$

Therefore, in the definition of the lightweight index according to Eq. (3.1), the external force F_0 can be replaced by material and geometry parameters as well as the maximum deflection:

$$M = \frac{F_0}{F_G} = \frac{Eh^2|u_{\max}|}{4\varrho g L^4} \quad \sim \frac{E}{\varrho}\,. \tag{3.11}$$

Based on this definition as a deformation criterion ($u_{\max}$), the lightweight indices of stainless steel (St), Al and Ti alloys are shown comparatively in Fig. 3.9. For this case, the lightest material (Al) has now the largest lightweight potential.

As a further alternative limit criterion, it may be required that the external load (force) does not exceed a predefined value ($F_{\max}$):

$$F_0 \overset{!}{=} F_{\max}\,. \tag{3.12}$$

With this condition, the definition of the lightweight index is given (see also Fig. 3.10):

$$M = \frac{F_0}{F_G} = \frac{|F_{\max}|}{\varrho A L g} \quad \sim \frac{1}{\varrho}\,. \tag{3.13}$$

Based on this assumption, the lightest material ($\rightarrow$ density) has the maximum lightweight potential.

The different formulations of the lightweight indices M due to the different limit conditions, that is

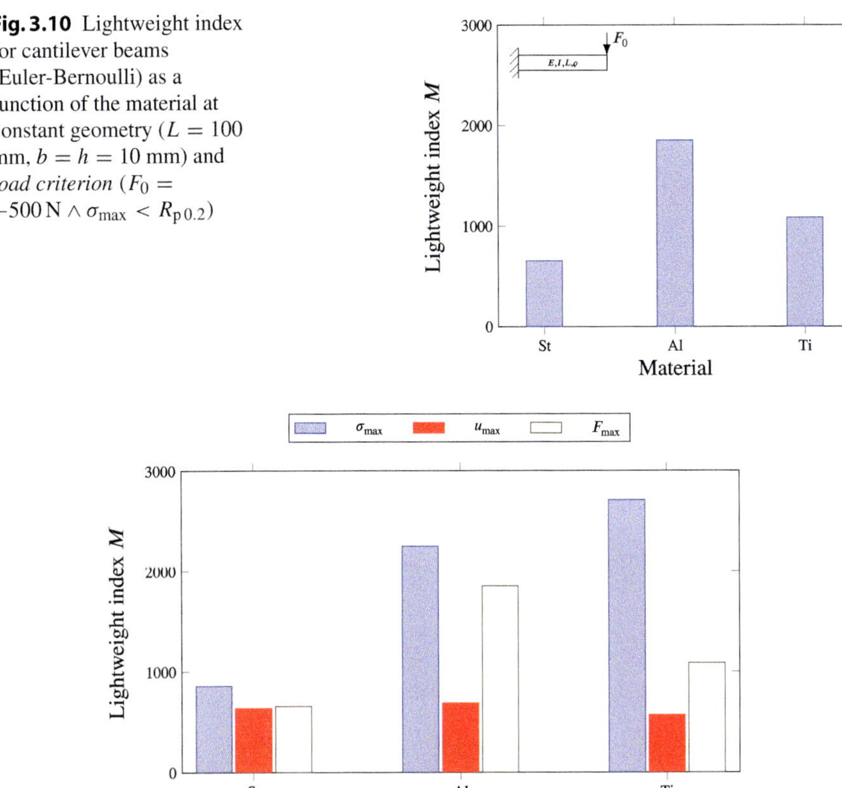

Fig. 3.10 Lightweight index for cantilever beams (Euler-Bernoulli) as a function of the material at constant geometry ($L = 100$ mm, $b = h = 10$ mm) and *load criterion* ($F_0 = -500\,\text{N} \wedge \sigma_{\max} < R_{\text{p}0.2}$)

Fig. 3.11 Comparison of lightweight indices for a cantilever beam (bending, Euler-Bernoulli) based on different limit criteria ($L = 100$ mm, $b = h = 10$ mm)

$$M(\sigma_{\max}) \sim \frac{R_{\text{p}0.2}}{\varrho}, \quad M(u_{\max}) \sim \frac{E}{\varrho}, \quad M(F_{\max}) \sim \frac{1}{\varrho}, \tag{3.14}$$

are shown comparatively in Fig. 3.11

The following example in Fig. 3.12 shows how a combined external load can be treated. For a load case with a transverse and axial force, the structural element must be considered as a combination of bar and beam.

Thus, due to the bending and compressive loading of the member, the two normal stress components must be superimposed to the total stress, see Fig. 3.13.

The maximum stress results here at the clamping point $x = 0$, at the lower surface ($z = -\frac{h}{2}$) to:

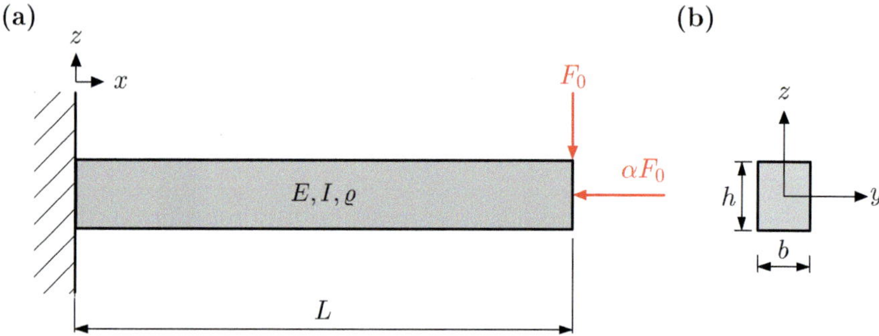

Fig. 3.12 Thin cantilever beam with combined shear and axial force: **(a)** boundary conditions and external loads; **(b)** cross section

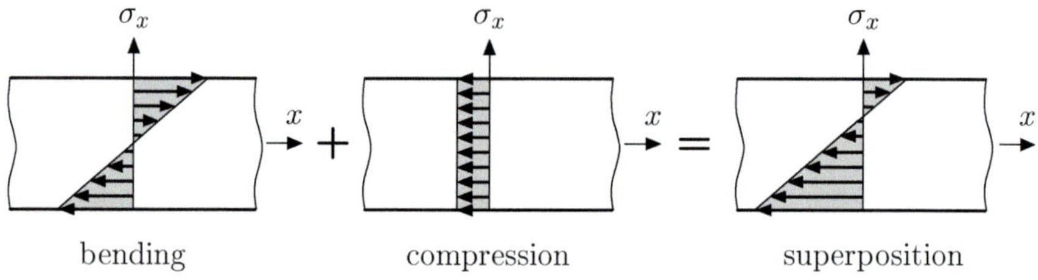

Fig. 3.13 Superposition of the normal stress components to the total stress

$$\sigma_{x,\max} = \sigma_x\!\left(x=0,\, z=-\tfrac{h}{2}\right) = \underbrace{\frac{M_y(x=0)}{I_y} \times \left(-\frac{h}{2}\right)}_{\text{bending}} + \underbrace{\frac{N_x(x=0)}{A}}_{\text{compression}}$$

$$= -\frac{F_0 L}{I_y} \times \frac{h}{2} - \frac{\alpha F_0}{A} = -F_0\left(\frac{Lh}{2I_y} + \frac{\alpha}{A}\right) \stackrel{!}{=} \left|R_{\mathrm{p0.2}}\right| . \qquad (3.15)$$

For the calculation of the lightweight index according to Eq. (3.1), the total force must first be determined by vector addition, see Fig. 3.14.

Fig. 3.14 Vector addition of
the force components to the
total force

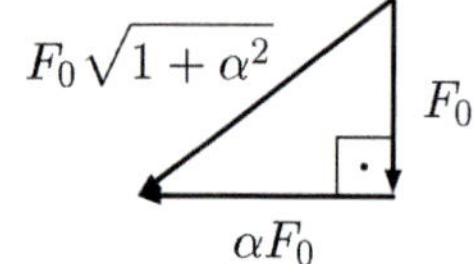

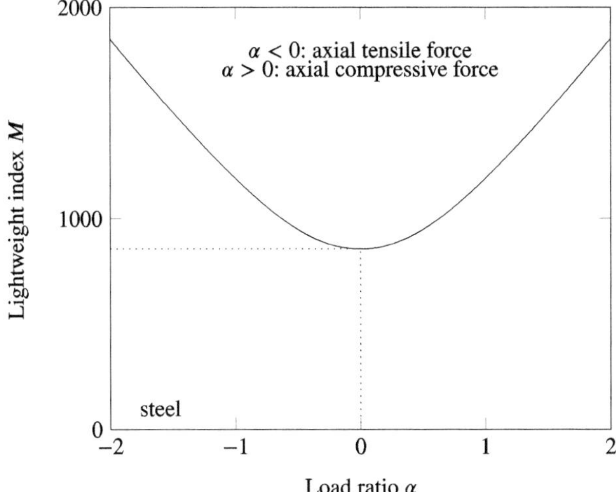

Fig. 3.15 Lightweight index for a cantilever beam with constant geometry ($L = 100$ mm, $b = h = 10$ mm) and *stress criterion* for different load ratios of axial and shear force. The dotted line relates to the reference case shown in Fig. 3.7

Thus, the lightweight index for combined load results (see also Fig. 3.15):

$$M = \frac{\sqrt{1 + \alpha^2}\, F_0}{F_G} = \frac{\sqrt{1 + \alpha^2}\, R_{p0.2}}{\left(\frac{6L^2}{h} + |\alpha|L\right)\varrho g} \quad \sim \quad \frac{R_{p0.2}}{\varrho}, \tag{3.16}$$

where the load ratio α allows a distinction between compression ($\alpha > 0$) and tension ($\alpha < 0$). However, Eq. (3.16) is valid for both cases due to the absolute value. Furthermore, it was assumed that the tensile and compressive yield stresses are identical.

In the following, the influence of shear force on the deformation and stress state is considered according to the Timoshenko beam theory. Assuming a stress distribution as shown in Fig. 2.16, it results that two different stress components are acting and a simple superposition as in Fig. 3.13 is not possible. In fact, an effective stress hypothesis is required which calculates a scalar value that can be compared to a uniaxial material characteristic. Assuming the hypothesis of von Mises, the maximum effective stress is given by Eq. (2.37) at the top or bottom side of the beam:

$$\sqrt{\left(\frac{6F_0 L}{bh^2}\right)^2 + 3\left(\frac{6F_0}{5bh}\right)^2} = R_{p0.2} \;\Rightarrow\; F_0 = \frac{R_{p0.2}}{\sqrt{\left(\frac{6L}{bh^2}\right)^2 + 3\left(\frac{6}{5bh}\right)^2}}. \tag{3.17}$$

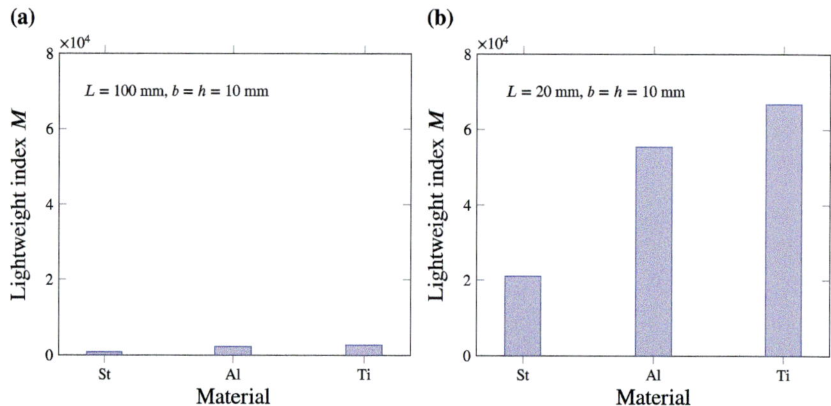

Fig. 3.16 Lightweight index for cantilever beams as a function of the material according to the Timoshenko theory and *stress criterion* ($\sigma_{\max} = R_{\mathrm{p}\,0.2}$): (**a**) $L = 100$ mm, $b = h = 10$ mm and (**b**) $L = 20$ mm, $b = h = 10$ mm

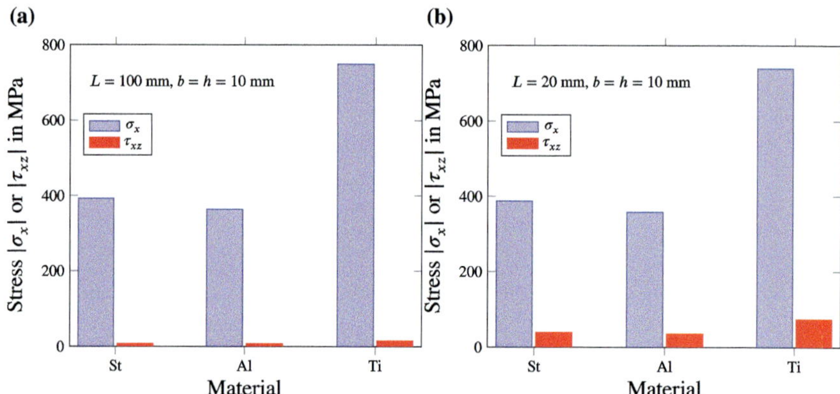

Fig. 3.17 Stress components for cantilever beams as a function of the material according to the Timoshenko theory and *stress criterion* ($\sigma_{\max} = R_{\mathrm{p}\,0.2}$): (**a**) $L = 100$ mm, $b = h = 10$ mm and (**b**) $L = 20$ mm, $b = h = 10$ mm

Thus, the lightweight index, assuming a stress criterion, results in:

$$M = \frac{F_0}{F_{\mathrm{G}}} = \frac{R_{\mathrm{p}0.2}}{\sqrt{\frac{36L^2}{h^2} + \frac{108}{24}} \times \varrho g L} \quad \sim \frac{R_{\mathrm{p}0.2}}{\varrho}. \tag{3.18}$$

Figure 3.16 shows the effect on the lightweight index for a slender ($L \gg b, h$) and compact beam ($L \sim b, h$) and Fig. 3.17 the corresponding ratios of normal and shear stress.

According to the procedure of the Timoshenko beam, the shear stress can also be considered according to the theory of Levinson. According to Fig. 2.19, a linearly distributed normal stress and a parabolic shear stress distribution is acting. Thus, the location of the maximum of the effective stress cannot be specified a priori here (if a stress criterion is to be used to calculate the lightweight index). For the beam configuration according to Fig. 3.1c, the bending moment and shear force distribution results in:

$$M_y(x) = F_0(x - L), \tag{3.19}$$

$$Q_z(x) = -F_0. \tag{3.20}$$

Thus, assuming the von Mises stress hypothesis according to Eq. (2.37) (see [4]) and the expressions for the normal and shear stresses according to Eq. (2.36), the effective stress can be stated as follows:

$$\sigma_{\text{eff}} = \sqrt{\sigma^2 + 3\tau^2} = \sqrt{\left(\frac{M_y}{I_y}z\right)^2 + 3\left(\frac{3Q_z}{2A}\frac{h^2 - 4z^2}{h^2}\right)^2} \tag{3.21}$$

$$= \sqrt{\left(\frac{F_0(x - L)}{I_y}z\right)^2 + 3\left(-\frac{3F_0}{2A}\frac{h^2 - 4z^2}{h^2}\right)^2}. \tag{3.22}$$

It follows from the above equation that the maximum occurs at the clamping point, i.e. $x = 0$. If we continue to consider a square cross-section with a side length h, one gets $I_y = \frac{h^4}{12} = \frac{Ah^2}{12}$. Thus, the effective stress results to:

$$\sigma_{\text{eff}} = \frac{F_0}{A}\sqrt{\left(\frac{12Lz}{h^2}\right)^2 + 3\left(\frac{3(h^2 - 4z^2)}{2h^2}\right)^2}, \tag{3.23}$$

or in a normalized representation:

$$\frac{\sigma_{\text{eff}}}{\frac{F_0}{A}} = \sqrt{\left(\frac{6L}{h}\left[\frac{z}{\frac{h}{2}}\right]\right)^2 + 3\left(\frac{3}{2}\left(1 - \left[\frac{z}{\frac{h}{2}}\right]^2\right)\right)^2}. \tag{3.24}$$

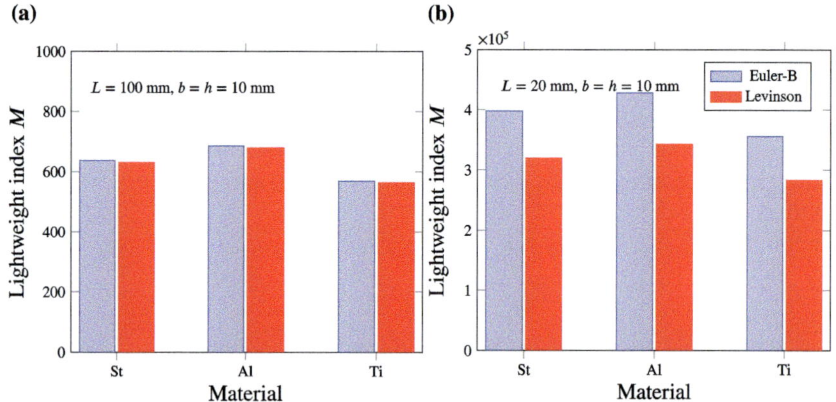

Fig. 3.18 Lightweight index for cantilever beams as a function of the material and different length ($L = 100$ mm or $L = 20$, $b = h = 10$ mm) according to the Levinson theory and *deformation criterion* ($u_{\mathrm{max}} = -1$ mm and $\sigma_{\mathrm{max}} < R_{\mathrm{p}0.2}$)

The evaluation of this relation for different height to length ratios is shown in Fig. 3.19. It can be seen that as the beam length reduces, the significance of the shear stress increases in relation to the normal stress. Although the shear stress remains constant, the magnitude of the normal stress (shorter lever to the clamping point) is reduced. The most important conclusion, however, is that the maximum of the effective stress occurs at the top or bottom side of the beam. Thus, the evaluation of the maximum lightweight index by means of a stress criterion would not give any difference to the theory for thin beams. More interesting, however, is the case with a deformation criterion. According to Eq. (2.35), the force for a given displacement for a square cross section (h) is given by:

$$F_0 = \frac{|u_{\mathrm{max}}|}{\frac{4L^3}{Eh^4} + \frac{3L}{2h^2 G}}. \tag{3.25}$$

Thus, the lightweight index can be given as follows:

$$M = \frac{|u_{\mathrm{max}}|}{\left(\frac{4L^4}{Eh^2} + \frac{3L^2}{2G}\right)\varrho g}. \tag{3.26}$$

It can be seen from Fig. 3.18 that for compact Levinson beams there are clear differences to the theory according to Euler-Bernoulli.

Fig. 3.19 Effective stress for a Levinson beam for different slenderness ratios: (**a**) slender with $\frac{h}{L} = 0.1$; (**b**) compact with $\frac{h}{L} = 0.6$; (**c**) compact with $\frac{h}{L} = 0.9$

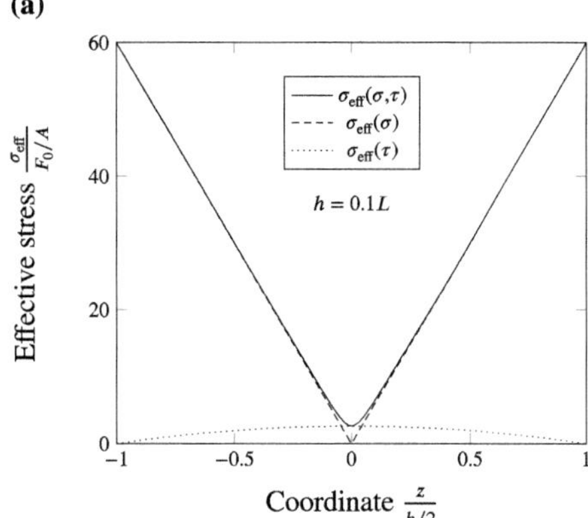

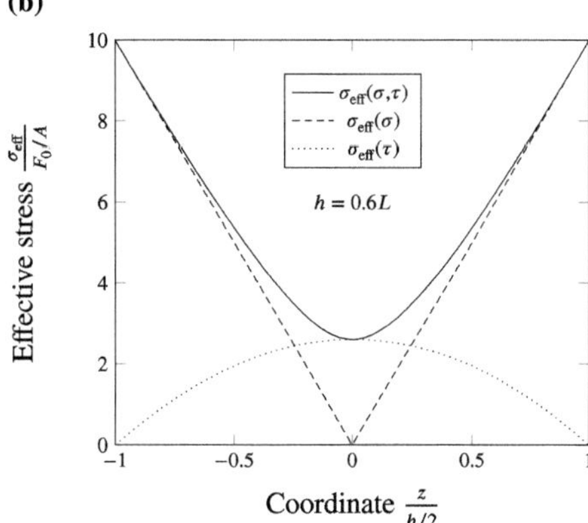

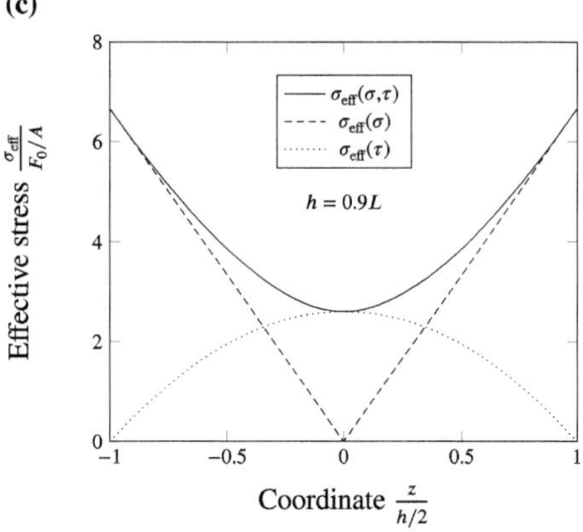

3.3   Evaluation Using the Specific Energy Absorption

However, the previous explanations also reveal some disadvantages in connection
with the definition of the lightweight index M (see Eq. (3.1)):

- Complicated load cases may not be possible to be summarized to a single resulting
  force;
- the consideration of distributed loads is not directly possible;
- the consideration of external bending moments or torsional moments is not possible;
- The consideration of plastic material behavior is difficult, since, for example, in
  the case of ideal plasticity, the deformation increases at a constant stress.

Therefore, an alternative approach will be introduced below. The specific energy
absorption (SEA) allows for the consideration of the above-mentioned limitations
and can be stated as follows:

$$SEA = \frac{\Pi}{m}, \tag{3.27}$$

where Π is the so-called total strain energy and m is the mass of the member. This
index number is often used to assess the crash behavior of lightweight structures
(see [2,7,8]). A descriptive definition of the *total strain energy* can be based on the
uniaxial tensile test as the area under the force-displacement diagram (see Fig. 3.20a).
It should be noted in this context that the area under the classical stress-strain diagram
(see Fig. 3.20b) represents the volume-specific strain energy, i.e. $\pi = \Pi/V$.
An easily interpreted graphic representation of the total strain energy—based on the
external force—is

$$\Pi = \frac{1}{2} F_0 u_0, \tag{3.28}$$

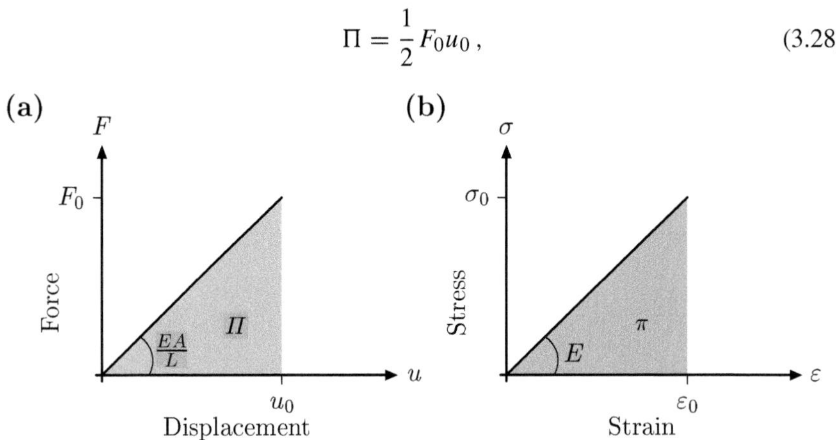

Fig. 3.20 Idealized data from a uniaxial tensile test: (**a**) force-displacement and (**b**) stress-strain
diagram

or in the following integral approach:

$$\Pi = \int_0^{u_0} F(u)\,\mathrm{d}u = \int_0^{u_0} \frac{EA}{L}u\,\mathrm{d}u = \frac{EA}{2L}u_0^2 = \frac{F_0^2 L}{2EA} = \frac{1}{2}F_0 u_0 \,. \tag{3.29}$$

Alternatively, the strain energy can be expressed based on the internal reactions, i.e. the normal force distribution, see Fig. 3.20b. For this purpose it is advisable to integrate the over the specific strain energy:

$$\mathrm{d}\Pi = \mathrm{d}\pi\,\mathrm{d}V = \sigma\,\mathrm{d}\varepsilon\,\mathrm{d}V = E\varepsilon\,\mathrm{d}\varepsilon\,\mathrm{d}V = \frac{E\varepsilon^2}{2}\,\mathrm{d}V = \frac{\sigma^2}{2E}(A\,\mathrm{d}x) = \frac{N^2}{2EA}\,\mathrm{d}x \,. \tag{3.30}$$

Similar derivations can be written for other simple modes of deformation of bar/beam members and the following cases can be distinguished for linear-elastic material behavior, see [6]:

- Tension or compression:

$$\Pi = \int_0^L \frac{N_x(x)^2}{2EA}\,\mathrm{d}x \,. \tag{3.31}$$

- Bending:

$$\Pi = \int_0^L \frac{M_y(x)^2}{2EI_y}\,\mathrm{d}x \,. \tag{3.32}$$

- Shear:

$$\Pi = \int_0^L \frac{Q_z(x)^2}{2GA_s}\,\mathrm{d}x = \int_0^L \frac{Q_z(x)^2}{2k_s GA}\,\mathrm{d}x \,. \tag{3.33}$$

- Torsion:

$$\Pi = \int_0^L \frac{M_x(x)^2}{2GI_p}\,\mathrm{d}x \,. \tag{3.34}$$

Thus, the total strain energy Π in a bar/beam-like structural member can be expressed as:

$$\Pi = \int_0^L \frac{N_x(x)^2}{2EA}\,\mathrm{d}x + \int_0^L \frac{M_y(x)^2}{2EI_y}\,\mathrm{d}x + \int_0^L \frac{Q_z(x)^2}{2GA_s}\,\mathrm{d}x + \int_0^L \frac{M_x(x)^2}{2GI_p}\,\mathrm{d}x \,, \tag{3.35}$$

Fig. 3.21 Internal reactions for a tensile bar

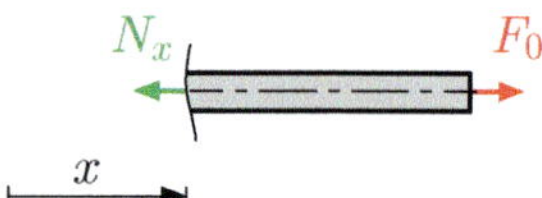

where N_x, M_y, Q_z, M_x represent the distributions of the internal reactions. It should be noted here that in Eq. (3.35), the shear area A_s and the shear correction factor k_s have been used, see [4] for details.

In the following, the example of the tensile rod according to Fig. 3.1a is reevaluated, taking into account the specific energy absorption (SEA). The internal normal force distribution follows from Fig. 3.21 as $N_x(x) = F_0$.

Thus, we can calculate the total strain energy Π as:

$$\Pi = \int_0^L \frac{N_x(x)^2}{2EA}\,dx = \frac{F_0^2}{2EA}\int_0^L dx = \frac{F_0^2 L}{2EA}, \tag{3.36}$$

whereby the specific energy absorption SEA can be written as:

$$SEA = \frac{\Pi}{m} = \frac{F_0^2 L}{2EA \times AL\varrho} = \frac{F_0^2}{2EA^2\varrho}. \tag{3.37}$$

Assuming again that the limit of the load-carrying capacity can be defined by the state where the tensile stress reaches a material characteristic (0.2-% strain limit), the external force is obtained from Eq. (3.2) as $F_0 = AR_{p0.2}$. Thus, the maximum specific energy absorption is obtained as:

$$SEA = \frac{A^2 R_{p0.2}^2}{2EA^2\varrho} = \frac{R_{p0.2}^2}{2E\varrho} \quad \sim \quad \frac{R_{p0.2}^2}{E\varrho}. \tag{3.38}$$

Based on Eq. (3.38) and the characteristic values according to Table 2.1, the specific energy absorption of stainless steel (St), Al and Ti alloys is shown in Fig. 3.22. Again, it is found that the highest energy absorption is achieved not by the lightest material (Al), but by the Ti alloy, thus giving the same trend as when using the lightweight index M, see Fig. 3.3.

As an alternative *limit criterion*, it may be required that the *maximum elongation* does not exceed a predefined value, i.e. $F_0 = \frac{EAu_{max}}{L}$. Thus, the specific energy absorption is obtained as:

$$SEA = \frac{1}{2EA^2\varrho}\frac{(EAu_{max})^2}{L^2} = \frac{u_{max}^2 E}{2L^2\varrho} \quad \sim \quad \frac{E}{\varrho}. \tag{3.39}$$

Based on the definition as deformation limit (u_{max}), the specific energy absorptions of stainless steel (St), Al and Ti alloys are shown comparatively in Fig. 3.23. For this case, the lightest material (Al) has now the largest lightweight potential, thus giving

Fig. 3.22 Specific energy absorption for a tensile rod as a function of the material at constant geometry ($L = 100$ mm, $b = h = 10$ mm) and *stress criterion*m

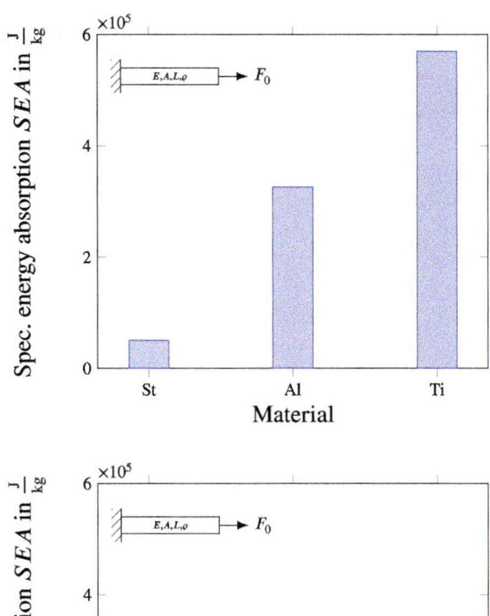

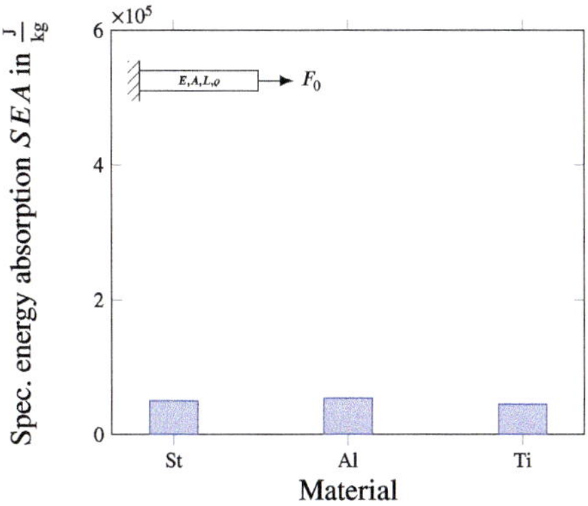

Fig. 3.23 Specific energy absorption for a tensile rod as a function of the material at constant geometry ($L = 100$ mm, $b = h = 10$ mm) and *deformation criterion* ($u_{\mathrm{max}} = 0.2$ mm and $\sigma_{\mathrm{max}} < R_{\mathrm{p0.2}}$)

the same trend as when using the lightweight index M, see Fig. 3.4. However, the values are at a significantly lower level.

As a further alternative *limit criterion*, it may be required that the *external load (force)* does not exceed a predefined value, i.e. $F_0 \overset{!}{=} F_{\mathrm{max}}$. Based on this definition, the specific energy absorption is obtained as (see Fig. 3.24):

$$SEA = \frac{F_{\mathrm{max}}^2}{2EA^2\varrho} \sim \frac{1}{E\varrho}. \tag{3.40}$$

In summary, it may be noted here that using the specific energy absorption SEA (see Figs. 3.22, 3.23 and 3.24) gives *the same trends* as for the lightweight index M (see Figs. 3.3, 3.4 and 3.5). It is important, however, that so far only linear-elastic material behavior has been considered.

In the following, *plastic material behavior* is considered based on the specific energy absorption[1]. A highly idealized stress-strain diagram is shown in Fig. 3.25. The initial yield stress is equated with the 0.2% yield stress ($R_{\mathrm{p0.2}}$). The corresponding strain is denoted $\varepsilon_{\mathrm{p0.2_t}}$. The elongation at break is ε_{A} and the total elongation at

---

[1] It is noted here again that this is difficult to achieve with the lightweight index M.

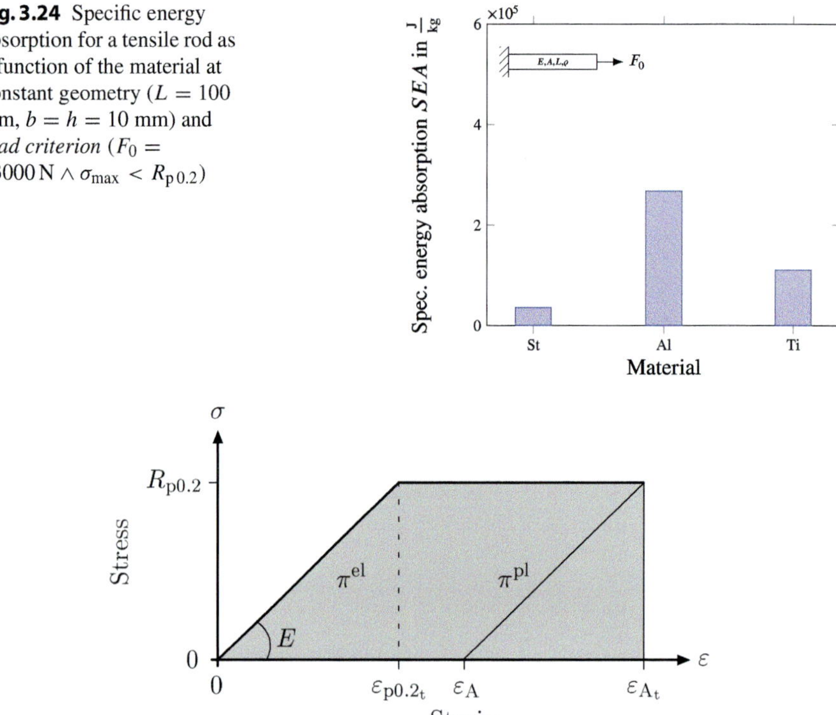

Fig. 3.24 Specific energy absorption for a tensile rod as a function of the material at constant geometry ($L = 100$ mm, $b = h = 10$ mm) and *load criterion* ($F_0 = 33000\,\mathrm{N} \wedge \sigma_{\max} < R_{\mathrm{p}\,0.2}$)

Fig. 3.25 Stress-strain diagram with elasto-plastic range in the case of ideal plasticity

break, i.e. with the elastic part, is denoted by $\varepsilon_{\mathrm{A}_t}$. Furthermore, it is assumed that a state immediately before fracture of the sample is considered. The specific strain energy can now be transformed into a purely elastic part π^{el} with $0 \leq \varepsilon \leq \varepsilon_{\mathrm{p}0.2_t}$ and a plastic part π^{el} with $\varepsilon_{\mathrm{p}0.2_t} \leq \varepsilon \leq \varepsilon_{\mathrm{A}_t}$.

Based on Eq. (3.30) and Fig. 3.25, the total specific strain energy can be written as

$$\pi = \pi^{\mathrm{el}} + \pi^{\mathrm{pl}} = \int_0^{\varepsilon_{\mathrm{p}0.2_t}} \sigma(\varepsilon)\mathrm{d}\varepsilon + R_{\mathrm{p}0.2} \times (\varepsilon_{\mathrm{A}_t} - \varepsilon_{\mathrm{p}0.2_t}), \qquad (3.41)$$

or as absolute strain energy:

$$\Pi = \Pi^{\mathrm{el}} + \Pi^{\mathrm{pl}} = \int_0^{L} \frac{N_x(x)^2}{2EA}\mathrm{d}x + R_{\mathrm{p}0.2} AL \times (\varepsilon_{\mathrm{A}_t} - \varepsilon_{\mathrm{p}0.2_t}). \qquad (3.42)$$

For the tensile bar shown in Fig. 3.1, the following relationships are given, assuming that the external force causes a reaching of the yield stress:

$$F_0 = N_x = R_{\text{p0.2}} A \,, \tag{3.43}$$

$$\varepsilon_{\text{p0.2}_t} = \frac{R_{\text{p0.2}}}{E} \,, \tag{3.44}$$

$$\varepsilon_{\text{A}_t} = \varepsilon_{\text{A}} + \frac{R_{\text{p0.2}}}{E} \,. \tag{3.45}$$

Thus, we can state the absolute strain energy as:

$$\Pi = \frac{R_{\text{p0.2}}^2 A^2 L}{2EA} + R_{\text{p0.2}} A L \times (\varepsilon_{\text{A}_t} - \varepsilon_{\text{p0.2}_t}) = R_{\text{p0.2}} A L \left(\frac{R_{\text{p0.2}}}{2E} + \varepsilon_{\text{A}} \right) , \tag{3.46}$$

or as index number:

$$SEA = \frac{R_{\text{p0.2}}}{\varrho} \left(\frac{R_{\text{p0.2}}}{2E} + \varepsilon_{\text{A}} \right) . \tag{3.47}$$

Based on Eq. (3.47) and the characteristic values according to Table 2.1, the specific energy absorption of stainless steel (St), Al and Ti alloys is shown comparatively in Fig. 3.26. Here, too, it is clear that the highest specific energy absorption is achieved not by the lightest material (Al), but by the Ti alloy. However, there are clearly different trends to those in the purely elastic range (see Fig. 3.3), because here the stainless steel performs much better.

It has already been noted that a pure *torsional load* cannot be considered by the lightweight index M. Therefore, the load case according to Fig. 3.1b is examined by means of the specific energy absorption SEA. The total strain energy for torsional loading can be calculated as follows:

Fig. 3.26 Specific energy absorption for a tension rod as a function of the material at constant geometry ($L = 100$ mm, $b = h = 10$ mm) and elasto-plastic material behavior (ideal plasticity)

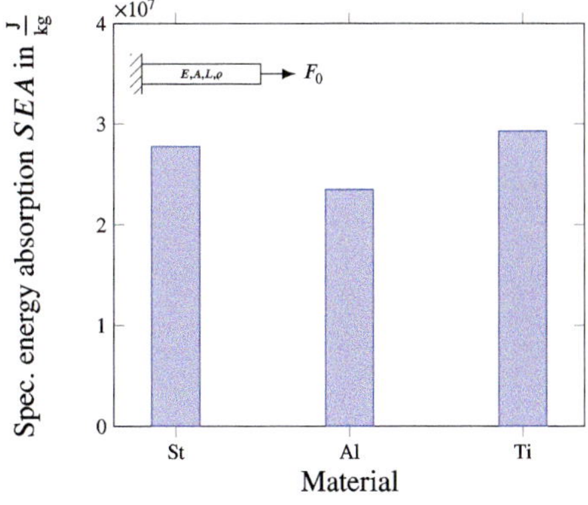

$$\Pi = \int_0^L \frac{M_x(x)^2}{2GI_\mathrm{p}}\, \mathrm{d}x = \frac{M_\mathrm{t}^2}{2GI_\mathrm{p}} \int_0^L \mathrm{d}x = \frac{M_\mathrm{t}^2 L}{2GI_\mathrm{p}}, \tag{3.48}$$

whereby the specific energy absorption can be written as:

$$SEA = \frac{M_\mathrm{t}^2 L}{2GI_\mathrm{p}} \times \frac{1}{AL\varrho} = \frac{64 M_\mathrm{t}^2}{\pi^2 h^6 G\varrho}. \tag{3.49}$$

In the last equation, the polar second moment of area was set to $I_\mathrm{p} = \frac{\pi h^4}{32}$. Assuming again a stress limit (0.2-% strain limit), Eq. (2.8) yields the moment $M_\mathrm{t} = \frac{R_{\mathrm{p}0.2}\pi h^3}{16\sqrt{3}}$. It was assumed that the shear yield stress can be calculated from the tensile yield stress using the equivalent stress hypothesis according to von Mises. Thus, the maximum specific energy absorption results in:

$$SEA = \frac{1}{12} \times \frac{R_{\mathrm{p}0.2}^2}{G\varrho} \quad \sim \quad \frac{R_{\mathrm{p}0.2}^2}{G\varrho}. \tag{3.50}$$

Based on the definition according to Eq. (3.50) and the characteristic values according to Table 2.1, the specific energy absorption of stainless steel (St), Al and Ti alloys is shown comparatively in Fig. 3.27. Here, too, it is clear that the highest energy absorption is achieved not by the lightest material (Al) but by the Ti alloy, which results in the same trend as when using the lightweight index M and the specific energy absorption SEA in the case of a tensile rod, see Fig. 3.7 and 3.22.

The following is the reevaluation of the example of the bending beam according to Fig. 3.1c, taking into account the specific energy absorption (SEA). The internal bending moment distribution is obtained as $M_y(x) = F_0(L - x)$.

Thus, we can express the total strain energy as:

Fig. 3.27 Specific energy absorption for a torsion bar as a function of the material at constant geometry ($L = 100$ mm, $b = h = 10$ mm) and *stress criterion*

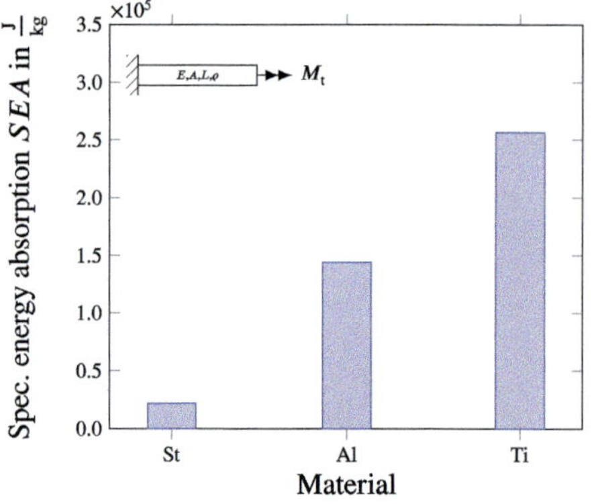

$$\Pi = \int\limits_0^L \frac{M_y(x)^2}{2EI_y}\,\mathrm{d}x = \frac{F_0^2}{2EI_y}\int\limits_0^L (L-x)^2\mathrm{d}x = \frac{F_0^2 L^3}{6EI_y}, \tag{3.51}$$

whereby the specific energy absorption can be written as follows:

$$SEA = \frac{\Pi}{m} = \frac{F_0^2 L^3}{6EI_y \times AL\varrho} = \frac{2F_0^2 L^2}{b^2 h^4 E\varrho}. \tag{3.52}$$

If one first assumes a stress limit value (0.2-% strain limit), the force results from Eq. (3.8) as $F_0 = \frac{bh^2}{6L}R_{p0.2}$. Thus, the maximum specific energy absorption is obtained as:

$$SEA = \frac{2L^2}{b^2 h^4 E\varrho} \times F_0^2 = \frac{R_{p0.2}^2}{18E\varrho} \quad \sim \quad \frac{R_{p0.2}^2}{E\varrho}. \tag{3.53}$$

Based on Eq. (3.53) and the characteristic values according to Table 2.1, the specific energy absorption of stainless steel (St), Al and Ti alloys is shown comparatively in Fig. 3.28. It can be seen that the highest lightweight potential is not achieved by the lightest material (Al), but by the Ti alloy. Thus, the same trends are obtained as in the case of the lightweight index M, see Fig. 3.7.

As an alternative limit criterion, it may be required that the maximum deflection does not exceed a predefined value u_{max}, i.e. $F_0 = \frac{Ebh^3|u_{max}|}{4L^3}$. Thus, the specific energy absorption is obtained as:

$$SEA = \frac{2L^2}{b^2 h^4 E\varrho} \times F_0^2 = \frac{h^2 u_{max}^2 E}{8L^4 \varrho} \quad \sim \quad \frac{E}{\varrho}. \tag{3.54}$$

Based on this definition as a deformation criterion (u_{max}), the specific energy absorptions of stainless steel (St), Al and Ti alloys are shown comparatively in Fig. 3.29.

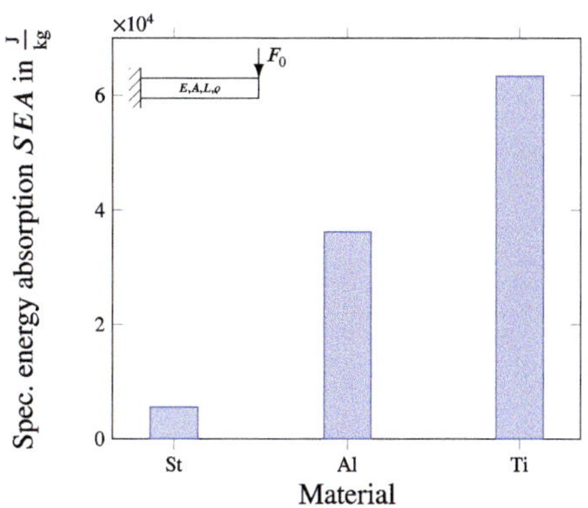

Fig. 3.28 Specific energy absorption for a beam as a function of the material at constant geometry ($L = 100$ mm, $b = h = 10$ mm) and *stress criterion*

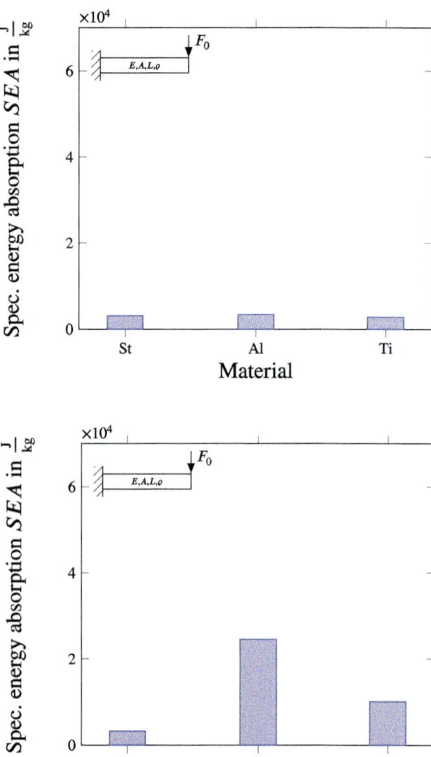

Fig. 3.29 Specific energy absorption for a beam as a function of the material at constant geometry ($L = 100$ mm, $b = h = 10$ mm) and *deformation criterion* ($u_{\max} = -1$ mm and $\sigma_{\max} < R_{\mathrm{p}0.2}$)

Fig. 3.30 Specific energy absorption for a beam as a function of the material at constant geometry ($L = 100$ mm, $b = h = 10$ mm) and *load criterion* ($F_0 = 500$ N $\wedge$ $\sigma_{\max} < R_{\mathrm{p}0.2}$)

For this case, the lightest material (Al) has now the largest lightweight potential, i.e. the same trend as in Fig. 3.9.

As a further alternative limit criterion, it may be required that the external load (force) does not exceed a predefined value, i.e. $F_0 \overset{!}{=} F_{\max}$. With this condition, the definition of the specific energy absorption is given (see also Fig. 3.30):

$$SEA = \frac{2F_{\max}^2 L^2}{E A^2 \varrho h^2} \quad \sim \quad \frac{1}{E\varrho} . \tag{3.55}$$

Finally, we will consider a beam with a combined load state, see Fig. 3.31. Since there is a load due to a single force and a single moment, the lightweight index according to Eq. (3.1) cannot be applied. Therefore, an estimation equation is derived by means of the specific energy absorption.

The bending moment distribution is obtained for this load case as, see Fig. 3.32:

$$M_y(x) = F_0(L - x) + M_0 , \tag{3.56}$$

or the integral over this moment:

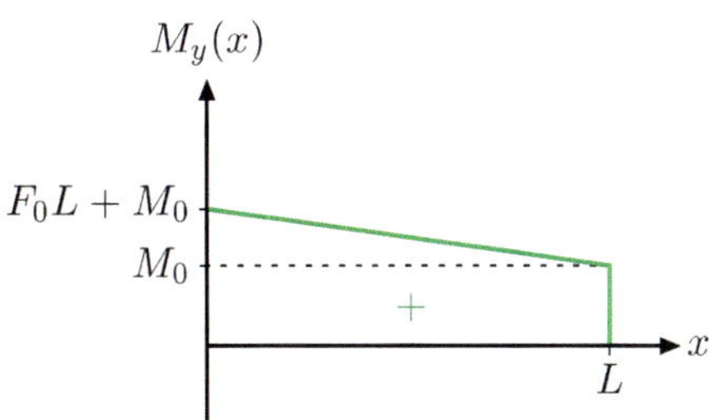

Fig. 3.31 Beam with combined loading

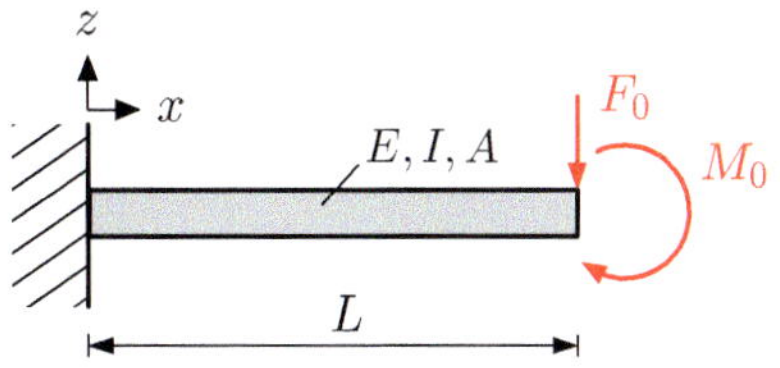

Fig. 3.32 Bending moment distribution for beam under combined loading

$$\int_0^L M_y^2(x)\,\mathrm{d}x = \int_0^L (F_0(L - x) + M_0)^2 \,\mathrm{d}x = \frac{1}{3} F_0^2 L^3 + F_0 M_0 L^2 + M_0^2 L \,. \quad (3.57)$$

Thus, based on Eq. (3.27) and (3.32), the specific energy absorption can be expressed as:

$$SEA = \frac{\Pi}{m} = \frac{\frac{1}{3} F_0^2 L^3 + F_0 M_0 L^2 + M_0^2 L}{2 E I_y A L \varrho} = \frac{1}{2} \times \frac{\frac{1}{3} F_0^2 L^2 + F_0 M_0 L + M_0^2}{E A I_y \varrho} \,.$$

$$(3.58)$$

3.4   Exercises

3.4.1   Knowledge Questions

- Explain the lightweight design concept based on material selection.
- State the definition equation of the lightweight index M and explain the quantities that occur.
- Name two external types of load on one-dimensional structural elements for which the application of the lightweight index M is not possible.
- The larger the lightweight index M, the ( ) larger / ( ) smaller the lightweight potential of a structure.
- State the advantages and disadvantages of the definition equation of the lightweight index M.
- Explain the use of a stress, deformation or load criterion in the context of the lightweight index M.

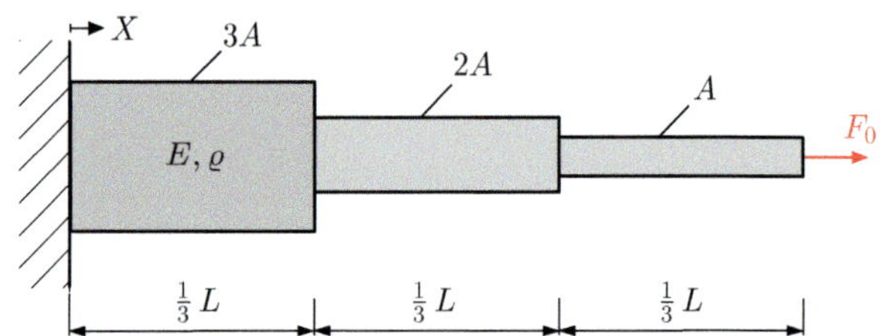

Fig. 3.33 Stepped tensile bar

- State the definition equation of the specific energy absorption SEA and explain the quantities that occur.
- What is the unit of the specific energy absorption SEA?
- How is the total strain energy energy Π calculated for a (a) tension bar, (b) Euler-Bernoulli beam and (c) torsion bar?
- The application of the lightweight index M and the specific energy absorption SEA result in ( ) the same / ( ) different tendencies when assessing the lightweight potential of a structure.
- Which index, i.e. M or SEA, can adequately take plastic material behavior into account when assessing the lightweight potential?

3.4.2   Calculation Problems

3.4.1 Finite element calculation of a stepped tensile bar

Calculate the lightweight index M for the tensile bar shown in Fig. 3.33 using a finite element approach. For this purpose, three linear bar elements of length $\frac{L}{3}$ are to be used. The entire rod is made of a homogeneous material with density ϱ and elastic modulus E.

3.4.2 Calculation of the lightweight index for a cantilever beam with constant distributed load

Calculate for the beam shown in Fig. 3.34 the general lightweight index for stainless steel (St), Al and Ti alloys under consideration of a stress criterion. The material properties can be taken from Table 2.1 as average values. Furthermore, assume for the calculation of the numerical values a constant geometry ($L = 100$ mm, $b = h = 10$ mm).

3.4.3 Calculation of the lightweight index for a simply supported beam with constant distributed load

Calculate for the beam shown in Fig. 3.35 the general lightweight index for stainless steel (St), Al and Ti alloys under consideration of a stress criterion. The material properties can be taken from Table 2.1 as average values. Furthermore, assume for the calculation of the numerical values a constant geometry ($L = 100$ mm, $b = h = 10$ mm).

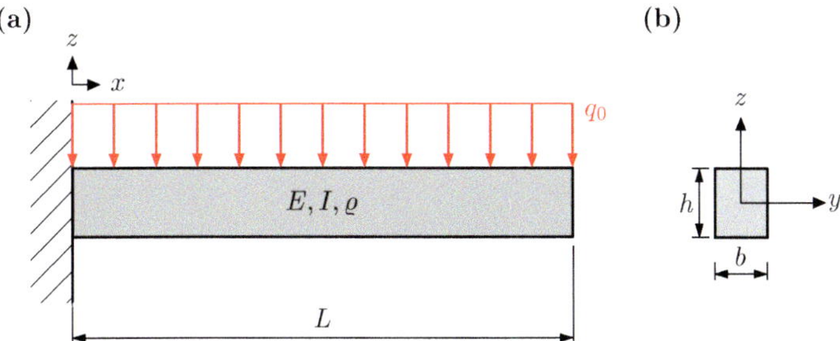

Fig. 3.34 Cantilever beam with constant distributed load: **(a)** general configuration; **(b)** cross section

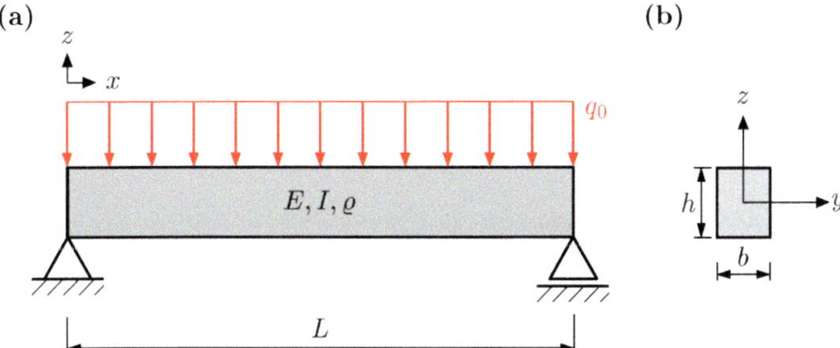

Fig. 3.35 Simply supported beam with constant distributed load: **(a)** general configuration; **(b)** cross section

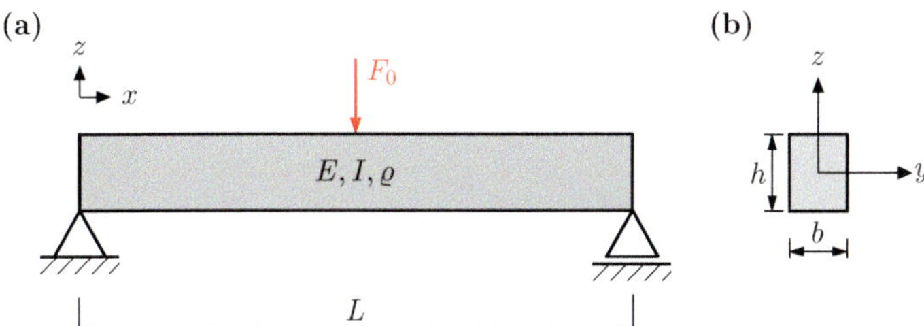

Fig. 3.36 Simply supported beam with point load: **(a)** general configuration; **(b)** cross section

3.4.4 Calculation of the lightweight index for a simply supported beam with point load

Calculate for the beam shown in Fig. 3.36 the general lightweight index for stainless steel (St), Al and Ti alloys under consideration of a stress criterion. The material properties can be taken from Table 2.1 as average values. Furthermore, assume for the calculation of the numerical values a constant geometry ($L = 100$ mm, $b = h = 10$ mm).

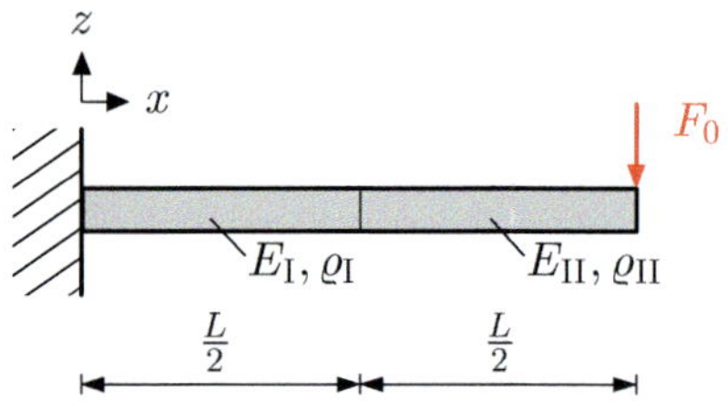

Fig. 3.37 Bimaterial cantilever beam loaded by a single force F_0

3.4.5 Calculation of the specific energy absorption for a cantilever beam with constant distributed load

Calculate for problem 3.4.2 the specific energy absorption.

3.4.6 Calculation of the specific energy absorption for a simply supported beam with constant distributed load

Calculate for problem 3.4.3 the specific energy absorption.

3.4.7 Calculation of the specific energy absorption for a simply supported beam with point load

Calculate for problem 3.4.4 the specific energy absorption.

3.4.8 Calculation of specific energy absorption for a bimaterial cantilever beam

Calculate for the bimaterial cantilever in Fig. 3.37 the specific energy absorption for a given force F_0. The geometric dimensions (I, L) and the material parameters (E_{I}, ϱ_{I} and E_{II}, ϱ_{II}) are assumed to be given. For simplification, a square cross-section (side length: a) can also be assumed.

References

1. Ashby, M.F.: Materials Selection in Mechanical Design. Butterworth-Heinemann, Burlington (2011)
2. Kim, H.-S.: New extruded multi-cell aluminum profile for maximum crash energy absorption and weight efficiency. Thin Wall. Struct. 40(4), 311–327 (2002). https://doi.org/10.1016/S0263-8231(01)00069-6
3. Klein, B.: Leichtbau-Konstruktion: Berechnungsgrundlagen und Gestaltung. Vieweg+Teubner, Wiesbaden (2009)
4. Öchsner, A.: Elasto-Plasticity of Frame Structure Elements: Modeling and Simulation of Rods and Beams. Springer, Berlin (2014)
5. Öchsner, A.: Theorie der Balkenbiegung: Einführung und Modellierung der statischen Verformung und Beanspruchung. Springer Vieweg, Wiesbaden (2016)
6. Öchsner, A.: A Project-Based Introduction to Computational Statics. Springer, Cham (2018)
7. Rezvani, M.J., Jahan, A.: Effect of initiator, design, and material on crashworthiness performance of thin-walled cylindrical tubes: A primary multi-criteria analysis in lightweight design. Thin Wall. Struct. 96, 169–182 (2015). https://doi.org/10.1016/j.tws.2015.07.026

8. Warrior, N.A., Turner, T.A., Robitaille, F., Rudd, C.D.: The effect of interlaminar toughening strategies on the energy absorption of composite tubes. Compos. Part A-Appl. S. 35(4), 431–437 (2004). https://doi.org/10.1016/j.compositesa.2003.11.001

Lightweight Design Concepts: Shape Optimization

4

Abstract

This chapter takes a closer look at the lightweight design in regard to shape optimization. The lightweight potential is mainly determined by adapting or optimizing the cross-section. The material of a component therefore initially remains unchanged. Using the lightweight index and the specific energy absorption, different configurations are assessed with regard to their lightweight potential.

4.1 Evaluation Using the Lightweight Index

The lightweight design concept shape optimization in the context of one-dimensional structural elements is about adapting or optimizing the cross-section to increase the lightweight potential. Thus, it is a purely geometrical problem. Let us consider in the following the example of a cantilever bending beam with a point load at the free end (see Fig. 3.1c) and optimize the structure based on a stress criterion. In general, the maximum stress in the cantilever beam is given by, see [6]:

$$\sigma_{x,\max} = \frac{M_y(x = 0)}{I_y} \times z_{\max},\tag{4.1}$$

where the cross sectional quantities I_y and $z_{\max}$ influence the maximum stress value. By increasing I_y the stress is reduced and thus the lightweight index should increase since the permissible stress is reached later. However, a change in I_y may also influence the maximum distance to the edge fiber ($z_{\max}$). Based on the definition of the axial second moment of area, i.e.

$$I_y = \int_A z^2 \mathrm{d}A,\tag{4.2}$$

it follows that the value is larger the further a surface element ($\mathrm{d}A$) is from the y-axis (which is located through the center of area), expressed by the distance (z).

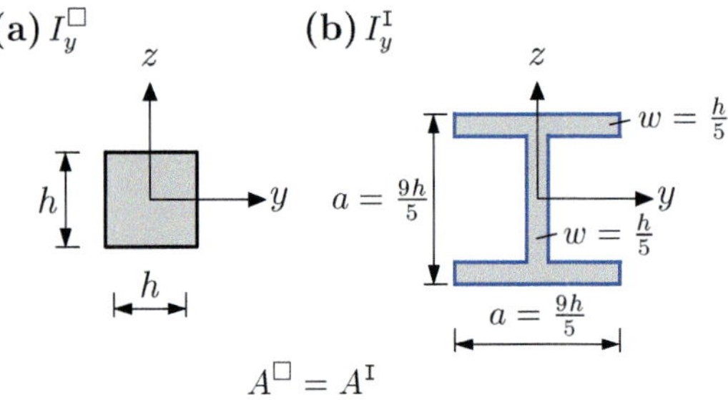

Fig. 4.1 Realization of the concept of lightweight design shapes based on $I_y^{\mathrm{I}} > I_y^{\square}$ and $A^{\square} = A^{\mathrm{I}}$: **a** Original cross section with $b = h$; **b** modified cross section

In the following, the axial second moment of area I_y is modified and optimized with the weight of the beam maintained constant (i.e. with $A^{\square} = A^{\mathrm{I}}$), see Fig. 4.1. The original quadratic cross section ($h \times h$) should be replaced by an I-profile with the same cross-sectional area.

The axial second moment of area (with $A^{\square} = A^{\mathrm{I}}$) can be calculated according to [1] as[1] :

$$I_y^{\square} = \frac{1}{12}hh^3 \,, \tag{4.3}$$

$$I_y^{\mathrm{I}} = \frac{w}{12}(a - 2w)^3 + 2\left(\frac{aw^3}{12} + aw\left(\frac{a}{2} - \frac{w}{2}\right)^2\right) \,. \tag{4.4}$$

These two equations can be further simplified under consideration of $w = \frac{h}{5}$ and $a = \frac{9h}{5}$:

$$I_y^{\square} = C^{\square}h^4 \quad \text{with} \quad C^{\square} = \frac{1}{12} \,, \tag{4.5}$$

$$I_y^{\mathrm{I}} = C^{\mathrm{I}}h^4 \quad \text{with} \quad C^{\mathrm{I}} = \frac{3817}{7500} \,. \tag{4.6}$$

Thus, the lightweight index is obtained with $z_{\max}^{\square} = \frac{h}{2}$ and $z_{\max}^{\mathrm{I}} = \frac{a}{2} = \frac{9h}{10}$ as (see also Fig. 4.2):

$$M^{\square} = 2 \times \frac{R_{\mathrm{p}0.2}}{\varrho g \frac{L^2}{h}} \times C^{\square} \,, \quad M^{\mathrm{I}} = \frac{10}{9} \times \frac{R_{\mathrm{p}0.2}}{\varrho g \frac{L^2}{h}} \times C^{\mathrm{I}} \,. \tag{4.7}$$

Alternatively, $I_y^{\square} \overset{!}{=} I_y^{\mathrm{I}}$ may be required. With $w = \frac{h}{5}$, we obtain $a = 1.055h$ (see Fig. 4.3) and the following mass ratio is finally obtained: $m^{\mathrm{I}} = 0.55m^{\square}$.

The comparison between the two profile sections shown in Figs. 4.1 and 4.3 was based on the assumption that the I-profile has the same height as width (see **b**)

---

[1] The axial second moment of area according to Eq. (4.4) can also alternatively be calculated using the basic shape of a rectangle: $I_y^{\mathrm{I}} = \frac{a^4}{12} - \frac{(a-w)(a-2w)^3}{12}$.

Fig. 4.2 Lightweight index for cantilever beams (Euler-Bernoulli) as a function of different axial second moments of area ($L = 100$ mm, $h = 10$ mm) and stress criterion. Same cross-sectional area ($A^{\square} = A^{\mathrm{I}}$) and thus same mass ($m^{\square} = m^{\mathrm{I}}$) assumed

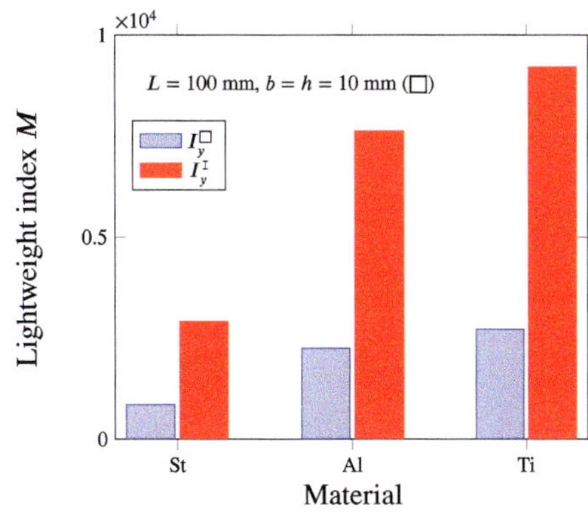

Fig. 4.3 Realization of the concept of lightweight design shapes based on $I^{\square} = I^{\mathrm{I}}$ and $A_y^{\square} > A_y^{\mathrm{I}}$: **a** Original cross section with $b = h$; **b** modified cross section with $a = 1.0554h$ and $w = \frac{h}{5}$

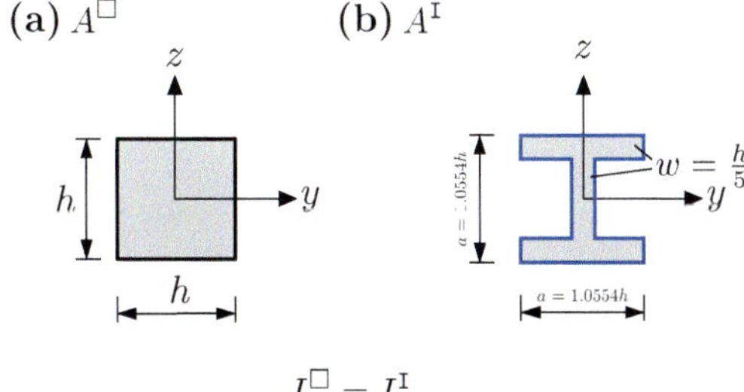

$$I^{\square} = I^{\mathrm{I}}$$

and furthermore the thicknesses of the flanges and the web are identical ($w = \frac{h}{5}$). A further improvement in the lightweight index can be achieved by considering the width-to-height ratio as a design variable (see Fig. 4.4c). However, it should still be assumed that the cross-sectional areas ($A^{\square} = A^{\mathrm{I}}$) and the flanges and web thicknesses ($w = \frac{h}{5}$) remain identical. Figure 4.4 also contains two limit cases, where either the web height becomes zero (see Fig. 4.4a) or the flange width is equated with the web thickness (see Fig. 4.4d).

For equal cross-sectional areas between a square cross section ($h \times h$) and an I-profile in Fig. 4.4c one gets

$$A^{\square} = h^2 = 2bw + w(a - 2w) = A^{\mathrm{I}}, \tag{4.8}$$

or rearranged with $w = \frac{h}{5}$:

$$b = \frac{27}{10}h - \frac{a}{2}. \tag{4.9}$$

For the I-profile, the axial second moment of area is still obtained as

$$I_y^{\mathrm{I}}(a, b) = \frac{w}{12}(a - 2w)^3 + 2\left(\frac{bw^3}{12} + bw\left(\frac{a}{2} - \frac{w}{2}\right)^2\right), \tag{4.10}$$

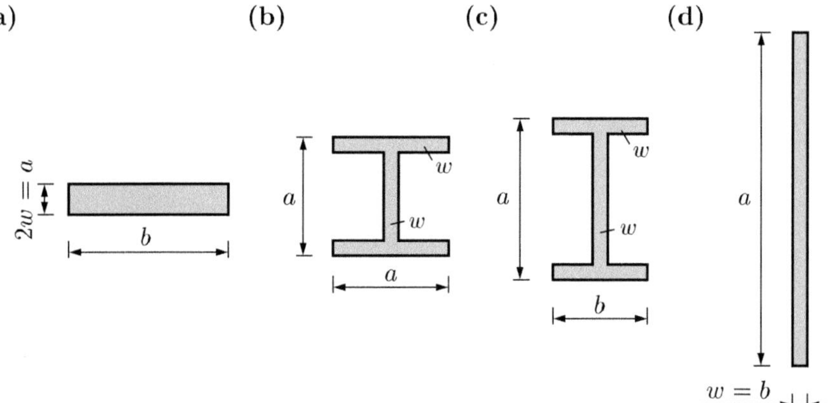

Fig. 4.4 Profiles with identical cross-sectional areas: **a** degenerated profile (web height $\rightarrow$ 0), **b** I-profile with the same width and height, **c** I-profile with different width-to -height ratio, **d** degenerated profile (flange width $b \rightarrow w$)

where by means of the degenerated profiles in Fig. 4.4 the following condition can be derived for the profile height:

$$2w = \frac{2h}{5} \le a \le 5h\,,\tag{4.11}$$

or as a condition for the profile width:

$$\frac{h}{5} \le b \le \frac{5h}{2}\,.\tag{4.12}$$

The last two equations can also be combined for the ratio of the design variables:

$$\underbrace{\frac{4}{25}}_{0.16} \le \frac{a}{b} \le 25\,.\tag{4.13}$$

The axial second moment of area according to Eq. (4.10) is shown in Fig. 4.5. It can be seen that there are large changes for small ratios $\frac{a}{b}$, but the functional values are for larger ratios converging to a constant value.

To evaluate the lightweight index M, it should be considered that the determination equation contains both, i.e. the axial second moment of area (see Fig. 4.5) as wells as the distance to the surface $z_{\max} = \frac{a}{2}$ (see Fig. 4.6) and both parameters change nonlinearly with the $\frac{a}{b}$ ratio:

$$M = \frac{F_0}{F_G} = \frac{R_{\mathrm{p}0.2}}{h^2 L^2 \varrho g} \times \frac{I_y^{\mathrm{I}}}{\frac{a}{2}}\,.\tag{4.14}$$

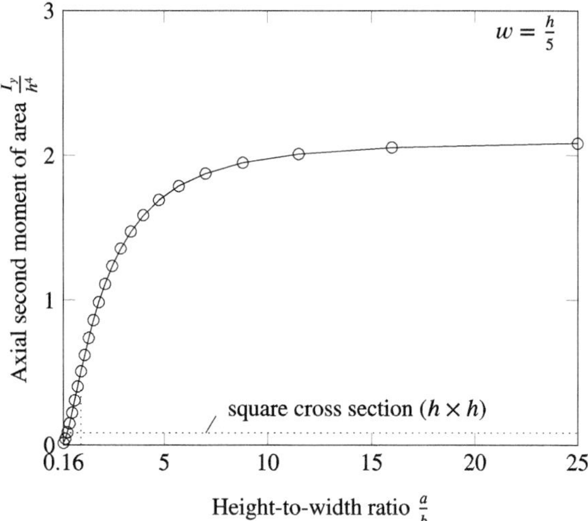

Fig. 4.5 Axial second moment of area for different I-profiles (see Fig. 4.4). Same cross sectional area $(A^{\square} = A^{\mathrm{I}})$ and thus same mass $(m^{\square} = m^{\mathrm{I}})$ assumed

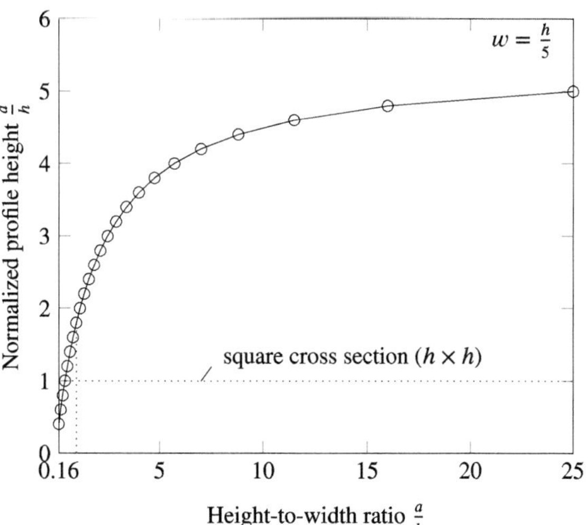

Fig. 4.6 Profile height for different I-profiles (see Fig. 4.4). Same cross sectional area $(A^{\square} = A^{\mathrm{I}})$ and thus same mass $(m^{\square} = m^{\mathrm{I}})$ assumed

The graphical representation of this lightweight index is shown in Fig. 4.7. It can be seen that a maximum value is obtained at $a \approx 5.7b$.

In regards to the diagrams shown in Figs. 4.5 and 4.7, it must be considered that there is no closed form $I_y = I_y(\frac{a}{b})$ and $M = M(\frac{a}{b})$. Thus, a tabular representation might be appropriate to generate pairs of values, see Table 4.1. Introducing in the equation for the second moment according to (4.10) the geometrical condition $w = \frac{h}{5}$ and replacing the width b based on (4.9), we obtain after a short calculation the

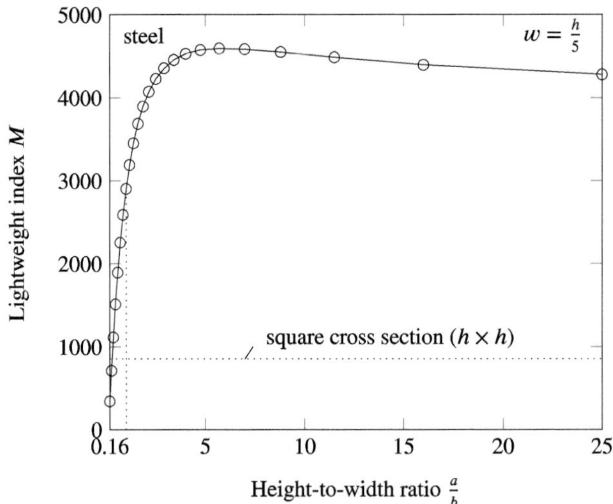

Fig. 4.7 Lightweight index for different I-profiles (see Fig. 4.4) and stress criterion ($\sigma_{max} = R_{p0.2}$). Same cross sectional area ($A^{\square} = A^{\mathrm{I}}$) and thus same mass ($m^{\square} = m^{\mathrm{I}}$) assumed ($L = 100\,\mathrm{mm}$, $h = 10\,\mathrm{mm}$)

following normalized representation:

$$\frac{I_y}{h^4} = \frac{1}{60}\left(\frac{a}{h} - \frac{2}{5}\right)^3 + \left(\frac{1}{750}\left(\frac{27}{10} - \frac{a}{2h}\right) + \frac{2}{5}\left(\frac{27}{10} - \frac{a}{2h}\right)\left(\frac{a}{2h} - \frac{1}{10}\right)^2\right).$$

(4.15)

The last relation is formulated as a function of the ratio $\frac{a}{h}$, which can be represented based on Eq. (4.9) as follows:

$$\frac{a}{h} = \left(\frac{h}{a}\right)^{-1} = \left(\frac{10}{27}\left(\frac{b}{a} + \frac{1}{2}\right)\right)^{-1}.$$

(4.16)

In the relation for the lightweight index according to Eq. (4.14), one should use finally for a graphical representation Eqs. (4.15) and (4.16). The pairs of values for the generation of the diagrams are summarized in Table 4.1.

Finally, the classic design of a profile, i.e. taking only a stress criterion into account, is discussed below. Using the values in Table 4.1, the maximum stress is

$$\sigma_{max} = \frac{M_{max}}{I} \times z_{max} = \frac{M_{max}\left(\frac{a}{h}\right) \times \frac{h}{2}}{\left(\frac{I}{h^4}\right)h^4},$$

(4.17)

or in normalized representation as (see Fig. 4.8):

$$\frac{\sigma_{max}}{\frac{M_{max}}{2h^3}} = \frac{\left(\frac{a}{h}\right)}{\left(\frac{I}{h^4}\right)}.$$

(4.18)

Table 4.1 Pairs of values for the generation of Figs. 4.5–4.7

$\frac{a}{b}$	$\frac{a}{h}$	$\frac{I_y}{h^4}$	M
0.16	0.4	0.013333	342.403095
0.25	0.6	0.041733	714.481124
0.347826	0.8	0.086933	1116.234089
0.454545	1.0	0.147333	1513.421678
0.571429	1.2	0.221333	1894.630457
0.70	1.4	0.307333	2254.968952
0.842105	1.6	0.403733	2591.991427
1.00	1.8	0.508933	2904.339139
1.176471	2.0	0.621333	3191.196842
1.375	2.2	0.739333	3452.045746
$\vdots$	$\vdots$	$\vdots$	$\vdots$
3.40	3.4	1.475333	4457.282639
4.00	3.6	1.587733	4530.373391
4.75	3.8	1.692933	4576.307467
5.714286	4.0	1.789333	4595.049531
7.00	4.2	1.875333	4586.570978
8.80	4.4	1.949333	4550.848404
11.50	4.6	2.009733	4487.862475
16.00	4.8	2.054933	4397.597080
25.00	5.0	2.083333	4280.038684

Here too, an optimal height-to-width ratio of $a \approx 5.7b$ is obtained, i.e. to achieve a minimum stress. Note that the diagram was created under the assumption of a constant mass. For the square reference cross-section ($h \times h$), the normal stress is:

$$\sigma_{\max} = \frac{M_{\max}}{I} \times \frac{h}{2} = \frac{M_{\max}}{\frac{1}{6}h^3} = \frac{M_{\max}}{\frac{1}{12} \times 2h^3}. \tag{4.19}$$

According to Eq. (4.1) it can also be seen that the stress is proportional to the internal bending moment. Thus, it may be useful to increase the axial second moment of area at the location of the largest bending moment (see Fig. 4.9a) and to make a reduction at the point of the minimum moment (see Fig. 4.9b).

According to Fig. 4.10 this results in a considerable enlargement of the lightweight index. However, it should be noted that in this example the cross section at $x = 0$ has been increased to $\frac{7h}{5}$, resulting in a significantly greater force F_0. Therefore, larger values of the lightweight index M also result.

In the previous example, the beam cross-section has been changed so that at the position of the maximum bending moment ($x = 0$) the stress just reached the limit value. However, in the following, we require that the stress at each point x of the beam just reaches the limit. Thus, the beam cross section along the beam axis

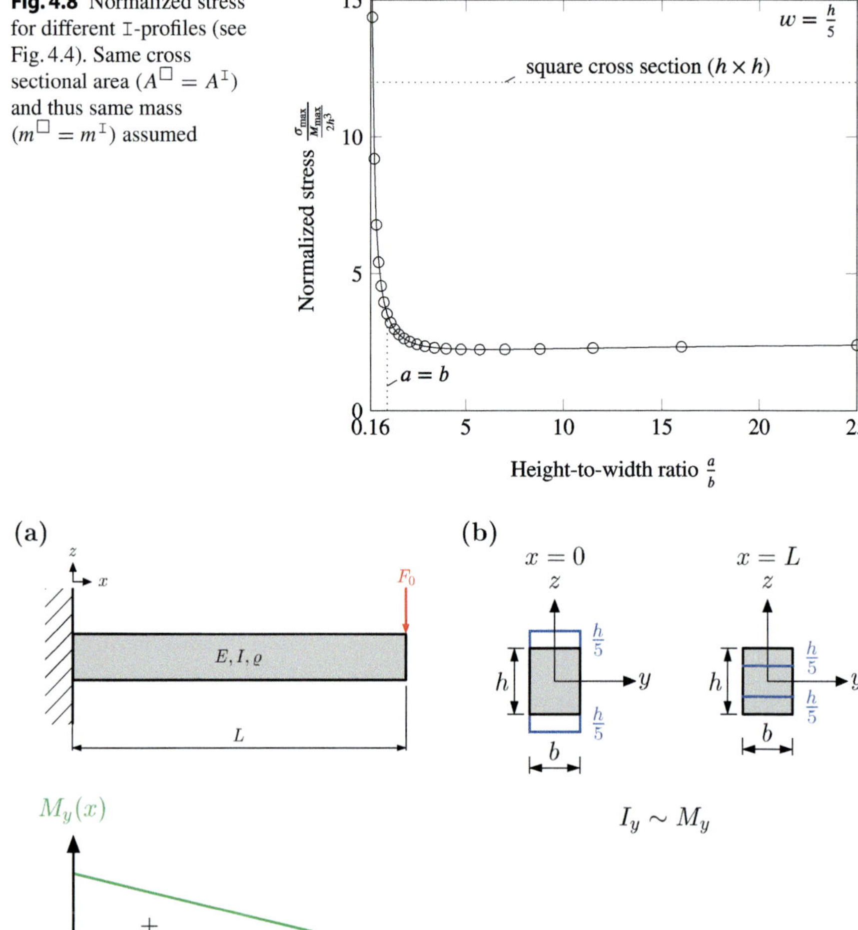

Fig. 4.8 Normalized stress for different I-profiles (see Fig. 4.4). Same cross sectional area ($A^\square = A^{\mathrm{I}}$) and thus same mass ($m^\square = m^{\mathrm{I}}$) assumed

Fig. 4.9 Realization of the concept of lightweight design shapes based on $I_y \sim M_y$: **a** beam configuration and moment distribution; **b** adaptation of cross section

is optimized. For comparison, we use the example with constant rectangular cross section ($b = h = 10\,\mathrm{mm}$) according to Fig. 4.2, i.e., the blue bars for the quadratic cross section. For the sake of simplicity, it is further assumed that the beam width (b) remains unchanged. If the limit condition is again assumed to be the maximum stress, that is,

$$\sigma_{x,\max}(x) = \frac{M_y(x)}{I_y(x)} \times \frac{h(x)}{2} = \frac{F_0 L\left(1-\frac{x}{L}\right)}{I_y(x)} \times \frac{h(x)}{2} = \frac{6F_0 L\left(1-\frac{x}{L}\right)}{bh(x)^2} \overset{!}{=} R_{\mathrm{p0.2}}, \quad (4.20)$$

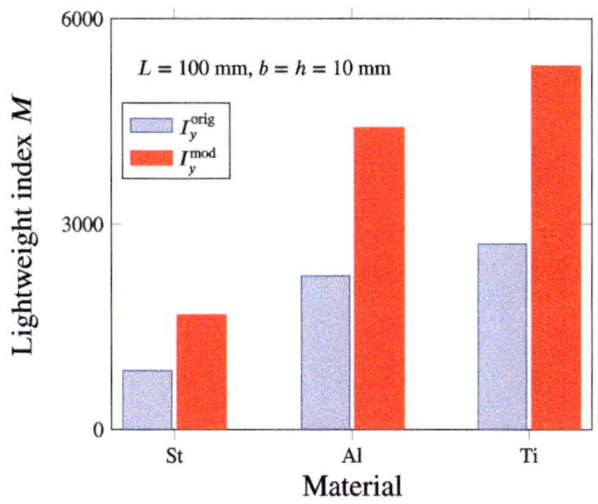

Fig. 4.10 Lightweight index for a cantilever beam made of different materials for $I_y = I_y(x)$ ($L = 100$ mm, $h = 10$ mm) and stress criterion

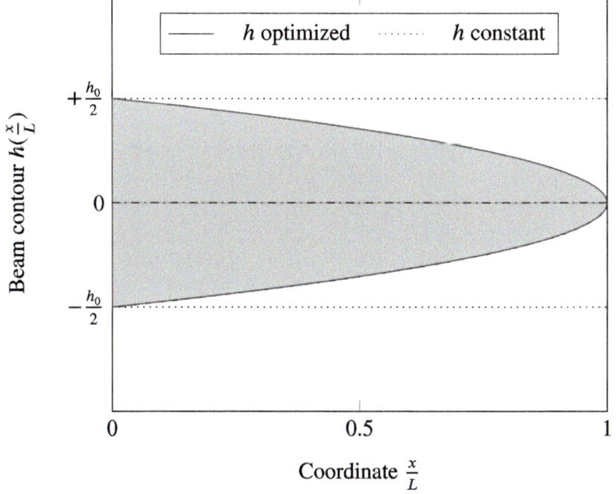

Fig. 4.11 Beam contour along the beam axis

the variable beam height is obtained (see also Fig. 4.11):

$$h(x) = \underbrace{\sqrt{\frac{6F_0 L}{b R_{\text{p0.2}}}}}_{h_0} \times \sqrt{1 - \frac{x}{L}} = h_0 \times \sqrt{1 - \frac{x}{L}}. \qquad (4.21)$$

Figure 4.12 shows that for the optimized beam the stress limit is reached at each point x.

The volume of the optimized beam is given for a small volume element as $\mathrm{d}V = bh(x)\mathrm{d}x$, or for the entire beam:

$$V = \int_{x=0}^{L} bh_0\sqrt{1 - \frac{x}{L}}\,\mathrm{d}x = \frac{2}{3}bh_0 L. \qquad (4.22)$$

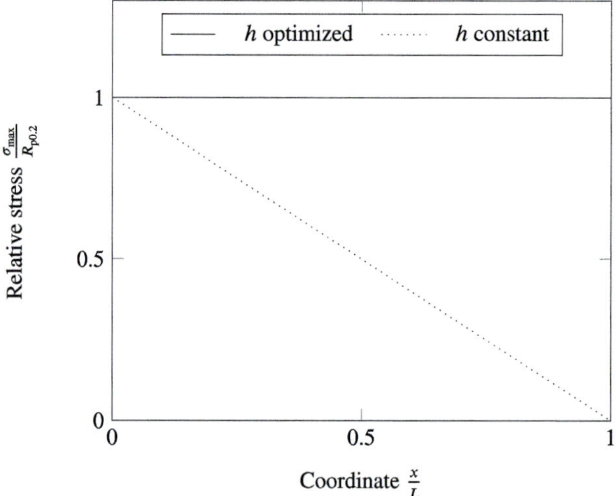

Fig. 4.12 Relative stress along the beam axis

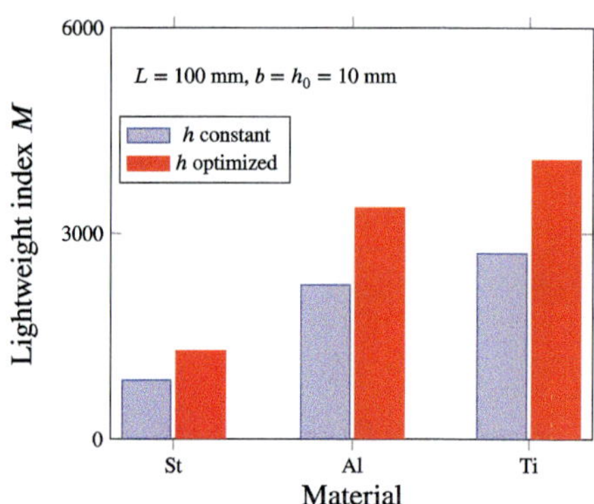

Fig. 4.13 Lightweight index
for cantilever beam made
from different materials for
$h = h(x)$ ($L = 100$ mm,
$h_0 = 10$ mm) and stress
criterion

Thus, the lightweight index can be calculated by means of Eq. (3.1), i.e. $M = \frac{F_0}{F_G}$,
see also Fig. 4.13:

$$M = \frac{F_0}{F_G} = \frac{\frac{R_{p0.2}bh_0^2}{6L}}{\varrho g \frac{2}{3} bh_0 L} = \frac{R_{p0.2}}{4\varrho g \frac{L^2}{h_0}} \,. \tag{4.23}$$

Finally, we will look again at the example in Fig. 3.1c, i.e. a cantilever beam clamped
on one side with a single force at the free end, taking the stress criterion into account.
However, in the following we will consider an I-profile that is to be optimized along
the entire beam axis, see Fig. 4.14. Compare the mass and lightweight index with a
beam of constant cross-section according to Fig. 4.14a.

Fig. 4.14 Beam with changing cross section along the axis: **a** location of support; **b** arbitrary position

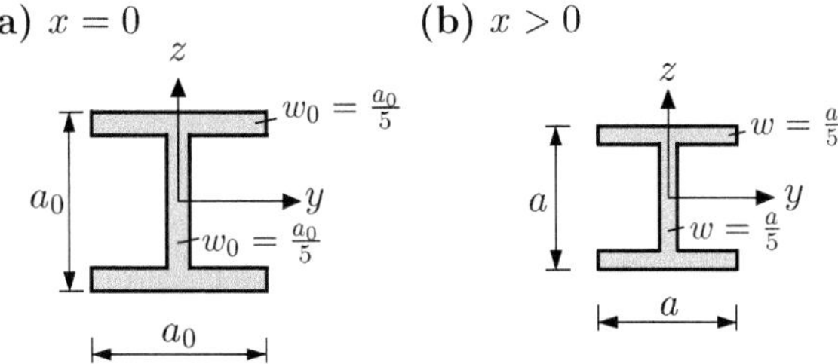

At the fixed support, i.e. at $x = 0$, the axial second moment of area $I_{y,0}^{\mathrm{I}}$ and the cross-sectional area A_0^{I} are:

$$I_{y,0}^{\mathrm{I}} = \frac{a_0^4}{12} - \frac{1}{12}(a_0 - w_0)(a_0 - 2w_0)^3 \overset{w_0 = \frac{a_0}{5}}{=} \frac{517}{7500}a_0^4 = C^{\mathrm{I}}a_0^4 , \tag{4.24}$$

$$A_0^{\mathrm{I}} = a_0^2 - (a_0 - w_0)(a_0 - 2w_0) = \frac{13}{25}a_0^2 . \tag{4.25}$$

At arbitrary location x the same expressions as in Eqs. (4.24)–(4.25) result, but without the index '0'. The beam should be optimized by allowing plastic flow to occur simultaneously at any point of the beam (and not only at the clamping point). The stress criterion can therefore be formulated as follows:

$$\sigma_{x,\mathrm{max}} = \frac{M_y(x)}{I_y^{\mathrm{I}}(x)} \times \frac{a(x)}{2} = \frac{F_0 L \left(1 - \frac{x}{L}\right)}{C^{\mathrm{I}}a(x)^4} \times \frac{a(x)}{2} \overset{!}{=} R_{\mathrm{p0.2}} , \tag{4.26}$$

which results in the function of the side dimension (see Fig. 4.15):

$$a(x) = \sqrt{3}\frac{F_0 L}{2C^{\mathrm{I}}R_{\mathrm{p0.2}}} \times \sqrt[3]{1 - \frac{x}{L}} = a_0 \times \sqrt[3]{1 - \frac{x}{L}} . \tag{4.27}$$

The volume of the optimized beam is given for a small volume element as $\mathrm{d}V = A(x)\mathrm{d}x$, or for the entire beam as:

$$V = \int_{x=0}^{L} \frac{13}{25}a(x)^2\mathrm{d}x = \frac{13}{25}a_0^2 \int_{x=0}^{L} \left(1 - \frac{x}{L}\right)^{\frac{2}{3}} \mathrm{d}x = \frac{39}{125}a_0^2 L . \tag{4.28}$$

Thus, the lightweight index can be calculated according to Eq. (3.1):

$$M^{\mathrm{I}} = \frac{F_0}{F_{\mathrm{G}}} = \frac{\frac{2C^{\mathrm{I}}R_{\mathrm{p0.2}}a_0^3}{L}}{\varrho g \frac{39}{125}a_0^2 L} = \frac{250}{39} \times \frac{C^{\mathrm{I}}R_{\mathrm{p0.2}}}{\varrho g \frac{L^2}{a_0}} . \tag{4.29}$$

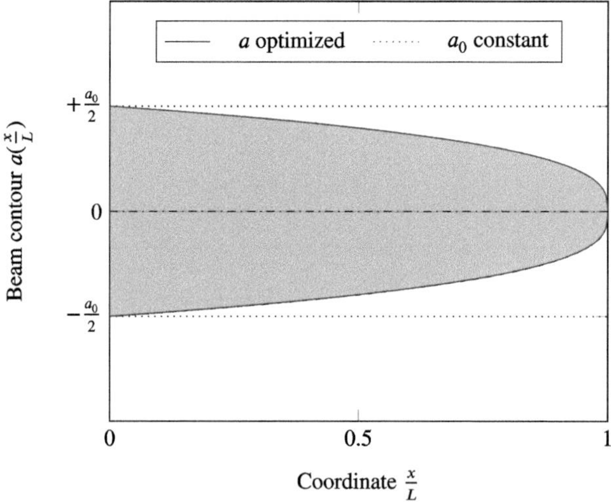

Fig. 4.15 Beam contour along the axis

The lightweight index for a beam with constant cross-section A_0 (see Fig. 4.14a) can also be determined accordingly:

$$M_0^{\text{I}} = \frac{50}{13} \times \frac{C^{\text{I}} R_{\text{p0.2}}}{\varrho g \frac{L^2}{a_0}}.$$

(4.30)

This gives the ratio of the lightweight indices as:

$$\frac{M^{\text{I}}}{M_0^{\text{I}}} = \frac{5}{3} \approx 1.667 .$$

(4.31)

The ratio of the masses is obtained as $\frac{m^{\text{I}}}{m_0^{\text{I}}} = \frac{3}{5} = 0.6$.

A comparison of the beam with the constant I-profile with a beam with a constant square cross-sectional area $(h \times h)$ gives for the same mass, i.e. $(A^{\text{I}} = A^{\square})$, the length ratio $h^2 = \frac{13}{25} a_0^2$. This result gives:

$$A^{\square} = h^2 = \frac{13}{25} a_0^2 ,$$

(4.32)

$$I_y^{\square} = \frac{1}{12} \left(\frac{13}{25}\right)^2 a_0^4 = \frac{169}{7500} a_0^4 .$$

(4.33)

This results in the following ratio of the lightweight indices for a stress criterion:

$$\frac{M_0^{\mathrm{I}}}{M^{\square}} = \frac{\frac{50}{13} \times C^{\mathrm{I}} \times \dfrac{R_{\mathrm{p0.2}}}{\varrho g \frac{L^2}{a_0}}}{\frac{1}{6} \times \sqrt{\frac{13}{25}} \times \dfrac{R_{\mathrm{p0.2}}}{\varrho g \frac{L^2}{a_0}}} \approx 2.206 \,. \tag{4.34}$$

4.2　Evaluation Using the Specific Energy Absorption

Let us reconsider the example with $I_y^{\square}$ and I_y^{I} (with $A^{\square} = A^{\mathrm{I}}$) after Fig. 4.1, i.e. a cantilever beam with a single shear force at the free end, treated under the consideration of a the stress criterion. The definition of specific energy absorption according to Eq. (3.27) requires that the strain energy is determined. The definition based on the internal reaction, see Eq. (3.32), yields the following relationship:

$$\Pi = \int_0^L \frac{M_y(x)^2}{2EI_y}\,\mathrm{d}x = \int_0^L \frac{F_0^2(L-x)^2}{2EI_y}\,\mathrm{d}x = \frac{F_0^2}{2EI_y}\int_0^L \left(L^2 - 2Lx + x^2\right)\mathrm{d}x \tag{4.35}$$
$$= \frac{F_0^2 L^3}{6EI_y}\,.$$

In the case of a stress criterion according to Eq. (4.1), the acting single force can be replaced by:

$$F_0 = \frac{R_{\mathrm{p0.2}} I_y}{L z_{\max}}\,. \tag{4.36}$$

Considering that the mass is given by $m = \varrho V = \varrho AL = \varrho Lh^2$, the specific energy absorption can be calculated as follows:

$$SEA = \frac{F_0^2 L^3}{6EI_y \times \varrho Lh^2} = \frac{R_{\mathrm{p0.2}}^2 I_y}{6E\varrho h^2 z_{\max}^2} \sim \frac{R_{\mathrm{p0.2}}^2}{E\varrho}\,. \tag{4.37}$$

Based on the expressions for the axial second moments of area according to Eqs. (4.5) and (4.6) and the maximum distance to the free surface $z_{\max}^{\square} = \frac{h}{2}$ and $z_{\max}^{\mathrm{I}} = \frac{a}{2} = \frac{9h}{10}$ (see also Fig. 4.2), one obtains:

$$SEA^{\square} = \frac{2}{3} \times \frac{R_{\mathrm{p0.2}}^2}{E\varrho} \times C^{\square}, \quad SEA^{\mathrm{I}} = \frac{50}{243} \times \frac{R_{\mathrm{p0.2}}^2}{E\varrho} \times C^{\mathrm{I}}\,. \tag{4.38}$$

The graphical representation of the specific energy absorption in Fig. 4.16 shows that the same trends are found as in the case of the lightweight index M (see Fig. 4.2). Using the I-profile with the same cross-sectional area results in a significant increase in the lightweight potential, with the highest value again achieved by the Ti alloy.

Fig. 4.16 Specific energy absorption for cantilever beam made of different materials for different second moments of area ($L = 100$ mm, $h = 10$ mm) and stress criterion. Same cross sectional area ($A^{\square} = A^{\mathrm{I}}$) and thus same mass ($m^{\square} = m^{\mathrm{I}}$) assumed

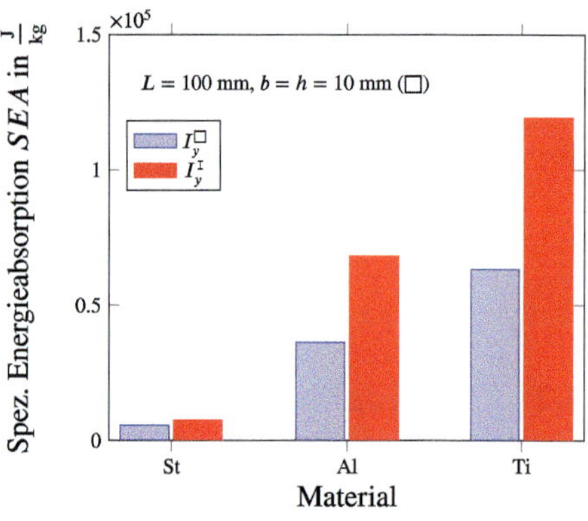

4.3   Consideration as a Classic Optimization Problem

In general, a classical optimization problem can be described (see [10, 11, 14]) by specifying an objective function F, e.g. the weight (mass) of a mechanical structure or the cost of a project, i.e.

$$F(X) = F(X_1, X_2, X_3, \ldots, X_n), \tag{4.39}$$

which is to be minimized under the following constraints

$$
\begin{aligned}
&g_j(X) \le 0 & &j = 1, m & &\text{inequality constraints}, & &(4.40)\\
&h_k(X) = 0 & &k = 1, l & &\text{equality constraints}, & &(4.41)\\
&X_i^{\min} \le X_i \le X_i^{\max} & &i = 1, n & &\text{explicit constraints}, & &(4.42)
\end{aligned}
$$

where the column matrix of n design variables, e.g. the geometric dimensions of the mechanical structure, can be represented as follows:

$$X = \begin{Bmatrix} X_1 \\ X_2 \\ X_3 \\ \vdots \\ X_n \end{Bmatrix}. \tag{4.43}$$

In the case of a one-dimensional objective function, i.e. $X \rightarrow X$, the following simplified set of equations results:

$$
\begin{aligned}
&F(X) & & & &\text{objective function}, & &(4.44)\\
&g_j(X) \le 0 & &j = 1, m & &\text{inequality constraints}, & &(4.45)\\
&h_k(X) = 0 & &k = 1, l & &\text{equality constraints}, & &(4.46)\\
&X^{\min} \le X \le X^{\max} & & & &\text{explicit constraints}. & &(4.47)
\end{aligned}
$$

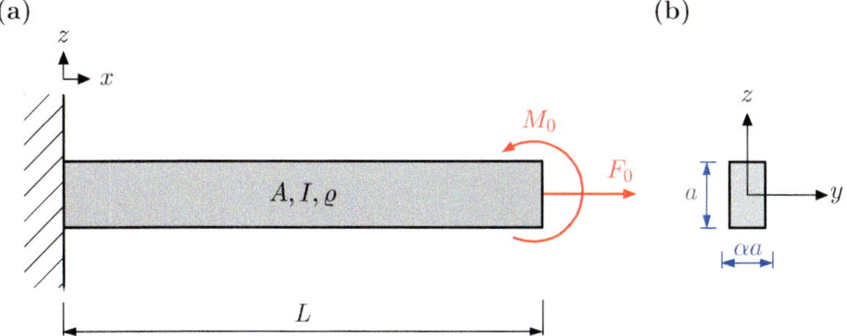

Fig. 4.17 Generalized beam fixed on one side under tension and moment loading: **a** General configuration; **b** beam cross section

In practice, numerical algorithms are used to determine the minimum of the objective function in the n-dimensional design space (see [8, 14]). In this description, however, only one- or two-dimensional design spaces, i.e. with a maximum of two design variables, are considered. In these simpler cases, the minimum can often be determined using analytical or graphical methods. The intention here is also to convey only basic relationships and ideas.

In the following, a simple example[2] with a single design variable (X) is considered to explain the basic approach to the optimization process in the case of *one* design variable. Figure 4.17 shows a generalized beam that is clamped at one side and is subjected to a single force F_0 in tension and a single moment M_0 in bending. The geometric cross-sectional dimensions of the beam, i.e. the width and the height (with a fixed ratio between height and width (α)), are to be optimized for minimal mass. As a restriction (g), it must be noted that no plastic deformations may occur. For this purpose, the beam can be considered thin. The beam length L, the external loads (F_0, M_0), the material properties ϱ and $R_{\mathrm{p0.2}}$, as well as the ratio α are known.

Due to the acting force and the moment, a composed normal stress state is obtained. This results in the condition that no plastic deformations may occur:

$$\sigma_{x,\max} = \frac{M_y}{I_y} \times z_{\max} + \frac{N}{A} \leq R_{\mathrm{p0.2}}\,, \tag{4.48}$$

$$= \frac{-M_0}{\frac{1}{12}\,a(\alpha a)^3} \times \left(-\frac{\alpha a}{2}\right) + \frac{F_0}{a \times \alpha a} \leq R_{\mathrm{p0.2}}\,, \tag{4.49}$$

$$\rightarrow \quad 6M_0 + \alpha a\,F_0 - \alpha^2 a^3 R_{\mathrm{p0.2}} \leq 0\,. \tag{4.50}$$

The mass of the beam, i.e. the objective function of the optimization problem, can be easily expressed using the following relation:

$$F = m = \varrho A L = \varrho L a(\alpha a)\,. \tag{4.51}$$

---

[2] The idea for this problem is taken from [12].

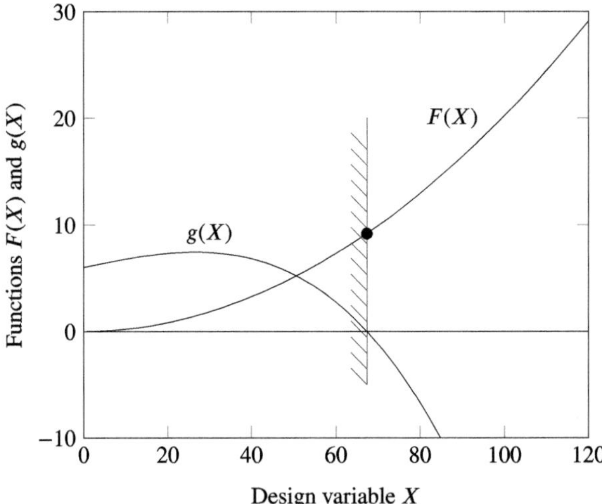

Fig. 4.18 Optimal design of a generalized beam fixed on one side under tension and moment loading: $L = 2500\,\text{mm}$, $F_0 = 90\,\text{kN}$, $M_0 = 1130\,\text{kNmm}$, $R_{\text{p0.2}} = 140\,\text{MPa}$, $\varrho = 2.7 \times 10^{-6}\,\text{kg/mm}^3$, $\alpha = 0.3$ (original set of Eqs. (4.52)–(4.53)). Exact solution for the minimum: $X_{\text{extr}} = 67.372\,\text{mm}$ (indicated by the • marker)

If one sets $X = a$ to generalize and normalizes the stress condition with M_0 for better graphical representation, the following set of equations results to describe the classical optimization problem:

$$F(X) = \varrho L \alpha X, \tag{4.52}$$

$$g(X) = 6 + \frac{F_0}{M_0} \times X - \frac{\alpha}{M_0 R_{\text{p0.2}}} \times X^3 \le 0. \tag{4.53}$$

From the graphical representation of the Eqs. (4.52)–(4.53) in Fig. 4.18 one can see that the permissible region must be to the right of the root of the inequality restriction. Thus, the optimal design for minimal mass is calculated using the following condition:

$$g(X_{\text{extr}}) \overset{!}{=} 0. \tag{4.54}$$

In the following, a simple example[3] with two design variables (X_1, X_2) is considered to explain the basic approach to the optimization process in the case of *two* design variables. Figure 4.19 shows a thin tube with a circular cross-section that is simply supported at both sides and is subjected to a compressive load by means of the force F_0. The geometric dimensions of the thin cross-section, i.e. the mean diameter d_{m} and the wall thickness s, are to be optimized for minimal mass. The restrictions (g_i) to be considered are that no global or local instabilities may occur, see Table 4.2.

---

[3] The idea for this problem is taken from [10].

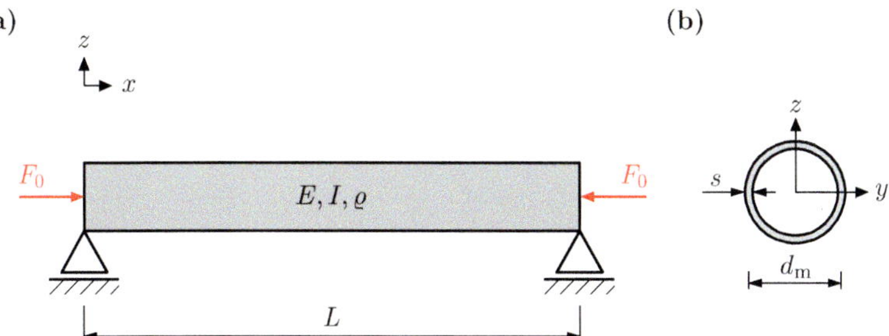

Fig. 4.19 Simply supported tube under compressive load: **a** General configuration; **b** tube cross section

Table 4.2 Instabilities of simply supported thin-walled tubes with a circular cross-section under axial load

	Schematic Representation	Description
(a)		Global instability: buckling
(b)		Local instability: wrinkling

The tube length L, the external load F_0 and the material properties E, v and ϱ can be considered as known.

The global instability failure due to buckling can be described for purely elastic material behavior using the buckling formula according to Euler (see Appendix A.1.3) as follows:

$$F_{cr} = \frac{\pi^2 E I}{L^2}. \tag{4.55}$$

To prevent buckling from occurring, the loading force F_0 must always be smaller than the buckling force F_{cr}.

The theoretical wrinkling stress (σ_{cr}) for a thin-walled tube with circular cross-section under axial pressure is for a simple support on both sides, see [1,3,13]:

$$\sigma_{cr} = \underbrace{\frac{1}{\sqrt{3(1 - v^2)}}}_{K} \times \frac{E s}{r_m}, \tag{4.56}$$

where the wrinkling coefficient for metallic materials with $v = 0.3$ is $K = 0.605$. Since the task involves a thin tube, the axial second moment of area I and the cross-sectional area A can be simplified according to problem 2.4.5 as follows:

$$I \approx \pi r_{\mathrm{m}}^3 s = \frac{\pi d_{\mathrm{m}}^3 s}{8}, \tag{4.57}$$

$$A \approx 2\pi r_{\mathrm{m}} s = \pi d_{\mathrm{m}} s. \tag{4.58}$$

With these approximations and the critical loads according to Eqs. (4.55) and (4.56) the two inequality constraints result in:

$$g_1 = F_0 - \frac{\pi^2 E I}{L^2} \leq 0, \tag{4.59}$$

$$= F_0 - \frac{\pi^3 E d_{\mathrm{m}}^3 s}{8 L^2} \leq 0, \tag{4.60}$$

or in relation to local wrinkling:

$$g_2 = \frac{F_0}{A} - 0.605 \times \frac{E s}{\frac{d_{\mathrm{m}}}{2}} \leq 0, \tag{4.61}$$

$$= F_0 - 1.21 \times E \pi s^2 \leq 0. \tag{4.62}$$

The objective function, i.e. the mass, can be simply expressed using the following relationship:

$$F = m = \varrho A L = \varrho L \pi d_{\mathrm{m}} s. \tag{4.63}$$

If we set $X_1 = d_{\mathrm{m}}$ and $X_2 = s$ to generalize, we get the following set of equations to describe a classical optimization problem:

$$F(X_1, X_2) = \varrho L \pi X_1 X_2, \tag{4.64}$$

$$g_1(X_1, X_2) = F_0 - \frac{\pi^3 E X_1^3 X_2}{8 L^2} \leq 0, \tag{4.65}$$

$$g_2(X_1, X_2) = F_0 - 1.21 \times E \pi X_2^2 \leq 0. \tag{4.66}$$

For graphical representation, it is advisable to formulate the three equations in the form $f(x)$ or in the current notation as $X_2(X_1)$:

$$X_2 = \frac{\text{const.}}{\varrho L \pi X_1}, \tag{4.67}$$

$$X_2 \geq \frac{8 F_0 L^2}{\pi^3 E X_1^3}, \tag{4.68}$$

$$X_2 \geq \sqrt{\frac{F_0}{1.21 E \pi}}. \tag{4.69}$$

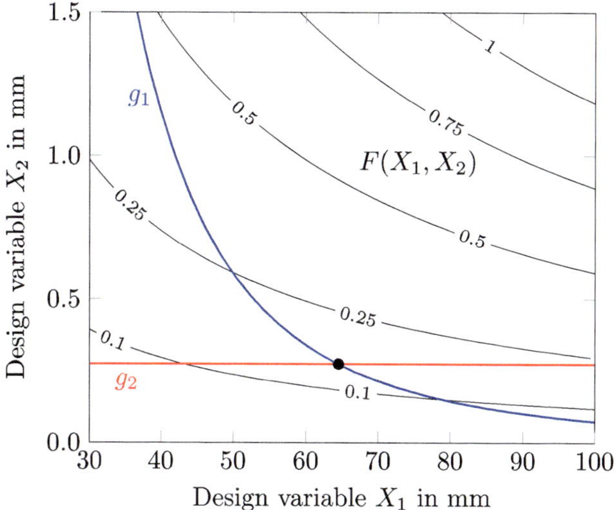

Design variable X_2 in mm (vertical axis)

Design variable X_1 in mm (horizontal axis)

Fig. 4.20 Optimal design of a tube with a circular cross section: $L = 1000$ mm, $F_0 = 20{,}000$ N, $E = 70{,}000$ MPa, $R_{p0.2} = 247$ MPa, $\varrho = 2.691 \times 10^{-6}$ kg/mm^3, (original set of Eqs. (4.64)–(4.66)). Exact solution for the minimum: $X_{1,\text{extr}} = 64.544$ mm, $X_{2,\text{extr}} = 0.274$ mm (indicated by the ● marker)

From the graphical representation of the Eqs. (4.64)–(4.66) in Fig. 4.20 one can see that the permissible range is above the red g_2 line and to the right of the blue g_1 curve. Thus, the optimal design for minimal mass is calculated as the intersection of the two inequality constraints:

$$X_{1,\text{extr}} = \left(\frac{8 F_0 L^2}{\pi^3 E X_2} \right)^{\frac{1}{3}}, \tag{4.70}$$

$$X_{2,\text{extr}} = \sqrt{\frac{F_0}{1.21 E \pi}}. \tag{4.71}$$

4.4 Exercises

4.4.1 Knowledge Questions

- Explain the lightweight design concept based on shape optimization.
- Name typical (technical) cross-sectional shapes to increase the bending stiffness of a beam.
- What equations make up the typical set of equations in an optimization problem?
- Name typical constraints that can occur in a classical optimization problem.
- Name some typical examples of the objective function in an optimization problem.

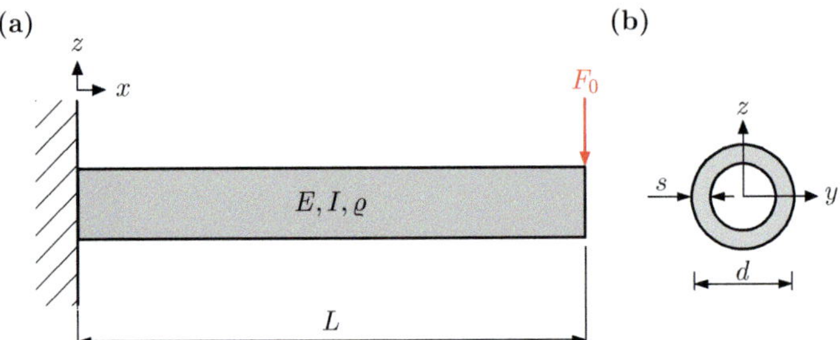

Fig. 4.21 Cantilever beam with point load: **a** General configuration; **b** beam cross section

- Name some typical examples of the design variable in a classical optimization problem.
- In the classical lectures on strength of materials, the cross-section of a bending beam is usually optimized at the point of maximum bending moment (stress). What other approaches are possible?
- What problems can occur if the beam cross-section is optimized proportionally to the bending moment?
- What is meant by an explicit constraint in the context of an optimization problem?
- What types of instabilities can be distinguished in a compression member?

4.4.2　Calculation Problems

4.4.1　Analysis of the lightweight index for a cantilever beam with end force and tubular cross section

For the cantilever beam with end force shown in Fig. 4.21, the lightweight index M is to be analyzed based on a stress criterion. To do this, first derive the general equation of the lightweight index as a function of d and s, i.e. $M = M(d, s)$. Then, the functions $F_0(d, s)$, $F_G(d, s)$ and $M(d, s) = F_0(d, s)/F_G(d, s)$ are to be sketched as a function of (a) s at $d = 10\,\mathrm{mm} = \mathrm{const.}$ and of (b) d at $s = 2\,\mathrm{mm} = \mathrm{const.}$

4.4.2　Implementation of the concept of lightweight design shapes: Optimization of a box profile

For a cantilever beam ($L = 100\,\mathrm{mm}$) with end force F_0, the cross section is to be optimized based on a stress criterion using the lightweight index. The original square cross-section ($h = 10\,\mathrm{mm}$) is to be modified and optimized by a box profile while maintaining the beam weight (i.e. with $A^{\boxtimes} = A^{\square}$), see Fig. 4.22.

The optimization is to be carried out for stainless steel with the material properties according to Table 2.1.

4.4.3　Optimization of a beam in bending along its axis

For a cantilevered beam in bending (bending stiffness $E \times I(x)$, length L) with a single force F_0 at the free end, optimize the tubular cross section along the beam axis x taking into account a stress criterion (0.2-% strain limit $R_{\mathrm{p}0.2}$). The cross-sectional

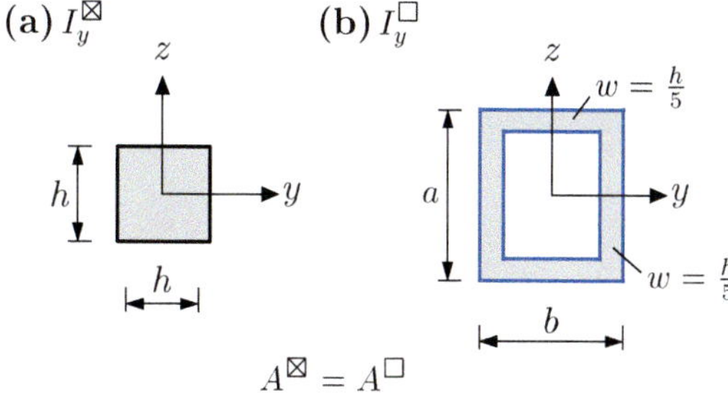

Fig. 4.22 Realization of the concept of lightweight design by shape optimization based on $I_y^{\square} > I_y^{\boxtimes}$ and $A^{\boxtimes} = A^{\square}$: **a** Original cross section with $b = h$; **b** modified cross section

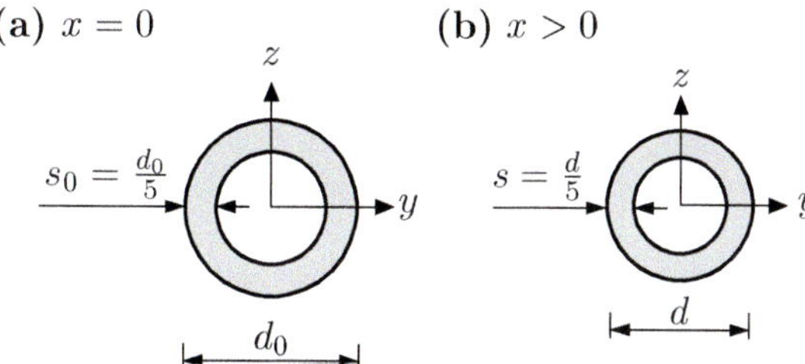

Fig. 4.23 Beam with changing cross section along the axis: **a** location of support; **b** arbitrary position

dimensions at the clamping point $x = 0$ (see Fig. 4.23a) are given by the outside diameter d_0 and the wall thickness $s_0 = \frac{d_0}{5}$. Determine:

a) the beam contour $d = d(x)$ along the beam axis (including sketch),
b) the ratio of the masses of the optimized cross-section with a beam of constant cross-section according to Fig. 4.23a,
c) the ratio of the lightweight index of the optimized cross section with a beam of constant cross-section according to Fig. 4.23a,
d) the ratio of the specific energy absorption of the optimized cross section with a beam of constant cross section according to Fig. 4.23a.

4.4.4 Optimization of a cantilevered beam loaded with two single forces of the same magnitude

For the cantilevered bending beam shown in Fig. 4.24 (bending stiffness $E \times I(x)$, length L) with two single forces F_0, optimize the tubular cross section along the beam axis x taking into account a stress criterion (0.2-% strain limit $R_{\mathrm{p}0.2}$). The cross-sectional dimensions at the clamping point $x = 0$ (see Fig. 4.23a) are given by the outside diameter d_0 and the wall thickness $s_0 = \frac{d_0}{5}$. The cross-sectional ratios at any location x are shown in Fig. 4.23b. Determine:

a) the beam contour $d = d(x)$ along the beam axis (including sketch),
b) the specific energy absorption SEA of the optimized cross section,
c) the problem in considering the lightweight index M in this case.

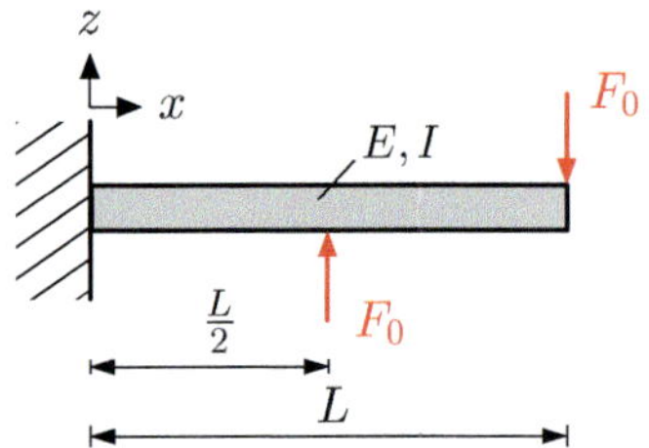

Fig. 4.24 Beam loaded with two single forces (same magnitude)

4.4.5 Optimization of a beam under 3-point bending: rectangular cross section

For a beam under symmetrical 3-point bending (bending stiffness $E \times I(x)$, length L, see Fig. 4.25a) which is loaded by a central single force F_0, optimize the rectangular cross section along the beam axis x taking into account a stress criterion (0.2-% strain limit $R_{p0.2}$), see Fig. 4.25b.

The cross-sectional dimensions at $x = L/2$ are given by the constant width b and the height h_0, and the cross-sectional ratios at any point $x \neq L/2$ are given by the height $h(x)$ to be optimized and the constant width b, see Fig. 4.25b.

Determine:

a) the beam contour $h = h(x)$ along the beam axis (including both courses of the functions and a basic sketch),
b) the lightweight index M of the optimized cross section. Compare this value with the lightweight index for a beam with a constant height h_0,
c) the specific energy absorption SEA of the optimized cross section. Compare this value with the specific energy absorption for a beam with a constant height h_0.

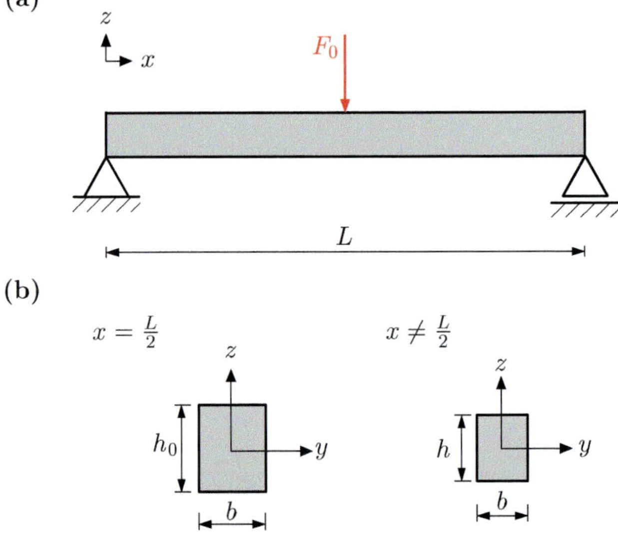

Fig. 4.25 Rectangular beam under 3-point bending: **a** General configuration and **b** cross sections

4.4.6 Optimization of a beam under 3-point bending: tubular cross section

For a beam under 3-point bending (bending stiffness $E \times I(x)$, length L, see Fig. 4.26a) which is loaded by a central single force F_0, optimize the tubular cross section along the beam axis x taking into account a stress criterion (0.2-% strain limit $R_{\mathrm{p0.2}}$), see Fig. 4.26b.

The cross-sectional dimensions in the middle at $x = L/2$ are given by the outer diameter d_0 and the wall thickness $s_0 = d_0/5$, and the cross-sectional ratios at any point $x \neq L/2$ are modified in the same ratio, see Fig. 4.26b. Determine:

a) the beam contour $d = d(x)$ along the beam axis (including sketch),
b) the specific energy absorption SEA of the optimized cross section. Compare this value with the specific energy absorption for a beam with a constant diameter d_0,
c) the lightweight index M of the optimized cross section. Compare this value with the lightweight index for a beam with a constant diameter d_0,
d) the problem of the technical realization of the contour $d = d(x)$.

4.4.7 Optimization of a beam under asymmetric 3-point bending

For a beam subjected to asymmetric 3-point bending (bending stiffness $E \times I(x)$, length L, see Fig. 4.27a) which is loaded by a single force F_0 at $x = b$, optimize the rectangular cross section along the beam axis x taking into account a stress criterion (0.2-% strain limit $R_{\mathrm{p0.2}}$), see Fig. 4.27b.

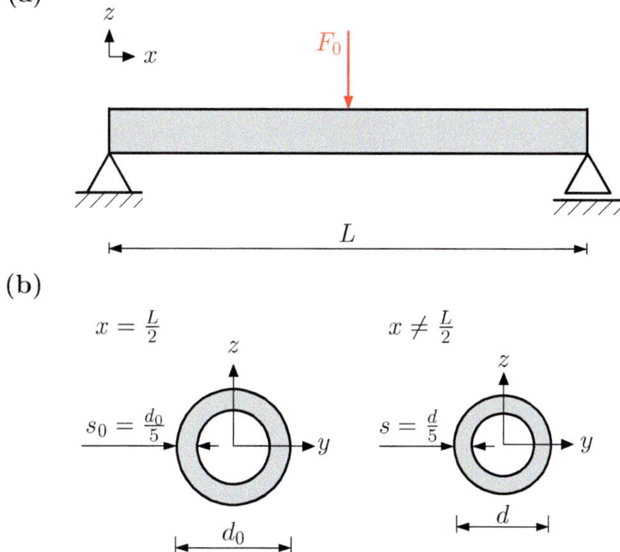

Fig. 4.26 Tubular beam under 3-point bending: **a** General configuration and **b** cross sections

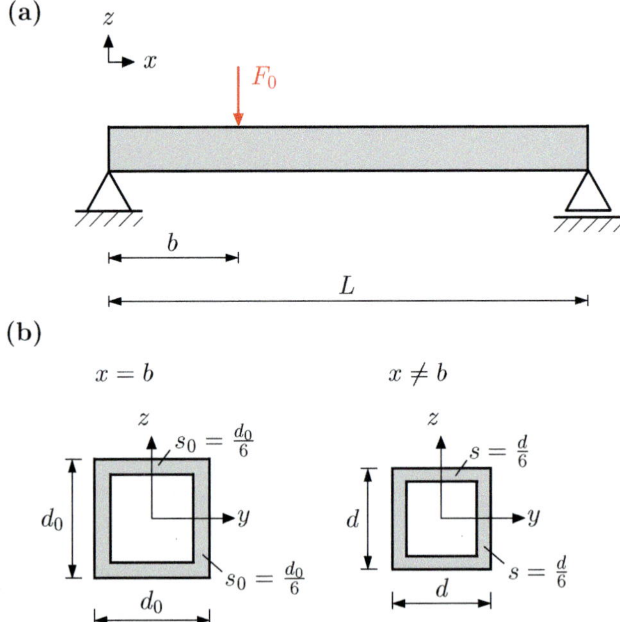

Fig. 4.27 Beam under asymmetrical 3-point bending: **a** General configuration and **b** cross sections

The cross-sectional dimensions at $x = b$ are given by the outer diameter d_0 and the wall thickness $s_0 = d_0/6$, and the cross-sectional ratios at any point $x \neq b$ are modified in the same ratio, see Fig. 4.27b. Determine:

a)   the beam contour $d = d(x)$ along the beam axis (including sketch),
b)   the lightweight index M of the optimized cross section. Compare this value with the lightweight index for a beam with a constant diameter d_0,
c)   the specific energy absorption SEA of the optimized cross section. Compare this value with the specific energy absorption for a beam with a constant diameter d_0,
d)   the problem of the technical realization of the contour $d = d(x)$.

Special values must be specified for $b = \frac{1}{4}L$, $\frac{1}{3}L$ and $\frac{1}{2}L$.

4.4.8 Optimization of a simply supported beam under constant distributed load: rectangular cross section

For a simply supported beam (bending stiffness $E \times I(x)$, length L, see Fig. 4.28a) which is loaded by a constant distributed load q_0, optimize the rectangular cross section along the beam axis x taking into account a stress criterion (0.2-% strain limit $R_{p0.2}$), see Fig. 4.28b.

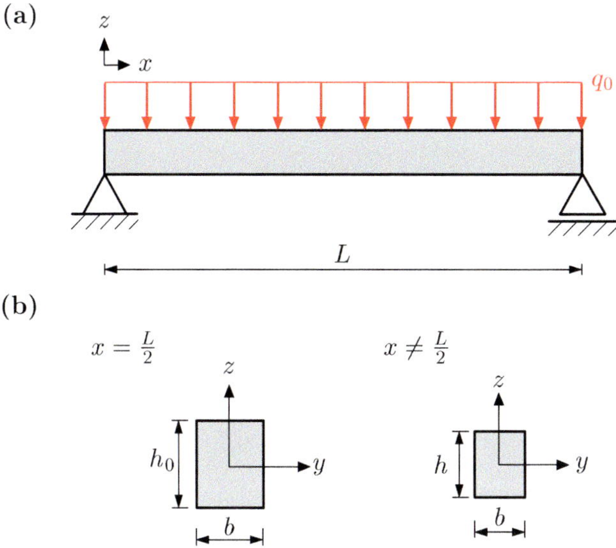

Fig. 4.28 Optimization of a simply supported beam under constant distributed load: **a** General configuration and **b** rectangular cross sections

The cross-sectional dimensions in the middle at $x = L/2$ are given by the height h_0 and the constant width b, and the cross-sectional ratios at any point $x \neq L/2$ are given by the height $h(x)$ and the constant width b, see Fig. 4.28b. Determine:

a) the optimized beam contour $h = h(x)$ along the beam axis x (including sketch),
b) the specific energy absorption SEA of the optimized cross section ($h(x) \times b$). Compare this value with the specific energy absorption for a beam with a constant cross section ($h_0 \times b$),
c) the problem of the technical realization of the contour $h = h(x)$.

Consider:

$$\int_0^L \left(-\left[\tfrac{x}{L}\right]^2 + \left[\tfrac{x}{L}\right]^1 \right)^{\frac{1}{2}} \mathrm{d}x = \frac{\pi L}{8} \quad \text{oder} \quad \int_0^L \left(-x^2 + Lx^1 \right)^{\frac{1}{2}} \mathrm{d}x = \frac{\pi L^2}{8}.$$

4.4.9 Optimization of a simply supported beam under constant distributed load: tubular cross section (numerical integration)

For a simply supported beam (bending stiffness $E \times I(x)$, length L, see Fig. 4.29a) which is loaded by a constant distributed load q_0, optimize the tubular cross section along the beam axis x taking into account a stress criterion (0.2-% strain limit $R_{\mathrm{p}0.2}$), see Fig. 4.29b.

The cross-sectional dimensions in the middle at $x = L/2$ are given by the outer diameter d_0 and the wall thickness $s_0 = d_0/6$, and the cross-sectional ratios at any point $x \neq L/2$ are modified in the same ratio, see Fig. 4.29b. Given are:

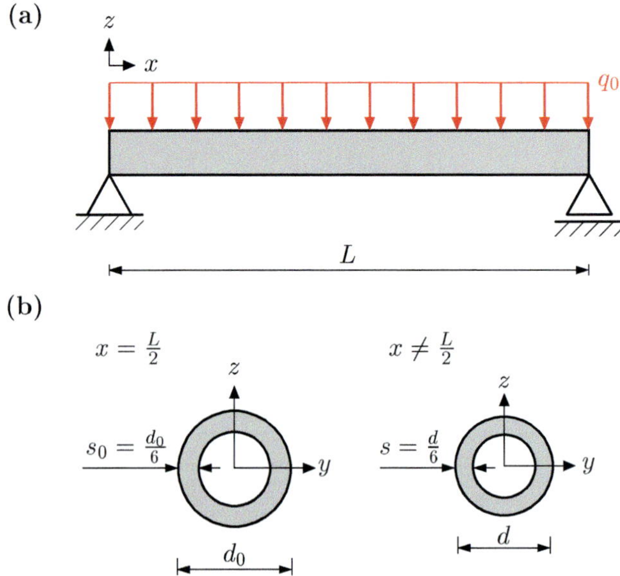

Fig. 4.29 Optimization of a simply supported beam under constant distributed load: **a** General configuration and **b** tubular cross sections

- Geometric dimension: $L = 2540\,\text{mm}$.
- Material properties of the beam: $E = 68{,}948\,\text{MPa}$, $\varrho = 2691\,\text{kg/m}^3$, $R_{\text{p0.2}} = 247\,\text{MPa}$.
- Load: $q_0 = 1.05\,\text{N/mm}$.

Determine:

a) the beam contour $d = d(x)$ along the beam axis (including sketch),
b) the specific energy absorption SEA of the optimized cross section. Compare this value with the specific energy absorption for a beam with a constant diameter d_0,
c) the problem of the technical realization of the contour $d = d(x)$.

4.4.10 Optimization of a beam with a square box profile along the longitudinal axis under the influence of a constant distributed load

For a simply supported beam (bending stiffness $E \times I(x)$, length L, see Fig. 4.30a) which is loaded by a constant distributed load q_0, optimize the rectangular cross section along the beam axis x taking into account a stress criterion (0.2-% strain limit $R_{\text{p0.2}}$), see Fig. 4.30b. The cross-sectional dimensions in the middle at $x = L/2$ are given by the height d_0, the width d_0 and the constant thickness $s_0 = d_0/6$, and the cross-sectional ratios at any point $x \neq L/2$ are given by the height $d(x)$, the width $d(x)$ and the thickness $s(x) = d(x)/6$, see Fig. 4.30b. *Consider the comments on integration at the end of the problem!*

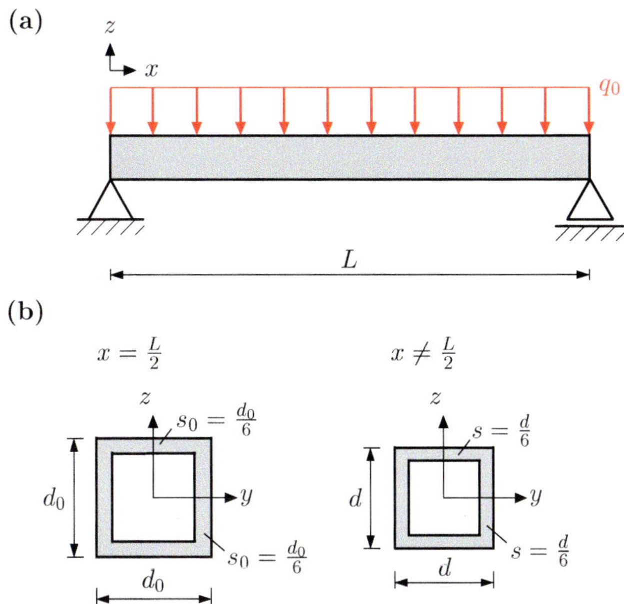

Fig. 4.30 Optimization of a simply supported beam under constant distributed load: **a** General configuration and **b** box-like cross sections

Determine:

a) The optimized beam contour $d = d(x)$ along the beam axis x (including sketch).
b) An approximate value for the lightweight index M_0 of the cross section with constant values (d_0, d_0, s_0).
c) An approximate value for the lightweight index M of the optimized cross section $(d(x), d(x), s(x))$.
d) The specific energy absorption SEA_0 of the cross-section with constant values (d_0, d_0, s_0).
e) The specific energy absorption SEA of the optimized cross section $(d(x), d(x), s(x))$.
f) What can be said about the technical realization of the contour $d = d(x)$?

Remarks:
$\int_0^L (-(x/L)^2 + x/L)^2 \mathrm{d}x \approx L/30$
$\int_0^L (-(x/L)^2 + x/L)^{2/3} \mathrm{d}x \approx 0.293342L.$

4.4.11 Optimization of a beam with a rectangular box profile along the longitudinal axis under the influence of a constant distributed load

For a simply supported beam (bending stiffness $E \times I(x)$, length L, see Fig. 4.31a) which is loaded by a constant distributed load q_0, optimize the rectangular cross section along the beam axis x taking into account a stress criterion (0.2-% strain

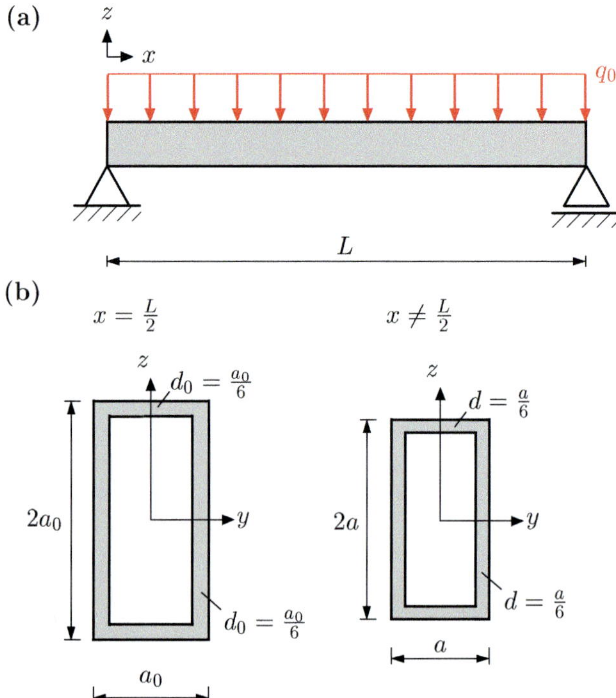

Fig. 4.31 Optimization of a simply supported beam under constant distributed load: **a** General configuration and **b** box-like cross sections

limit $R_{p0.2}$), see Fig. 4.31b. The cross-sectional dimensions in the middle at $x = L/2$ are given by the outer dimensions $2a_0 \times a_0$ and the constant thickness $d_0 = a_0/6$, and the cross-sectional ratios at any point $x \neq L/2$ are given by the outer dimensions $2a(x) \times a(x)$ and the thickness $d(x) = a(x)/6$, see Fig. 4.31b. *Consider the comments on integration at the end of the problem!*
Determine:

(a)   The optimized beam contour $a = a(x)$ along the beam axis x (including sketch).
(b)   An approximate value for the lightweight index M_0 of the cross section with constant value $(2a_0, a_0, d_0)$.
(c)   An approximate value for the lightweight index M of the optimized cross-section $(2a(x), a(x), d(x))$.
(d)   The specific energy absorption SEA_0 of the cross section with constant values $(2a_0, a_0, d_0)$.
(e)   The specific energy absorption SEA of the optimized cross section $(2a(x), a(x), d(x))$.
(f)   What can be said about the technical realization of the contour $a = a(x)$?

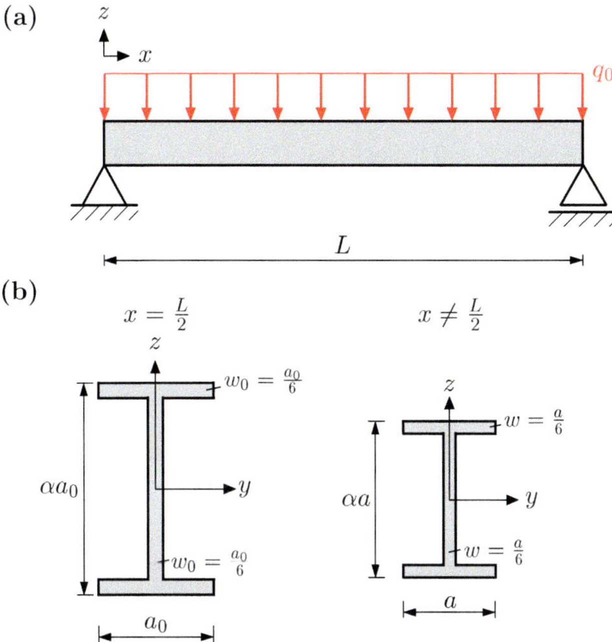

Fig. 4.32 Optimization of a simply supported beam under constant distributed load: **a** General configuration and **b** I cross sections

Remarks:

$$\int_0^L (-(x/L)^2 + x/L)^2 \, \mathrm{d}x \approx L/30$$

$$\int_0^L (-(x/L)^2 + x/L)^{2/3} \, \mathrm{d}x \approx 0.293342 L.$$

4.4.12 Optimization of a beam with I profile along the longitudinal axis under the influence of a constant distributed load

For a simply supported beam (bending stiffness $E \times I(x)$, length L, see Fig. 4.32a) which is loaded by a constant distributed load q_0, optimize the cross-section along the beam axis x taking into account a stress criterion (0.2-% strain limit $R_{\mathrm{p}0.2}$), see Fig. 4.32b. The cross-sectional dimensions in the middle at $x = L/2$ are given by the outer dimensions $\alpha a_0 \times a_0$ and the constant thickness $w_0 = a_0/6$. The cross-sectional ratios at any point $x \neq L/2$ are given by the external dimensions $\alpha a(x) \times a(x)$ and the thickness $w(x) = a(x)/6$, see Fig. 4.32b. *Consider the comments on integration at the end of the problem!*

Determine for $\alpha = 2$ and alternatively for $\alpha = 3$:

(a) The optimized beam contour $a = a(x)$ along the beam axis x (including sketch).
(b) An approximate value for the lightweight index M_0 of the cross-section with constant values ($\alpha a_0, a_0, w_0$).

(c)   An approximate value for the lightweight index M of the optimized cross-section $(\alpha a(x), a(x), w(x))$.

(d)   The specific energy absorption SEA_0 of the cross section with constant values $(\alpha a_0, a_0, w_0)$.

(e)   The specific energy absorption SEA of the optimized cross section with variable values $(\alpha a(x), a(x), w(x))$.

(f)   What can be said about the technical realization of the contour $a = a(x)$?

Remarks:
$\int_0^L (-(x/L)^2 + x/L)^2 \mathrm{d}x \approx L/30$
$\int_0^L (-(x/L)^2 + x/L)^{2/3} \mathrm{d}x \approx 0.293342L.$

4.4.13 Optimization of a beam with I-profile along the longitudinal axis and the height-to-width ratio under the influence of a constant distributed load

For problem 4.4.12, determine the optimal α to maximize the specific energy absorption SEA of the optimized cross-section.

4.4.14 Optimization of a beam along the longitudinal axis under the influence of two different single forces

For a cantilever beam (bending stiffness $E \times I(x)$, length L) which is loaded by two different single forces, optimize the box profile along the beam axis x taking into account a stress criterion (0.2-% strain limit $R_{p0.2}$), see Fig. 4.33a.

The cross-sectional dimensions at the fixed support at $x = 0$ (see Fig. 4.33b) are given by the outer diameter d_0 and the wall thickness $s_0 = d_0/5$. The cross-sectional ratios at any point $x > 0$ are shown in Fig. 4.33b. Determine:

Fig. 4.33 Cantilever beam loaded by two different point loads: **a** General configuration and **b** cross sections

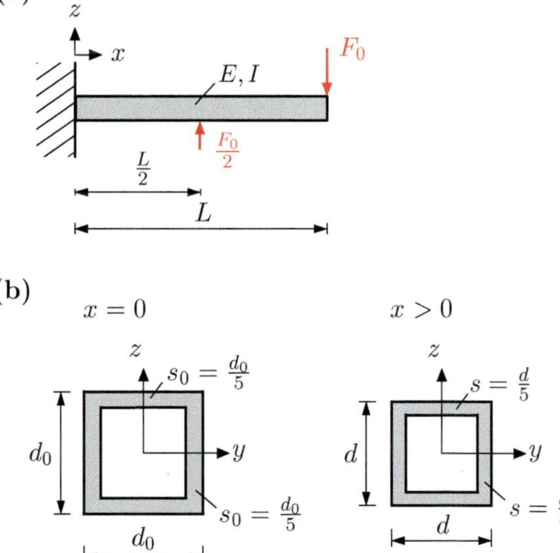

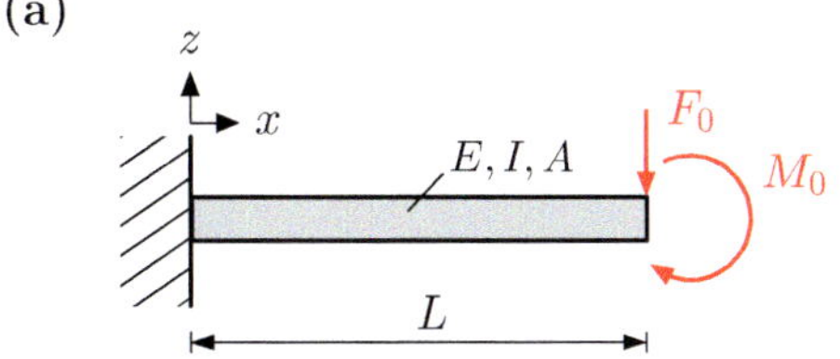

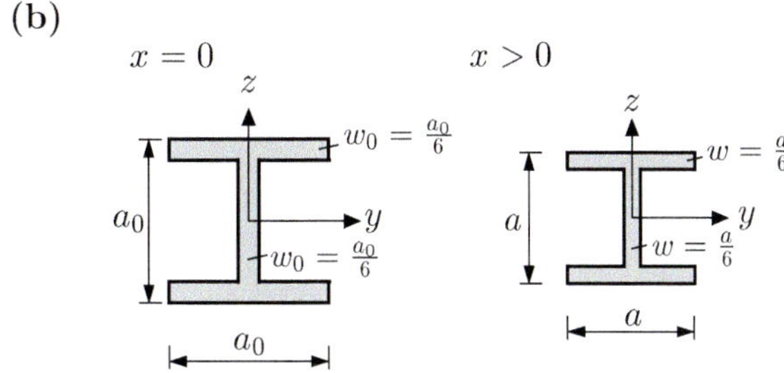

Fig. 4.34 Cantilever beam under the influence of a single force and a single moment: **a** General configuration and **b** I cross sections

a) the beam contour $d = d(x)$ along the beam axis (including sketch),
b) the specific energy absorption SEA of the optimized cross section.

4.4.15 Optimization of a beam along the longitudinal axis under the influence of a single force and a single moment

For a cantilever beam (bending stiffness $E \times I(x)$, length L) loaded by a single force F_0 and a single moment $M_0 = \frac{F_0 L}{6}$, optimize the I-profile along the beam axis x taking into account a stress criterion (0.2-% strain limit $R_{\mathrm{p0.2}}$), see Fig. 4.34a. The cross-sectional dimensions at the fixed support at $x = 0$ (see Fig. 4.34b) are given by the external dimensions a_0 and the wall thicknesses $w_0 = \frac{a_0}{6}$. The cross-sectional ratios at any point $x > 0$ are shown in Fig. 4.34b. Determine:

a) the beam contour $a = a(x)$ along the beam axis (including sketch),
b) the specific energy absorption SEA of the optimized cross section,
c) compare the specific energy absorption SEA of the optimized cross-section with the specific energy absorption for a beam with a constant cross section (a_0 and w_0).

4.4.16 Optimization of a cantilever beam along the longitudinal axis under the influence of a constant distributed load

For a cantilever beam (bending stiffness $E \times I(x)$, length L) subjected to a constant distributed load (q_0), optimize the box section along the beam axis x taking into account a stress criterion (0.2-% strain limit $R_{\mathrm{p0.2}}$), see Fig. 4.35a. The cross-sectional dimensions at the fixed support $x = 0$ (see Fig. 4.35b) are given by the external dimension a_0 and the wall thickness $d_0 = \frac{a_0}{6}$. The cross-sectional ratios at any point $x > 0$ are shown in Fig. 4.35b.

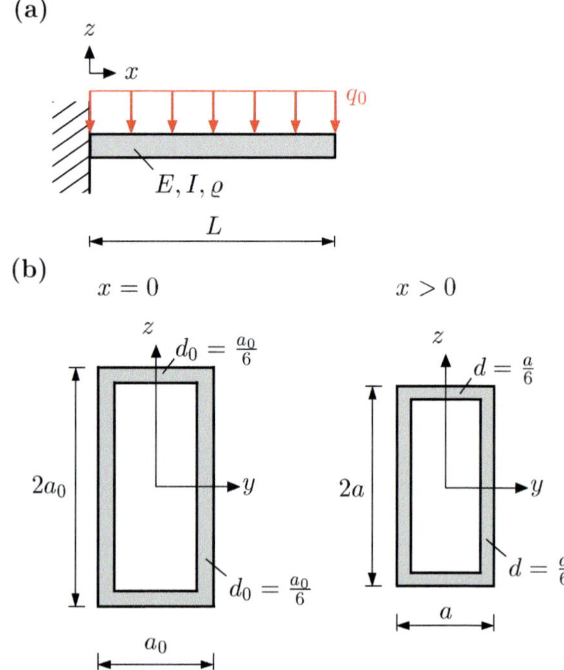

Fig. 4.35 Cantilever beam under the influence of a constant distributed load: **a** General configuration and **b** box-like cross sections

First consider a beam with a constant cross section (see Fig. 4.35b (left)) and calculate:

a) the cross-sectional area A_0, the axial second moment of area I_0 and the mass m_0,
b) the specific energy absorption SEA_0 taking into account a stress criterion (0.2-% strain limit $R_{\mathrm{p0.2}}$).

The optimized beam will now be considered. To do this, determine:

c) the optimized beam dimension $a(x)$,
d) the cross-sectional area $A(x)$, the axial second moment of area $I(x)$ and the mass m,
e) the specific energy absorption SEA taking into account a stress criterion (0.2-% strain limit $R_{\mathrm{p0.2}}$).

Finally, calculate the ratios m/m_0 and SEA/SEA_0. What would an external load have to look like so that the 'optimized' beam has the same cross section at every point x?

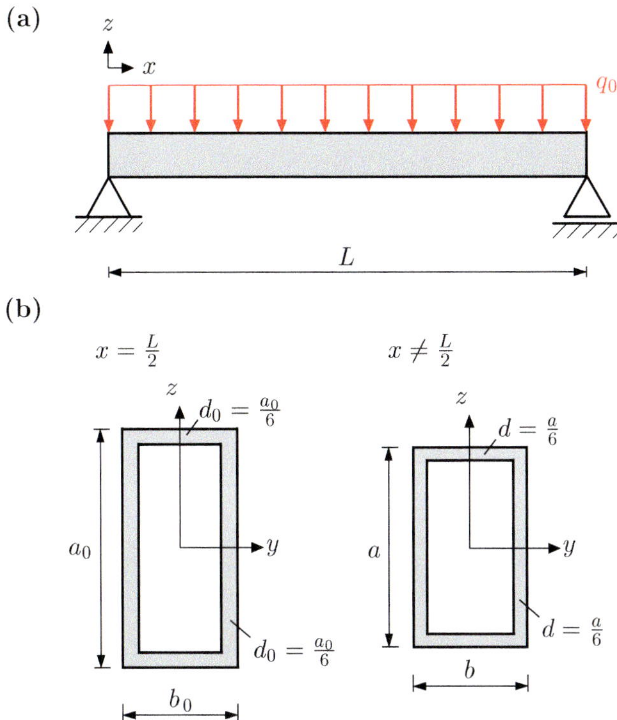

Fig. 4.36 Optimization of a simply supported beam under constant distributed load: **a** General configuration and **b** box-like cross sections

4.4.17 Optimization of a simply supported beam under constant distributed load: box cross section

For a simply supported beam (bending stiffness $E \times I(x)$, length L, see Fig. 4.36a) which is loaded by a constant distributed load q_0, optimize the rectangular cross section along the beam axis x taking into account a stress criterion (0.2-% strain limit $R_{p0.2}$), see Fig. 4.36b. The cross-sectional dimensions in the middle at $x = L/2$ are given by the height a_0, the width $b_0 = a_0/2$ and the constant thickness $d_0 = a_0/6$, and the cross-sectional ratios at any point $x \neq L/2$ are given by the height $a(x)$, the width $b(x) = a(x)/2$ and the thickness $d(x) = a(x)/6$, see Fig. 4.36b.

Determine the optimized beam contour $a = a(x)$ along the beam axis x (including sketch).

4.4.18 Optimization of a simply supported bending under constant distributed load: box cross section. Numerical calculation of the specific energy absorption

For problem 4.4.17 determine the specific energy absorption taking into account a stress criterion (0.2-% strain limit $R_{p0.2}$). The numerical calculation according to Appendix A.1.5 is to be carried out for the reference cross section and the optimized cross section.

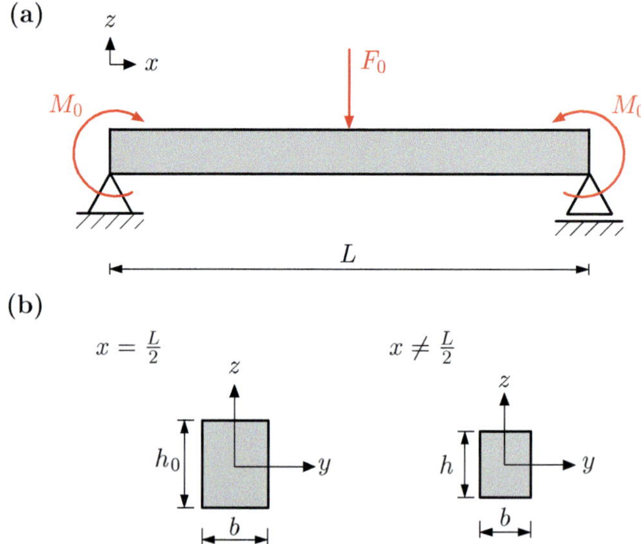

Fig. 4.37 Beam problem: **a** General configuration; **b** rectangular cross sections

4.4.19 Optimization of a simply supported beam under the influence of a single moment and a single force

For a simply supported beam under symmetric 3-point bending with superimposed moment load (bending stiffness $E \times I(x)$, length L, see Fig. 4.37a) which is loaded by a single force F_0 at $x = L/2$ and single moments $M_0 = \frac{F_0 L}{2}$ at $x = 0$ and $x = L$, optimize the rectangular cross-section along the beam axis x taking into account a stress criterion (0.2-% strain limit $R_{p0.2}$), see Fig. 4.37b.

The cross-sectional dimensions at $x = L/2$ are given by the constant width b and the height h_0, and the cross-sectional ratios at any point $x \neq L/2$ are given by the height $h(x)$ to be optimized and the constant width b, see Fig. 4.37b.

Determine:

a) the beam contour $h = h(x)$ along the beam axis (including sketch),
b) the specific energy absorption SEA of the optimized cross section. Compare this value with the specific energy absorption for a beam with a constant height h_0.

4.4.20 Optimal design of a cantilever beam (classical optimization problem)

Consider a cantilever beam fixed at one end as shown in Fig. 4.38. The beam is loaded by a single force F_0 and has constant material (E, ϱ) and geometric (I) properties along its longitudinal axis. The material is isotropic and homogeneous and the beam theory for thin beams (Euler-Bernoulli) can be used for this example.

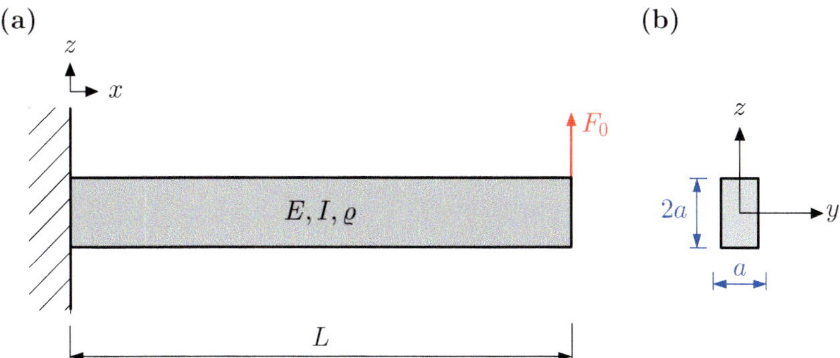

Fig. 4.38 **a** General configuration of the beam problem; **b** beam cross section

Given are:

- Geometric dimension: $L = 2540\,\text{mm}$.
- Material properties of the beam: Elastic modulus $E = 68{,}948\,\text{MPa}$, mass density $\varrho = 2691\,\text{kg/m}^3$, 0.2-% strain limit $R_{\text{p0.2}} = 247\,\text{MPa}$.
- Load: $F_0 = 2667\,\text{N}$.

Determine the optimized cross-sectional dimension a under the condition that the effective stress does not exceed the initial yield stress. Furthermore, a maximum deflection of $u_z(L) = r_1 L$ with $r_1 = 0.03$ should not be exceeded. Based on a graphical representation of the objective function and the corresponding inequality constraints, the optimum is to be determined analytically.

4.4.21 Optimal design of a compression strut (classical optimization problem)

A compression strut is considered according to Fig. 4.39. The strut is loaded by a single force F_0 and has constant material (E, ϱ) and geometric (I, A) properties along its longitudinal axis. Furthermore, the material can be considered isotropic and homogeneous.

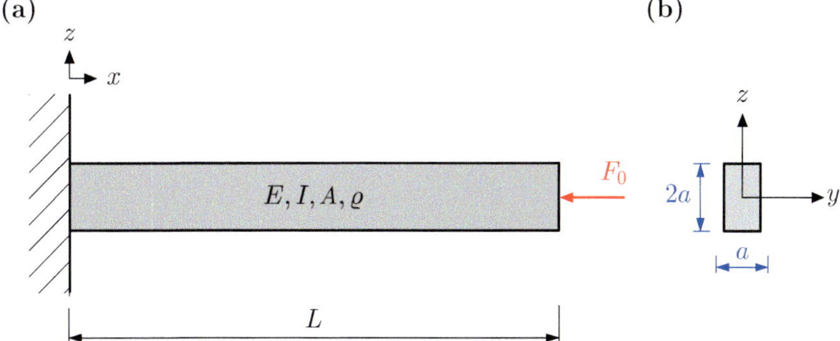

Fig. 4.39 **a** General configuration of the compression strut problem; **b** cross section

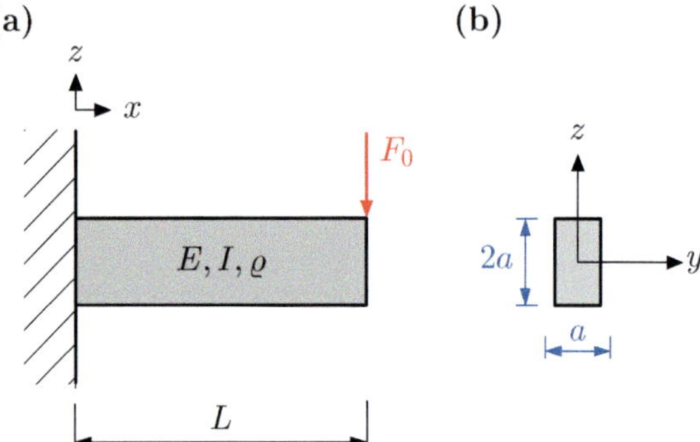

Fig. 4.40 a General configuration of the short cantilever beam problem; **b** cross section

Given are:

- Geometric dimension: $L = 1000\,\text{mm}$.
- Material properties of the strut: Elastic modulus $E = 70{,}000\,\text{MPa}$, mass density $\varrho = 2691\,\text{kg/m}^3$, 0.2-% strain limit $R_{\text{p}0.2} = 247\,\text{MPa}$.
- Load: $F_0 = 2667\,\text{N}$.

Determine the optimized cross-sectional dimension a under the condition that the effective stress does not exceed the initial yield stress. Furthermore, no buckling should occur in the elastic region. The optimum is to be determined analytically based on a graphical representation of the objective function and the corresponding inequality constraints.

4.4.22 Optimal design of a short cantilever beam (classical optimization problem)

Consider a short cantilever beam as shown in Fig. 4.40. The beam is loaded by a single force F_0 and has constant material (E, ϱ) and geometric (I) properties along its longitudinal axis. The material is isotropic and homogeneous and the beam theory for thick beams (i.e., consideration of the acting shear stress) can be used for this example.

Given are:

- Geometric dimension: $L = 846.33\,\text{mm}$.
- Material properties of the beam: elastic modulus $E = 68{,}948\,\text{MPa}$, mass density $\varrho = 2691\,\text{kg/m}^3$, 0.2-% strain limit $R_{\text{p}0.2} = 247\,\text{MPa}$.
- Load: $F_0 = 2667\,\text{N}$.

Determine the optimized cross-sectional dimension a under the condition that the acting normal and shear stresses do not exceed the corresponding initial yield stress. The initial shear yield stress can be approximated using the Tresca hypothesis. The normal stress distribution can be assumed to be linear and the shear stress distribution to be parabolic over the beam height. Based on a graphical representation of the

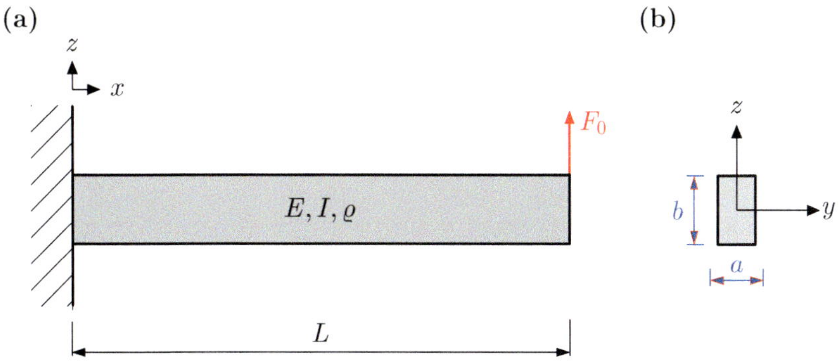

Fig. 4.41 **a** General configuration of the beam problem; **b** beam cross section

objective function and the corresponding inequality restrictions, the optimum is to be determined analytically.

4.4.23 Optimal design of a cantilever beam: constant rectangular cross section (classical optimization problem)

Consider a cantilever beam clamped at one end as shown in Fig. 4.41. The beam is loaded by a single force F_0 and has constant material (E, ϱ) and geometric (I) properties along its longitudinal axis. The material is isotropic and homogeneous and the beam theory for thin beams (Euler-Bernoulli) can be used for this problem.
Given are:

- Geometric dimension: $L = 2540\,\mathrm{mm}$.
- Material properties of the beam: elastic modulus $E = 68{,}948\,\mathrm{MPa}$, mass density $\varrho = 2691\,\mathrm{kg/m^3}$, 0.2-% strain limit $R_{\mathrm{p0.2}} = 247\,\mathrm{MPa}$.
- Load: $F_0 = 2667\,\mathrm{N}$.

Determine the optimized cross-sectional dimensions a and b under the condition that the acting normal stress does not exceed the initial yield stress. Furthermore, a maximum deflection of $u_z(L) = r_1 L$ with $r_1 = 0.03$ should not be exceeded. Finally, the height-to-width ratio should be limited to $b \le 20a$ in order to avoid instabilities. Based on a graphical representation of the objective function and the corresponding inequality constraints, the optimum is to be determined analytically. Compare the optimized geometry with the results from problem 4.4.20.

4.4.24 Optimal design of a simply supported beam: constant rectangular cross section (classical optimization problem)

Consider a simply supported beam as shown in Fig. 4.42. The beam is loaded in the middle by a single force F_0 and has constant material (E, ϱ) and geometric (I) properties along its longitudinal axis. The material is isotropic and homogeneous and the beam theory for thin beams (Euler-Bernoulli) can be used for this problem.
Given are:

- Geometric dimension: $L = 2540$ mm.
- Material properties of the beam: elastic modulus $E = 68{,}948$ MPa, mass density $\varrho = 2691$ kg/m^3, 0.2-% strain limit $R_{\text{p0.2}} = 247$ MPa, shear yield strength $\tau_{\text{p}} = R_{\text{p0.2}}/2$.
- Load: $F_0 = 2667$ N.

Determine the optimized cross-sectional dimensions a and b under the condition that the acting normal and shear stresses do not exceed the corresponding initial yield stresses. Furthermore, a maximum deflection of $u_z(L) = r_1 L$ with $r_1 = 0.03$ or $r'_1 = 0.003$ should not be exceeded. Finally, the height-to-width ratio (aspect ratio) should be limited to $b \leq 20a$ in order to avoid instabilities. Based on a graphical representation of the objective function and the corresponding inequality constraints, the optimum is to be determined analytically.

4.4.25 Optimal design of a compression strut: constant rectangular cross section (classical optimization problem)

A compression strut is considered according to Fig. 4.43. The strut is loaded by a single force F_0 and has constant material (E, ϱ) and geometric (I, A) properties along its longitudinal axis. Furthermore, the material can be considered isotropic and homogeneous.

Given are:

- Geometric dimension: $L = 1000$ mm.
- Material properties of the rod: elastic modulus $E = 70{,}000$ MPa, mass density $\varrho = 2691$ kg/m^3, 0.2-% strain limit $R_{\text{p0.2}} = 247$ MPa.
- Load: $F_0 = 2667$ N.

Determine the optimized cross-sectional dimensions a and b under the condition that the effective stress does not exceed the initial yield stress. Furthermore, no buckling should occur in the elastic range. Based on a graphical representation of the objective function and the corresponding inequality constraints, the optimum is to

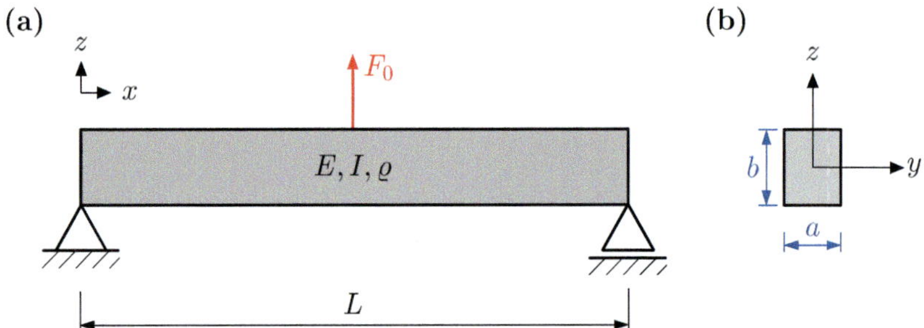

Fig. 4.42 **a** General configuration of a simply supported beam under 3-point bending; **b** cross section

be determined analytically. Compare the optimized geometry with the results from problem 4.4.21.

4.4.26 Optimal design of a short cantilever beam: constant rectangular cross section (classical optimization problem)

Consider a short, cantilevered bending beam as shown in Fig. 4.44. The beam is loaded by a single force F_0 and has constant material (E, ϱ) and geometric (I) properties along its longitudinal axis. The material is isotropic and homogeneous and the beam theory for thick beams (i.e., consideration of the acting shear stress) can be used for this example.

Given are:

- Geometric dimension: $L = 846.33\,\text{mm}$.
- Material properties of the beam: elastic modulus $E = 68{,}948\,\text{MPa}$, mass density $\varrho = 2691\,\text{kg/m}^3$, 0.2-% strain limit $R_{\text{p0.2}} = 247\,\text{MPa}$.
- Load: $F_0 = 2667\,\text{N}$.

Determine the optimized cross-sectional dimensions a and b under the condition that the acting normal and shear stresses do not exceed the corresponding initial

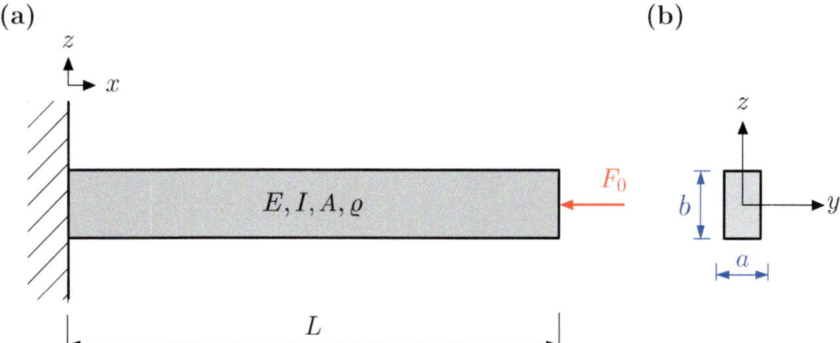

Fig. 4.43 a General configuration of the compression strut problem; **b** cross section

Fig. 4.44 a General configuration of the short cantilever beam problem; **b** cross section

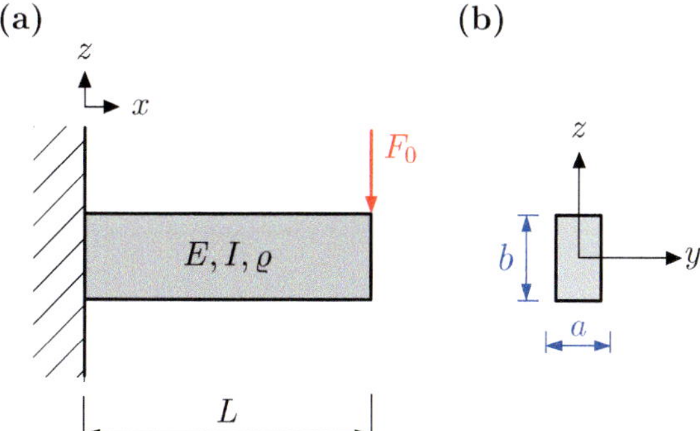

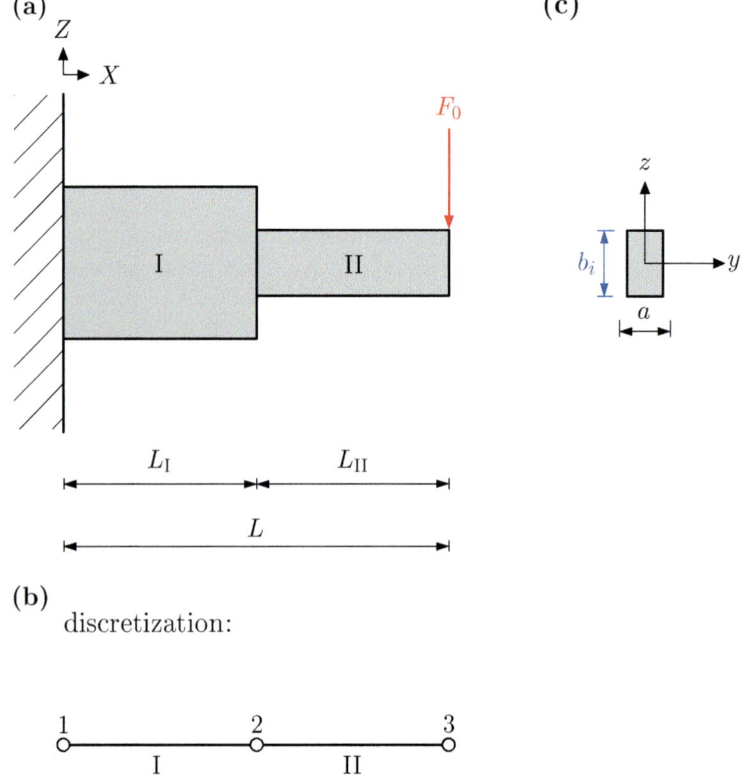

Fig. 4.45 Cantilever stepped beam with two sections: **a** General configuration, **b** discretization, and **c** cross section of element i

yield stresses. The initial shear yield stress can be approximated using the Tresca hypothesis. The normal stress distribution can be assumed to be linear and the shear stress distribution to be parabolic over the beam height. Furthermore, the height-to-width ratio should be limited to $b \leq 20a$ in order to avoid instabilities. Based on a graphical representation of the objective function and the corresponding inequality constraints, the optimum is to be determined analytically. Compare the optimized geometry with the results from problem 4.4.22.

4.4.27 Optimal design of a stepped cantilever beam with two sections (classical optimization problem)

Consider a cantilever beam that is fixed on one side as shown in Fig. 4.45 and loaded at the right-hand end by a single force F_0 in the negative Z-direction. The beam is divided into two sections of length $L_\mathrm{I} = L_\mathrm{II} = L/2$. The material of each section is the same, i.e. $E_\mathrm{I} = E_\mathrm{II} = E$, but the cross sections are different in both sections. Optimize the two cross sections, i.e. the heights b_i $(i = \mathrm{I}, \mathrm{II})$, whereby the width of the two sections (a) remains identical.

Given are:

- Geometric dimensions: $L = 2540\,\text{mm}$, $a = 45.16\,\text{mm}$.
- Material properties of the beam: elastic modulus $E = 68{,}948\,\text{MPa}$, mass density $\varrho = 2691\,\text{kg/m}^3$, 0.2-% strain limit $R_{\text{p}0.2} = 247\,\text{MPa}$.
- Load: $F_0 = 2667\,\text{N}$.

Determine the optimized cross-sectional dimensions b_i under the condition that the acting normal and shear stresses do not exceed the corresponding initial yield stresses. The initial shear yield stress can be approximated using the Tresca hypothesis. The normal stress distribution can be assumed to be linear and the shear stress distribution to be parabolic over the beam height. Furthermore, a maximum deflection of $|u_z(L)| = r_1 L$ with $r_1 = 0.06$ should not be exceeded. Finally, the height-to-width ratio should be limited to $b \leq 20a$ to avoid instabilities. Based on a graphical representation of the objective function and the corresponding inequality constraints, the optimum is to be determined analytically. Use a finite element approach ('hand calculation') to determine the deformations and internal reactions, see [4,7,9].

4.4.28 Optimal design of a simply supported beam with three sections (classical optimization problem)

Consider a simply supported beam as shown in Fig. 4.46, which is loaded in the middle by a single force F_0 in the negative Z-direction. The beam is divided into three sections of length $L_\text{I} = L_\text{II} = L_\text{III} = L/3$. The material of each section is the same, i.e. $E_\text{I} = E_\text{II} = E_\text{III} = E$, but the middle cross section has different dimensions than the two outer sections ($\text{I} = \text{III}$). Optimize the cross sections, i.e. the heights b_i ($i = \text{I, II, III}$), whereby the width of the three sections (a) remains identical. For simplification, consider the symmetry of the problem.

Given are:

- Geometric dimensions: $L = 2540\,\text{mm}$, $a = 45.16\,\text{mm}$.
- Material properties of the beam: elastic modulus $E = 68{,}948\,\text{MPa}$, mass density $\varrho = 2691\,\text{kg/m}^3$, 0.2-% strain limit $R_{\text{p}0.2} = 247\,\text{MPa}$.
- Load: $F_0 = 2667\,\text{N}$.

Determine the optimized cross-sectional dimensions b_i under the condition that the acting normal and shear stresses do not exceed the corresponding initial yield stresses. The initial shear yield stress can be approximated using the Tresca hypothesis. The normal stress distribution can be assumed to be linear and the shear stress distribution to be parabolic over the beam height. Furthermore, a maximum deflection of $|u_z(L/2)| = r_1 L$ with $r_1 = 0.01$ should not be exceeded. Finally, the height-to-width ratio should be limited to $b \leq 20a$ to avoid instabilities. Based on a graphical representation of the objective function and the corresponding inequality constraints, the optimum is to be determined analytically. Use a finite element approach ('hand calculation') to determine the deformations and internal reactions, see [4,7,9].

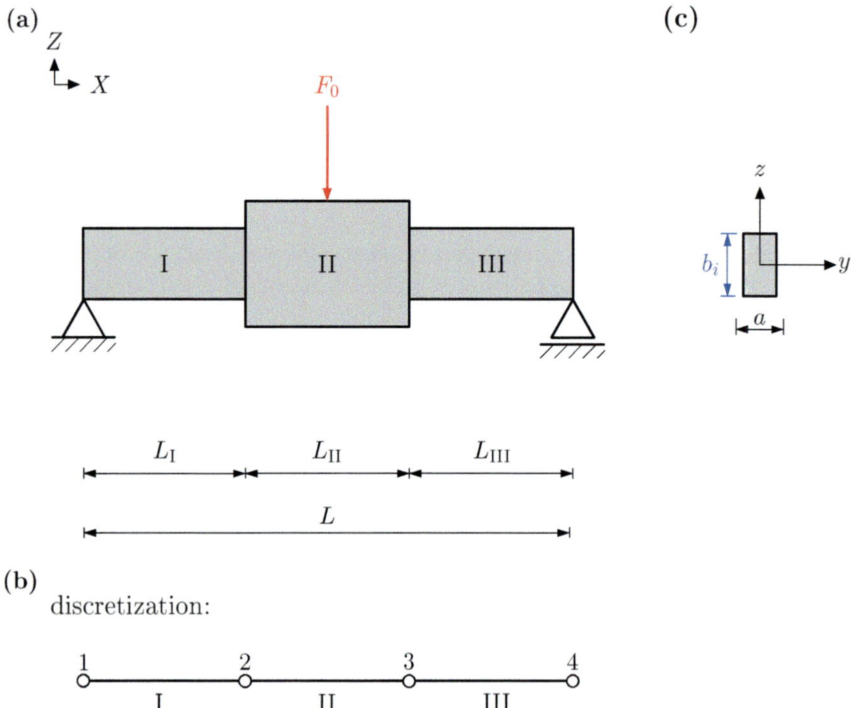

Fig. 4.46 Simply supported beam with three sections: **a** General configuration, **b** discretization, and **c** cross section of element i

References

1. Altenbach, H.: Holzmann/Meyer/Schumpich Technische Mechanik Festigkeitslehre. Springer Vieweg, Wiesbaden (2016)
2. Boresi, A.P., Schmidt, R.J.: Advanced Mechanics of Materials. John Wiley & Sons, New York (2003)
3. Bruhn, E.F.: Analysis and Design of Flight Vehicle Structures. Jacobs Publishing, Indianapolis (1973)
4. Javanbakht, Z., Öchsner, A.: Computational Statics Revision Course. Springer, Cham (2018)
5. Öchsner, A.: Elasto-Plasticity of Frame Structure Elements: Modeling and Simulation of Rods and Beams. Springer, Berlin (2014)
6. Öchsner, A.: Computational Statics and Dynamics: An Introduction Based on the Finite Element Method. Springer, Singapore (2016)
7. Öchsner, A.: A Project-Based Introduction to Computational Statics. Springer, Cham (2018)
8. Öchsner, A., Makvandi, R.: Numerical Engineering Optimization: Application of the Computer Algebra System Maxima. Springer Nature, Cham (2020)
9. Öchsner, A.: Computational Statics and Dynamics: An Introduction Based on the Finite Element Method. Springer, Singapore (2023)
10. Rothwell, A.: Optimization Methods in Structural Design. Springer, Cham (2017)

11. Schumacher, A.: Optimierung mechanischer Strukturen: Grundlagen und industrielle Anwendungen. Springer Vieweg, Berlin (2013)
12. Spillers, W.R., MacBain, K.M.: Structural Optimization. Springer, New York (2009)
13. Timoshenko, S.P., Gere, J.M.: Theory of Elastic Stability. McGraw-Hill, Singapore (1963)
14. Vanderplaats, G.N.: Numerical Optimization Techniques for Engineering Design. Vanderplaats Research & Development, Colorado Springs (1999)

Sandwich Structures: Basic Mechanical Load Cases

Abstract

In this chapter, the basic mechanical load cases for one-dimensional sandwich beams under static loading conditions are treated in detail. The cases of tension/compression, bending and shear are considered. For each case the corresponding stress distributions are derived. The cases of general sandwich structures and the special cases of thin face sheets and/or soft core are investigated.

5.1 Introductory Remarks

In the case of material and form lightweight design, an attempt is made to achieve an optimal lightweight potential through a suitable choice and combination of different materials and their distribution in the cross section. Typical representatives of this come from the class of composite materials in their most general definition, with sandwich elements being considered in more detail below, see [1,7]. A typical sandwich element is shown in Fig. 5.1

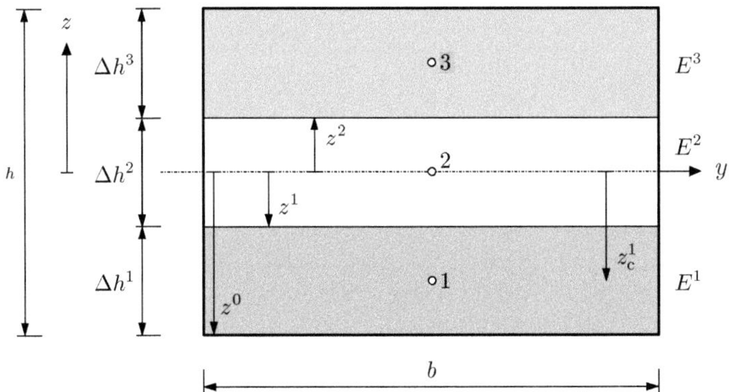

Fig. 5.1 Structure of a sandwich element: 1, 3: top layer (skin, face), 2: core layer

© The Author(s), under exclusive license to Springer Fachmedien Wiesbaden GmbH, part of Springer Nature 2025
A. Öchsner, *Lightweight Design*,
https://doi.org/10.1007/978-3-658-48162-9_5

Fig. 5.2 Investigated load cases: **(a)** bending, **(b)** tensile or compressive loading, and **(c)** shear loading

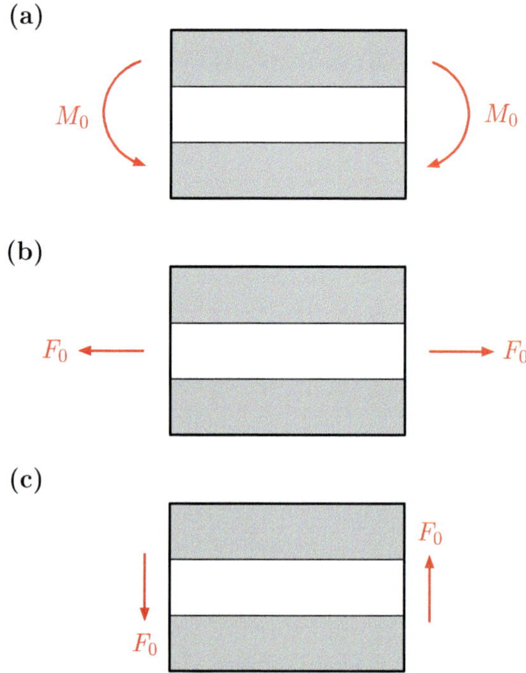

Here, two top layers, which often consist of the same material, are connected by a core and thus kept at a defined distance. It is also noted that the top layers and the core usually have different tasks, see [3].

In the following, three different load cases are examined in more detail, see Fig. 5.2. For simplification, the beam theory according to Euler-Bernoulli (see Sect. 2.2.1) and the bar theory (see Sect. 2.1) are used as a basis. Thus, a generalized, i.e. a combination of beam and bar member, composite beam is considered.

5.2   Bending Load

The average bending stiffness $\overline{EI}_y$ for such a composite with three layers results from [5,6] as follows:

$$\overline{EI}_y = \sum_{k=1}^{3} E^k \left[\tfrac{1}{12} b \left(\Delta h^k \right)^3 + b \Delta h^k \left(z_\text{c}^k \right)^2 \right] = \sum_{k=1}^{3} E^k I_y^k , \qquad (5.1)$$

where z_c^k is the vertical distance (i.e. in the z-direction) of the center of area of the partial body k to the total center of area. The layer thickness Δh can also be calculated using $\Delta h^k = z^{k-1} - z^k$. In Eq. (5.1), the total second moment of area was calculated using the fraction with regard to the center of area of the partial body and the additional contribution from the parallel-axis theorem ('Steiner' fraction). Thus, from Eq. (5.1) for a homogeneous element of width b and height h, $\overline{EI}_y = EI_y = E\frac{1}{12}bh^3$ would result. In the case of a transverse strain constraint of the cover layers (e.g. by the core or a significant dimension in the y-direction, i.e. $b \approx L$), instead of E the modified modulus of elasticity $E \rightarrow E/(1 - v^2)$ can be used, see [7].

The stress distribution in the sandwich can be described using the following modified approach (see [5,6]):

$$\sigma_{x,k}(z) = \frac{M_y E^k}{\overline{EI}_y} \times z .$$

(5.2)

According to Fig. 5.3a, there are discontinuities in stress at the transition from one material to the next in the case of different stiffnesses (E^k). In contrast to this, however, the strain is assumed without these discontinuities, i.e. a linear progression through the origin of the coordinate system, see Fig. 5.3b.

The derivation of Eq. (5.2) is briefly presented in the following. Assuming that each layer k has the same curvature κ , the stress in each layer can be given as follows:

$$\sigma_{x,k} = -E^k z \frac{\mathrm{d}^2 u_z(x)}{\mathrm{d}x^2} = +E^k z \kappa .$$

(5.3)

The internal bending moment results from integration over the stress distribution using:

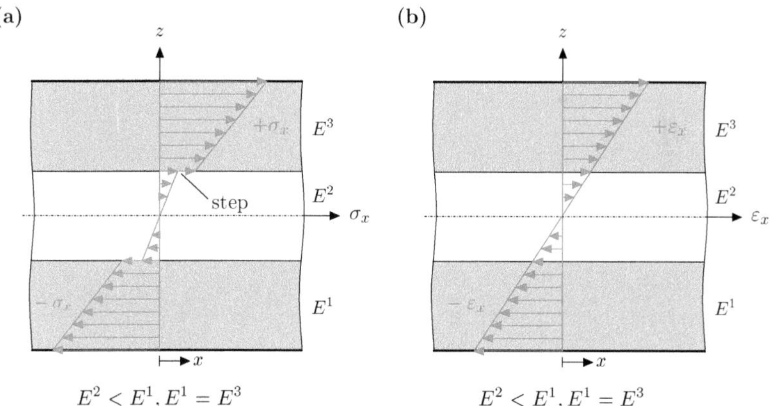

Fig. 5.3 Course of the (a) normal stress component σ_x and (b) strain component ε_x under bending load

$$M_y = \int_A z\sigma_x \mathrm{d}A = \int_{A_1} z\sigma_{x,1}\mathrm{d}A_1 + \int_{A_2} z\sigma_{x,2}\mathrm{d}A_2 + \int_{A_3} z\sigma_{x,3}\mathrm{d}A_3 \tag{5.4}$$

$$= \int_{A_1} z^2 E^1 \kappa \mathrm{d}A_1 + \int_{A_2} z^2 E^2 \kappa \mathrm{d}A_2 + \int_{A_3} z^2 E^3 \kappa \mathrm{d}A_3 \tag{5.5}$$

$$= \left(E^1 \int_{A_1} z^2 \mathrm{d}A_1 + E^2 \int_{A_2} z^2 \mathrm{d}A_2 + E^1 \int_{A_3} z^2 \mathrm{d}A_3 \right) \kappa \tag{5.6}$$

$$= \left(E^1 I_y^1 + E^2 I_y^2 + E^3 I_y^3 \right) \kappa = \overline{EI}_y \kappa . \tag{5.7}$$

Using Eq. (5.3), the stress relationship (5.2) results from the last equation.

In the following, different sandwich structures with different material and geometry combinations are compared in relation to the lightweight potential. The basic configuration for numerous examples, i.e. cantilever with end shear force, is again as in Fig. 3.1. A homogeneous aluminum beam is now used as a reference (see Fig. 5.4, configuration (1)). By using a foam core and then reducing the thickness of the face sheets (with the same external dimensions of the cross section), an increase in the lightweight potential can be achieved (see Fig. 5.4, configurations (2) and (3)). A configuration in which the core layer was completely neglected is given as a limiting case (see Fig. 5.4, configuration (4)). It should be noted here that the calculation of the lightweight index was carried out with a limit value as the maximum external load.

An alternative design concept can be realized by using carbon fiber-reinforced plastic (CFRP) (characteristic values for CFRP according to Klein [5]: $\varrho = 1.50 \times$

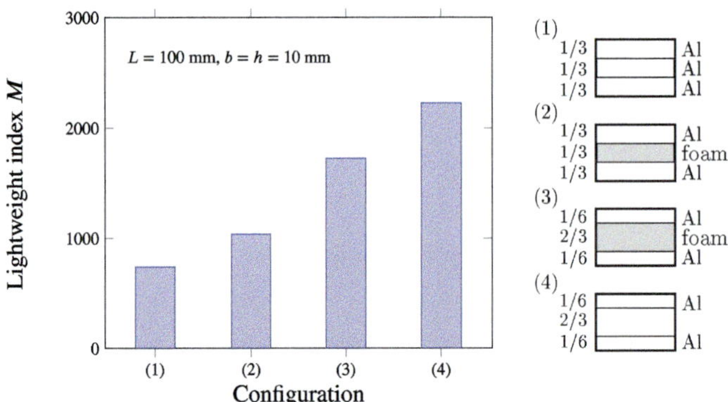

Fig. 5.4 Lightweight index for different configurations of sandwich elements ($L = 100$ mm, $b = h = 10$ mm) and load criterion for bending load case ($F_0 = -200$ N $\wedge$ $\sigma_{\max} < R_{p0.2}$). Characteristic values of the Al foam according to [5]: $\varrho = 0.4 \times 10^{-6}\ \frac{\mathrm{kg}}{\mathrm{mm}^3}$, $E = 2500$ MPa, $R_{p0.2}^{\mathrm{t}} = 4$ MPa, $R_{p0.2}^{\mathrm{c}} = 6$ MPa

10^{-6} kg/mm^3, $E = 120.000$ MPa, $R^{\mathrm{t}} = 1700$ MPa; unidirectional layer (UD) with fiber volume fraction of $\phi_{\mathrm{F}} = 0.55$). For case (4) according to Fig. 5.4 this now results in $M_{\mathrm{CFRP}} = 4077$.

5.3 Tensile/Compressive Load

In the following, the load case of tension or compression (see Fig. 5.2b) is considered in more detail, although instabilities such as buckling or wrinkling are not yet taken into account here. Under the influence of an external axial force, the stress and strain distributions shown in Fig. 5.5 result. It is important to assume that all layers are perfectly connected and that the strain in each individual layer is equal to the total strain, see Fig. 5.5b.

The internal normal force results from integration over the stress distribution using:

$$N_x = \int_A \sigma_x \mathrm{d}A = \int_{A_1} \sigma_{x,1} \mathrm{d}A_1 + \int_{A_2} \sigma_{x,2} \mathrm{d}A_2 + \int_{A_3} \sigma_{x,3} \mathrm{d}A_3 \tag{5.8}$$

$$= E_1 \varepsilon_x \Delta h^1 b + E_2 \varepsilon_x \Delta h^2 b + E_3 \varepsilon_x \Delta h^3 b \tag{5.9}$$

$$= \left(E_1 \Delta h^1 b + E_2 \Delta h^2 b + E_3 \Delta h^3 b \right) \varepsilon_x = \overline{EA} \varepsilon_x . \tag{5.10}$$

The average tensile stiffness $\overline{EA}$ is thus:

$$\overline{EA} = \sum_{k=1}^{3} E^k \Delta h^k b . \tag{5.11}$$

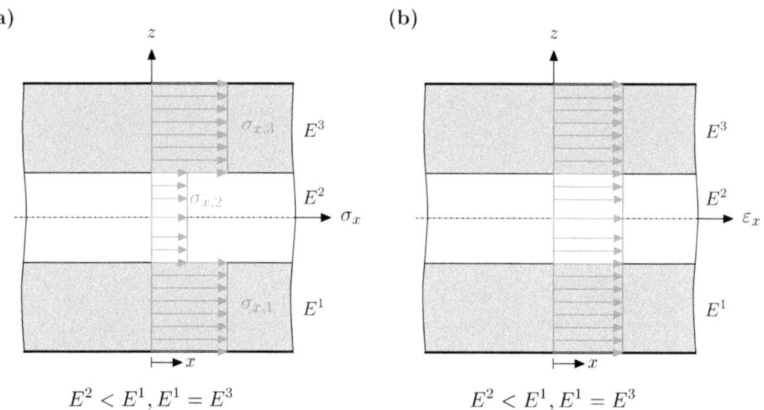

Fig. 5.5 Course of the **(a)** normal stress component σ_x and **(b)** strain component ε_x under tensile load

Substituting in Eq. (5.10) the total strain using Hooke's law for a layer and considering that the internal normal force is equal to the external force F_0, i.e.

$$N_x = F_0 = \overline{EA}\,\frac{\sigma_k}{E^k},\tag{5.12}$$

the relationship for the stress in the k-th layer is obtained as:

$$\sigma_{x,k} = \frac{F_0 E^k}{\overline{EA}}.\tag{5.13}$$

5.4   Shear Load

In the following, the shear load case (see Fig. 5.2c) is considered in more detail. In general, the shear stress distribution can be calculated according to Eq. (A.6) by integration over the normal stress distribution (or more precisely, the corresponding gradient).

For the top layers (see layers 1 and 3 in Fig. 5.1) this results, as an example, for layer 3 ($\frac{\Delta h^2}{2} \leq z \leq \frac{\Delta h^2}{2} + \Delta h^3$) in:

$$\tau_{zx,3}(z) = \int_{z}^{\frac{\Delta h^2}{2}+\Delta h^3} \frac{d\sigma_{x,3}(x)}{dx}\,dz' + c_3.\tag{5.14}$$

With the stress gradient

$$\frac{d\sigma_{x,3}(x)}{dx} = \frac{d}{dx}\left(\frac{M_y(x)E^3}{EI_y}z\right) = \frac{E^3 z}{EI_y}\frac{dM_y(x)}{dx} = \frac{E^3 Q_z(x)}{EI_y}z\tag{5.15}$$

the shear stress distribution results in:

$$\tau_{zx,3}(z) = \frac{E^3 Q_z(x)}{EI_y}\int_{z}^{\frac{\Delta h^2}{2}+\Delta h^3} z'\,dz' + c_3 = \frac{E^3 Q_z(x)}{2EI_y}\left[\left(\frac{\Delta h^2}{2}+\Delta h^3\right)^2 - z^2\right] + c_3.\tag{5.16}$$

The constant of integration c_3 can be determined using the condition that no shear stresses occur at the free edge, i.e. $\tau_{zx,3}(\Delta h^2/2 + \Delta h^3) = 0$, as $c_3 = 0$. This finally gives the shear stress distribution in layer 3:

$$\tau_{zx,3}(z) = \frac{E^3 Q_z(x)}{2\overline{EI_y}}\left[\left(\frac{\Delta h^2}{2} + \Delta h^3\right)^2 - z^2\right], \tag{5.17}$$

or in layer 1:

$$\tau_{zx,1}(z) = \frac{E^1 Q_z(x)}{2\overline{EI_y}}\left[\left(\frac{\Delta h^2}{2} + \Delta h^1\right)^2 - z^2\right]. \tag{5.18}$$

For the core layer (see layer 2 in Fig. 5.1) we get $\left(-\frac{\Delta h^2}{2} \leq z \leq \frac{\Delta h^2}{2}\right)$:

$$\tau_{zx,2}(z) = \int_z^{\frac{\Delta h^2}{2}} \frac{\mathrm{d}\sigma_{x,2}(x)}{\mathrm{d}x}\mathrm{d}z' + c_2. \tag{5.19}$$

With the stress gradient

$$\frac{\mathrm{d}\sigma_{x,2}(x)}{\mathrm{d}x} = \frac{E^2 Q_z(x)}{\overline{EI_y}}z \tag{5.20}$$

the shear stress distribution results in:

$$\tau_{zx,2}(z) = \frac{E^2 Q_z(x)}{\overline{EI_y}}\int_z^{\frac{\Delta h^2}{2}} z'\mathrm{d}z' + c_2 = \frac{E^2 Q_z(x)}{2\overline{EI_y}}\left[\left(\frac{\Delta h^2}{2}\right)^2 - z^2\right] + c_2. \tag{5.21}$$

The integration constant c_2 results from the transition condition for the stress τ_{zx} between layers 2 and 3, i.e. $\tau_{zx,2}(z = \Delta h^2/2) = \tau_{zx,3}(z = \Delta h^2/2)$, to

$$c_2 = \frac{E^3 Q_z(x)}{2\overline{EI_y}}\left[\left(\frac{\Delta h^2}{2} + \Delta h^3\right)^2 - \left(\frac{\Delta h^2}{2}\right)^2\right] = \frac{E^3 Q_z(x)}{2\overline{EI_y}}\left[(\Delta h^2 + \Delta h^3)\Delta h^3\right]. \tag{5.22}$$

Thus, the shear stress distribution in the core layer (layer '2') results in:

$$\tau_{zx,2}(z) = \frac{Q_z(x)}{2\overline{EI_y}}\left[E^2\left(\left(\frac{\Delta h^2}{2}\right)^2 - z^2\right) + E^3\left(\Delta h^2 + \Delta h^3\right)\Delta h^3\right]. \tag{5.23}$$

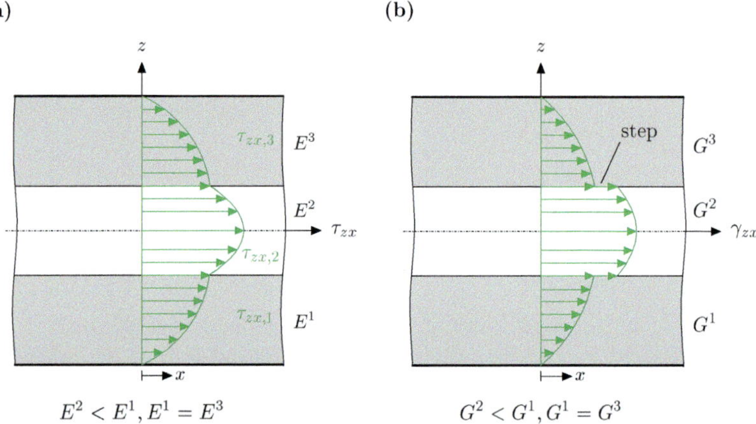

Fig. 5.6 Course of the (a) shear stress component τ_{xz} and the (b) shear strain γ_{xz} under shear load

The distribution of the shear stress over the height of the sandwich element is shown in Fig. 5.6a. Different parabolas form the distribution here.

The distribution of the shear strain shows, because $\gamma_{zx}(z) = \frac{\tau_{zx}}{G_{zx}}$ with different shear moduli in layers 2 and 3 (= 1), discontinuities, i.e. steps at the layer transitions, see Fig. 5.6b.

5.5   Technical Sandwich

For the further explanations, a more technical configuration of a sandwich is considered, see Fig. 5.7. The face sheets of the symmetrical structure are generally significantly thinner than the core layer. Furthermore, the core layer is usually softer than the face sheets[1]. In addition, homogeneous and isotropic materials that deform linearly and elastically are assumed for all layers.

As tasks[2] of the face sheets in a technical sandwich, the following functions can be cited:

- They absorb practically the entire bending moment.
- They absorb practically the entire axial tensile or compressive load.

The core, on the other hand, has the following tasks:

- It practically absorbs the entire lateral force (shear).
- Fixing, supporting and stabilizing the face sheets in their mutual position (i.e. vertical spacing of the face sheets, avoidance of the face sheets sliding relative to

---

[1] A composite element with $\Delta h^C < \Delta h^F$, $G^C < G^F$ is called an anti-sandwich, see [2].
[2] The following subchapters will explain certain facts in more detail.

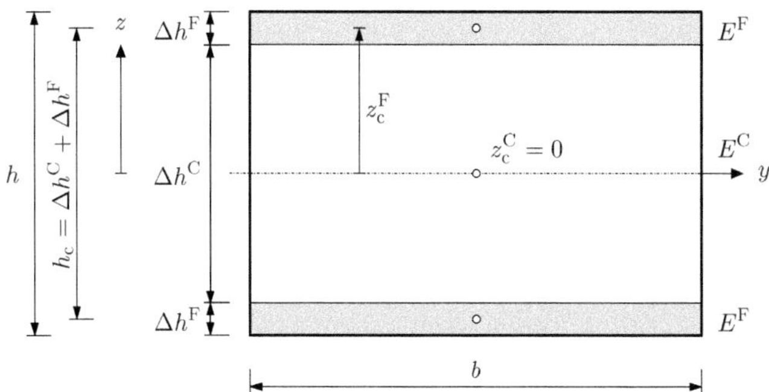

Fig. 5.7 Technical sandwich element: symmetrical structure with thin face sheets (F) and soft core (C)

each other, ensuring the evenness—as a stabilizing elastic bedding—of the face sheets to avoid buckling/wrinkling).

In many cases, the tensile and shear-resistant assemblage between the face sheets and the core is created by gluing[3] . However, self-forming adhesion after the foaming of the core, welding, nailing, screwing or dowelling can also be found in technical applications, see [7].

5.5.1 Bending Load

For this case, according to Eq. (5.1), the average bending stiffness for the configuration according to Fig. 5.7 can be set as follows:

$$\overline{EI}_y = 2E^F\left[\frac{1}{12}b(\Delta h^F)^3 + b\Delta h^F(z_c^F)^2\right] + E^C\frac{1}{12}b(\Delta h^C)^3 \tag{5.24}$$

$$= \frac{E^F b(\Delta h^F)^3}{6} + \frac{E^F b\Delta h^F(\Delta h^C + \Delta h^F)^2}{2} + \frac{E^C b(\Delta h^C)^3}{12} \tag{5.25}$$

$$= \underbrace{\frac{E^F b(\Delta h^F)^3}{6}}_{\overline{EI}_{y,\mathrm{F}\square}} + \underbrace{\frac{E^F b\Delta h^F(h_c)^2}{2}}_{\overline{EI}_{y,\mathrm{FSt}}} + \underbrace{\frac{E^C b(\Delta h^C)^3}{12}}_{\overline{EI}_{y,\mathrm{C}\square}} . \tag{5.26}$$

The average bending stiffness is therefore made up of three parts: Part $\overline{EI}_{y,\mathrm{F}\square}$, which describes the bending stiffness with regard to the partial center of area of the face sheet, part $\overline{EI}_{y,\mathrm{FSt}}$, which describes the Steiner part (i.e., the additional part from the parallel axis theorem) of the face sheets with regard to the overall center of area

---

[3] The thickness of the adhesive layer is usually neglected when considering the different layers.

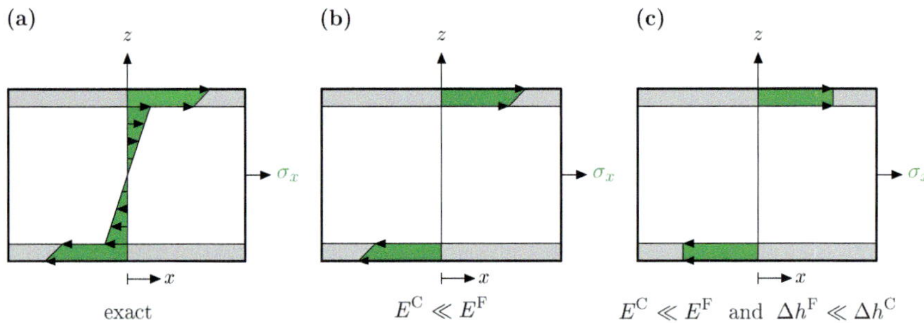

Fig. 5.8 Normal stress distribution due to bending load: (**a**) exact, (**b**) soft core, (**c**) soft core and thin face sheets

and the part $\overline{EI}_{y,\mathrm{C}\square}$, which relates to the bending stiffness of the core of the overall center of area.[4,5]

This results in stress distribution in the core $\left(-\frac{\Delta h^{\mathrm{C}}}{2} \leq z \leq \frac{\Delta h^{\mathrm{C}}}{2}\right)$ according to Eq. (5.2)

$$\sigma_{x,\mathrm{C}}(z) = \frac{M_y(x)E^{\mathrm{C}}}{\dfrac{E^{\mathrm{F}}b(\Delta h^{\mathrm{F}})^3}{6} + \dfrac{E^{\mathrm{F}}b\Delta h^{\mathrm{F}}(h_{\mathrm{c}})^2}{2} + \dfrac{E^{\mathrm{C}}b(\Delta h^{\mathrm{C}})^3}{12}} \times z, \tag{5.27}$$

or in the upper face sheet $\left(\frac{\Delta h^{\mathrm{C}}}{2} \leq z \leq \frac{\Delta h^{\mathrm{C}}}{2} + \Delta h^{\mathrm{F}}\right)$ or in the lower face sheet $\left(-\frac{\Delta h^{\mathrm{C}}}{2} - \Delta h^{\mathrm{F}} \leq z \leq -\frac{\Delta h^{\mathrm{C}}}{2}\right)$:

$$\sigma_{x,\mathrm{F}}(z) = \frac{M_y(x)E^{\mathrm{F}}}{\dfrac{E^{\mathrm{F}}b(\Delta h^{\mathrm{F}})^3}{6} + \dfrac{E^{\mathrm{F}}b\Delta h^{\mathrm{F}}(h_{\mathrm{c}})^2}{2} + \dfrac{E^{\mathrm{C}}b(\Delta h^{\mathrm{C}})^3}{12}} \times z. \tag{5.28}$$

The graphical representation of these two curves is shown in Fig. 5.8a.

For certain geometry or material pairings, the bending stiffness can be further simplified according to Eq. (5.26). First, let us examine for what ratio $h_{\mathrm{c}}/\Delta h^{\mathrm{F}}$ is the bending stiffness of the face sheets $(\overline{EI}_{y,\mathrm{F}\square})$ neglectable compared to the Steiner fraction $(\overline{EI}_{y,\mathrm{F}\mathrm{St}})$. If a limit value of 1% is set, the result from Eq. (5.26) is:

---

[4] In the case of the core, this is identical to the partial center of area of the core.
[5] The symbol "$\square$" stands for a part in relation to a center of area. The abbreviation "St" for a "Steiner fraction" (resulting from the parallel axis theorem).

$$\frac{\frac{E^{\mathrm{F}} b (\Delta h^{\mathrm{F}})^3}{6}}{\frac{E^{\mathrm{F}} b \Delta h^{\mathrm{F}} (h_{\mathrm{c}})^2}{2}} \leq 0.01 \,, \tag{5.29}$$

$$\Leftrightarrow \quad \frac{(\Delta h^{\mathrm{F}})^2}{3 (h_{\mathrm{c}})^2} \leq 0.01 \,, \tag{5.30}$$

$$\Rightarrow \quad \frac{h_{\mathrm{c}}}{\Delta h^{\mathrm{F}}} \geq \sqrt{\frac{1}{0.03}} = 5.77 \,. \tag{5.31}$$

If one also considers the relationship $h_{\mathrm{c}} = \Delta h^{\mathrm{C}} + \Delta h^{\mathrm{F}}$ in the last equation, the following limit value condition results:

$$\frac{\Delta h^{\mathrm{C}}}{\Delta h^{\mathrm{F}}} \geq 4.77 \,. \tag{5.32}$$

Thus, for thin face sheets, i.e. $\Delta h^{\mathrm{F}} \ll \Delta h^{\mathrm{C}}$, the part of the bending stiffness of the face sheets $(\overline{EI}_{y,\mathrm{F}\square})$ can be neglected compared to the Steiner fraction $(\overline{EI}_{y,\mathrm{F}\mathrm{St}})$:

$$\overline{EI}_y \approx \underbrace{\frac{E^{\mathrm{F}} b \Delta h^{\mathrm{F}} (h_{\mathrm{c}})^2}{2}}_{\overline{EI}_{y,\mathrm{F}\mathrm{St}}} + \underbrace{\frac{E^{\mathrm{C}} b (\Delta h^{\mathrm{C}})^3}{12}}_{\overline{EI}_{y,\mathrm{C}\square}} \,. \tag{5.33}$$

A further simplification can be made by using

$$\frac{h_{\mathrm{c}}}{\Delta h^{\mathrm{C}}} \approx 1 \,. \tag{5.34}$$

According to

$$\frac{h_{\mathrm{c}}}{\Delta h^{\mathrm{C}}} = \frac{\Delta h^{\mathrm{C}} + \Delta h^{\mathrm{F}}}{\Delta h^{\mathrm{C}}} = 1 + \underbrace{\frac{\Delta h^{\mathrm{F}}}{\Delta h^{\mathrm{C}}}}_{< 0.01} \approx 1 \,, \tag{5.35}$$

this further simplification is permissible if

$$\frac{\Delta h^{\mathrm{F}}}{\Delta h^{\mathrm{C}}} < 0.01 \quad \text{or} \quad \frac{\Delta h^{\mathrm{C}}}{\Delta h^{\mathrm{F}}} > 100 \tag{5.36}$$

holds. Thus, the classification of sandwich structures shown in Table 5.1 can also be made in relation to the thickness ratio of core to face sheet[6].

---

[6] In the following, however, no distinction is made between 'thin' and 'very thin' face sheets.

Table 5.1 Classification of sandwich structures with regard to the thickness ratio of core to face sheet, see [1]

Description	Thickness ratio
Thick face sheets	$\dfrac{\Delta h^{C}}{\Delta h^{F}} < 4.77$
Thin face sheets	$100 \geq \dfrac{\Delta h^{C}}{\Delta h^{F}} \geq 4.77$
Very thin face sheets	$\dfrac{\Delta h^{C}}{\Delta h^{F}} > 100$

Next, it is examined for which condition the bending stiffness of the core $(\overline{EI}_{y,\mathrm{C}\square})$ can be neglected compared to the bending stiffness of the Steiner fraction $(\overline{EI}_{y,\mathrm{F^{St}}})$. If a limit value of 1% is also set here, the result from Eq. (5.26) is:

$$\frac{\frac{E^{C}b(\Delta h^{C})^{3}}{12}}{\frac{E^{F}b\Delta h^{F}(h_{c})^{2}}{2}} \leq 0.01 \,, \tag{5.37}$$

$$\frac{6E^{F}\Delta h^{F}(h_{c})^{2}}{E^{C}(\Delta h^{C})^{3}} \geq 100 \,. \tag{5.38}$$

Thus, for soft cores[7], i.e. $E^{C} \ll E^{F}$, the bending stiffness of the core can be neglected compared to the Steiner fraction:

$$\overline{EI}_{y} \approx \underbrace{\frac{E^{F}b(\Delta h^{F})^{3}}{6}}_{\overline{EI}_{y,\mathrm{F}\square}} + \underbrace{\frac{E^{F}b\Delta h^{F}(h_{c})^{2}}{2}}_{\overline{EI}_{y,\mathrm{F^{St}}}} \,. \tag{5.39}$$

From this, Eq. (5.2) approximately gives the stress distribution in the core $\left(-\frac{\Delta h^{C}}{2} \leq z \leq \frac{\Delta h^{C}}{2}\right)$ to

$$\sigma_{x,\mathrm{C}}(z) \approx \frac{M_{y}(x)E^{C}}{\overline{EI}_{y}} \times z \tag{5.40}$$

$$= \frac{M_{y}(x)E^{C}}{\dfrac{E^{F}b(\Delta h^{F})^{3}}{6} + \dfrac{E^{F}b\Delta h^{F}(h_{c})^{2}}{2}} \times z$$

$$= \underbrace{\frac{E^{C}}{E^{F}}}_{\ll 1} \times \frac{M_{y}(x)}{\dfrac{b(\Delta h^{F})^{3}}{6} + \dfrac{b\Delta h^{F}(h_{c})^{2}}{2}} \times z \approx 0 \,, \tag{5.41}$$

---

[7] Taking typical values for applications of $\frac{\Delta h^{F}}{\Delta h^{C}} = 0.02\ldots 0.1$ and $\frac{h_{c}}{\Delta h^{C}} \approx 1$, the resulting stiffness relationships are $\frac{E^{F}}{E^{C}} = 833\ldots 167$, see [1].

or in the upper face sheet $\left(\frac{\Delta h^{C}}{2} \leq z \leq \frac{\Delta h^{C}}{2} + \Delta h^{F}\right)$ or in the lower face sheet $\left(-\frac{\Delta h^{C}}{2} - \Delta h^{F} \leq z \leq -\frac{\Delta h^{C}}{2}\right)$:

$$\sigma_{x,\mathrm{F}}(z) \approx \frac{M_y(x)E^{\mathrm{F}}}{\overline{EI}_y} \times z \tag{5.42}$$

$$= \frac{M_y(x)}{\dfrac{b(\Delta h^{\mathrm{F}})^3}{6} + \dfrac{b\Delta h^{\mathrm{F}}(h_{\mathrm{c}})^2}{2}} \times z . \tag{5.43}$$

The graphical representation of these two curves is shown in Fig. 5.8b.

Both simplifications, i.e. thin face sheets and soft core can also be combined, so that for $\Delta h^{\mathrm{F}} \ll \Delta h^{\mathrm{C}}$ and $E^{\mathrm{C}} \ll E^{\mathrm{F}}$ the total average stiffness is only determined by the Steiner fraction of the face sheets:

$$\overline{EI}_y \approx \underbrace{\frac{E^{\mathrm{F}}b\Delta h^{\mathrm{F}}(h_{\mathrm{c}})^2}{2}}_{\overline{EI}_{y,\mathrm{FSt}}} . \tag{5.44}$$

From this, Eq. (5.2) approximately gives the stress distribution in the core $\left(-\frac{\Delta h^{C}}{2} \leq z \leq \frac{\Delta h^{C}}{2}\right)$ to

$$\sigma_{x,\mathrm{C}}(z) \approx 0 , \tag{5.45}$$

or in the upper face sheet $\left(\frac{\Delta h^{C}}{2} \leq z \leq \frac{\Delta h^{C}}{2} + \Delta h^{F}\right)$ or in the lower face sheet $\left(-\frac{\Delta h^{C}}{2} - \Delta h^{F} \leq z \leq -\frac{\Delta h^{C}}{2}\right)$:

$$\sigma_{x,\mathrm{F}}(z) \approx \frac{M_y(x)E^{\mathrm{F}}}{\overline{EI}_y} \times z = \frac{M_y(x)}{\dfrac{b\Delta h^{\mathrm{F}}(h_{\mathrm{c}})^2}{2}} \times z \tag{5.46}$$

$$= \frac{2M_y(x)}{b\Delta h^{\mathrm{F}}(\Delta h^{\mathrm{C}} + \Delta h^{\mathrm{F}})^2} \times z = \frac{2M_y(x)}{b\Delta h^{\mathrm{F}}(h^{\mathrm{C}})^2 \left(1 + \underbrace{\frac{\Delta h^{\mathrm{F}}}{\Delta h^{\mathrm{C}}}}_{\ll 1}\right)^2} \times z$$

$$= \frac{2M_y(x)}{b\Delta h^{\mathrm{F}}(h^{\mathrm{C}})^2} \times z . \tag{5.47}$$

Since the change in the normal stress in the face sheet with the z-coordinate occurs over a very small thickness Δh^{F}, the course of the function can also be approximated by the function value in the middle of the face sheet:

Table 5.2 Formulations of the average bending stiffness $\overline{EI}_y = \sum\limits_{k=1}^{3} E^k I_y^k$

Formulation of $\overline{EI}_y$	Remark
general case	
$\underbrace{\dfrac{E^F b(\Delta h^F)^3}{6}}_{\overline{EI}_{y,F\square}} + \underbrace{\dfrac{E^F b\Delta h^F(h_c)^2}{2}}_{\overline{EI}_{y,FSt}} + \underbrace{\dfrac{E^C b(\Delta h^C)^3}{12}}_{\overline{EI}_{y,C\square}}$	with bending stiffness of the face sheets and the core as well as Steiner's share of the face sheets
thin face sheets	
$\underbrace{\dfrac{E^F b\Delta h^F(h_c)^2}{2}}_{\overline{EI}_{y,FSt}} + \underbrace{\dfrac{E^C b(\Delta h^C)^3}{12}}_{\overline{EI}_{y,C\square}}$	with bending stiffness of the core and Steiner's share of the face sheets
soft core	
$\underbrace{\dfrac{E^F b(\Delta h^F)^3}{6}}_{\overline{EI}_{y,F\square}} + \underbrace{\dfrac{E^F b\Delta h^F(h_c)^2}{2}}_{\overline{EI}_{y,FSt}}$	with bending stiffness and Steiner's share of the face sheets
thin face sheets and soft core	
$\underbrace{\dfrac{E^F b\Delta h^F(h_c)^2}{2}}_{\overline{EI}_{y,FSt}}$	only Steiner's share of the face sheets

$$\sigma_{x,F}(z) \approx \sigma_{x,F}\left(z = \frac{\Delta h^C}{2} + \frac{\Delta h^F}{2}\right) = \frac{2M_y(x)}{b\Delta h^F(\Delta h^C)^2} \times \frac{\Delta h^C + \Delta h^F}{2} \tag{5.48}$$

$$= \frac{M_y(x)}{b\Delta h^F(\Delta h^C)^2} \times \Delta h^C\left(1 + \underbrace{\frac{\Delta h^F}{\Delta h^C}}_{\ll 1}\right) = \frac{M_y(x)}{b\Delta h^F \Delta h^C}. \tag{5.49}$$

The graphical representation of the curves according to Eqs. (5.45) and (5.49) shows Fig. 5.8c.

The various expressions of the average bending stiffness of the considered cases are shown in Table 5.2 for comparison.

At the end of this section, some universal design curves should be derived. For a technical sandwich according to Fig. 5.7 under bending load (assume a cantilever with end shear force) derive normalized design curves depending on the normalized core thickness $\frac{\Delta h^C}{\Delta h^F}$. The normalization should be done with the limit without core ($\Delta h^C = 0$). The following relationships are sought:

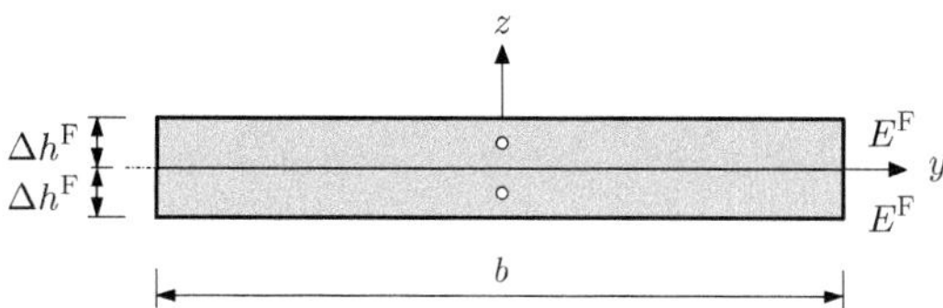

Fig. 5.9 Limit case of a technical sandwich element without a core layer

- average bending stiffness,
- strength (max. force) according to stress criterion,
- mass,
- lightweight index.

The derived relationships should be compared with the special case for thin face sheets and a soft core.

The limit case to be considered for the normalization is shown in Fig. 5.9. For this limit case, the average bending stiffness $\overline{EI_{y}}_{,\mathrm{ref}}$ is:

$$\overline{EI_{y}}_{,\mathrm{ref}} = E^{\mathrm{F}} \frac{1}{12} b \left(2\Delta h^{\mathrm{F}}\right)^{3} = \frac{2}{3} E^{\mathrm{F}} b \left(\Delta h^{\mathrm{F}}\right)^{3} . \tag{5.50}$$

Based on the requirement that the maximum stress must not exceed the 0.2% initial yield stress, the maximum force that can be endured can be determined:

$$\sigma_{x,\max} = \frac{F_{0} L}{\frac{2}{3} b (\Delta h^{\mathrm{F}})^{3}} \times \Delta h^{\mathrm{F}} \overset{!}{=} R^{\mathrm{F}}_{\mathrm{p0.2}} . \tag{5.51}$$

This results in the maximum bearable force:

$$F_{0,\mathrm{ref}} = \frac{2b \left(\Delta h^{\mathrm{F}}\right)^{2} R^{\mathrm{F}}_{\mathrm{p0.2}}}{3L} . \tag{5.52}$$

The mass results from density and volume:

$$m_{\mathrm{ref}} = \varrho^{\mathrm{F}} V = \varrho^{\mathrm{F}} L A = 2\varrho^{\mathrm{F}} L b \Delta h^{\mathrm{F}} . \tag{5.53}$$

Finally, the lightweight index of the reference configuration is:

$$M_{\mathrm{ref}} = \frac{F_{0,\mathrm{ref}}}{m_{\mathrm{ref}} \times g} = \frac{R^{\mathrm{F}}_{\mathrm{p0.2}}}{3\varrho^{\mathrm{F}} g \frac{L^{2}}{\Delta h^{\mathrm{F}}}} . \tag{5.54}$$

The average bending stiffness of the sandwich with three layers results from Eq. (5.26) as follows:

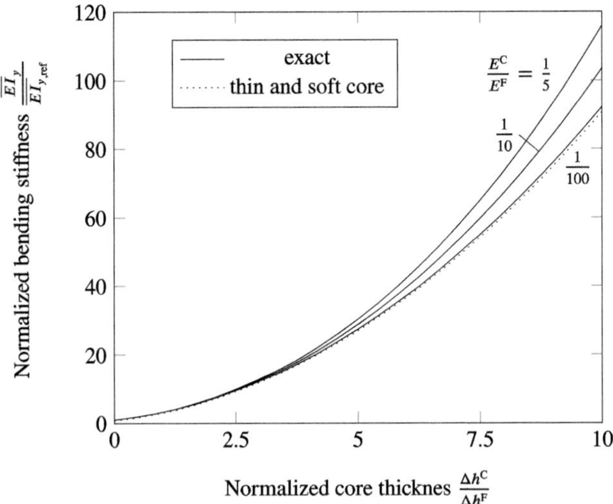

Fig. 5.10 Normalized average bending stiffness for sandwich elements

$$\overline{EI}_y = \underbrace{\frac{E^F b (\Delta h^F)^3}{6}}_{\overline{EI}_{y,\mathrm{F}\square}} + \underbrace{\frac{E^F b \Delta h^F (h_c)^2}{2}}_{\overline{EI}_{y,\mathrm{FSt}}} + \underbrace{\frac{E^C b (\Delta h^C)^3}{12}}_{\overline{EI}_{y,\mathrm{C}\square}} \tag{5.55}$$

$$= \frac{2}{3} E^F b \left[\frac{(\Delta h^F)^3}{4} + \frac{3 \Delta h^F (\Delta h^C + \Delta h^F)^2}{4} + \frac{E^C (\Delta h^C)^3}{8 E^F} \right]$$

$$= \underbrace{\frac{2}{3} E^F b (\Delta h^F)^3}_{\overline{EI}_{y,\mathrm{ref}}} \left[\frac{1}{4} + \frac{3}{4} \left(\frac{\Delta h^C}{\Delta h^F} + 1 \right)^2 + \frac{1}{8} \left(\frac{E^C}{E^F} \right) \left(\frac{\Delta h^C}{\Delta h^F} \right)^3 \right], \tag{5.56}$$

or in the normalized representation (see Fig. 5.10):

$$\frac{\overline{EI}_y}{\overline{EI}_{y,\mathrm{ref}}} = \left[\frac{1}{4} + \frac{3}{4} \left(\frac{\Delta h^C}{\Delta h^F} + 1 \right)^2 + \frac{1}{8} \left(\frac{E^C}{E^F} \right) \left(\frac{\Delta h^C}{\Delta h^F} \right)^3 \right]. \tag{5.57}$$

For thin face sheets and soft cores, this relationship can be significantly simplified:

$$\frac{\overline{EI}_y}{\overline{EI}_{y,\mathrm{ref}}} \approx \frac{3}{4} \left(\frac{\Delta h^C}{\Delta h^F} + 1 \right)^2. \tag{5.58}$$

The mass of the sandwich is composed of the contribution of the face sheets (see Eq. (5.53)) and the core:

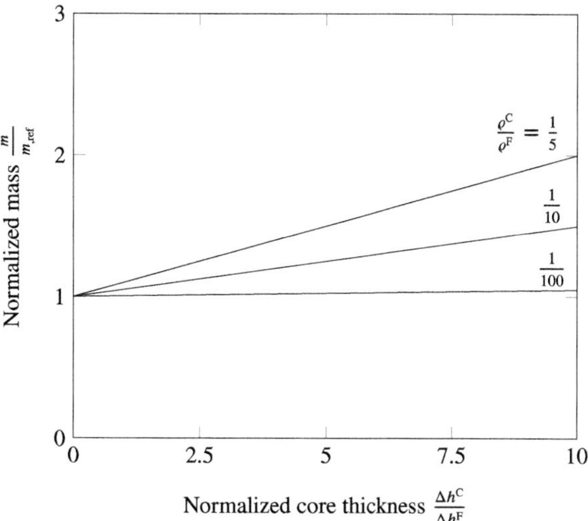

Fig. 5.11 Normalized mass for sandwich elements

or in the normalized representation (see Fig. 5.11):

$$m = \varrho^{\mathrm{F}} A^{\mathrm{F}} L + \varrho^{\mathrm{C}} A^{\mathrm{C}} L = \underbrace{2\varrho^{\mathrm{F}} L b \Delta h^{\mathrm{F}}}_{m_{\mathrm{ref}}} \left(1 + \frac{1}{2}\frac{\varrho^{\mathrm{C}}}{\varrho^{\mathrm{F}}}\frac{\Delta h^{\mathrm{C}}}{\Delta h^{\mathrm{F}}}\right), \qquad (5.59)$$

$$\frac{m}{m_{\mathrm{ref}}} = 1 + \frac{1}{2}\frac{\varrho^{\mathrm{C}}}{\varrho^{\mathrm{F}}}\left(\frac{\Delta h^{\mathrm{C}}}{\Delta h^{\mathrm{F}}}\right). \qquad (5.60)$$

The maximum force that can be endured under a stress criterion[8] results from Eq. (5.2) to:

$$\sigma_{x,\mathrm{F,max}} = \frac{M_{y,\mathrm{max}} E^{\mathrm{F}}}{EI_y} \times z_{\mathrm{max}} = \frac{F_0 L E^{\mathrm{F}}}{EI_y} \times \left(\frac{\Delta h^{\mathrm{C}}}{2} + \Delta h^{\mathrm{F}}\right) \overset{!}{=} R^{\mathrm{F}}_{\mathrm{p0.2}}. \qquad (5.61)$$

This results in the maximum force:

$$F_0 = \frac{R^{\mathrm{F}}_{\mathrm{p0.2}}}{L E^{\mathrm{F}}} \times \frac{1}{\dfrac{\Delta h^{\mathrm{C}}}{2} + \Delta h^{\mathrm{F}}} \times \overline{EI_y} \qquad (5.62)$$

$$= \frac{R^{\mathrm{F}}_{\mathrm{p0.2}}}{L E^{\mathrm{F}}} \times \frac{1}{\dfrac{\Delta h^{\mathrm{C}}}{2} + \Delta h^{\mathrm{F}}} \times \frac{2}{3} E^{\mathrm{F}} b \left(\Delta h^{\mathrm{F}}\right)^3 \times$$

---

[8] Here, for the sake of simplicity, it is assumed that the failure occurs in the face sheet. Core failure should be excluded at this point.

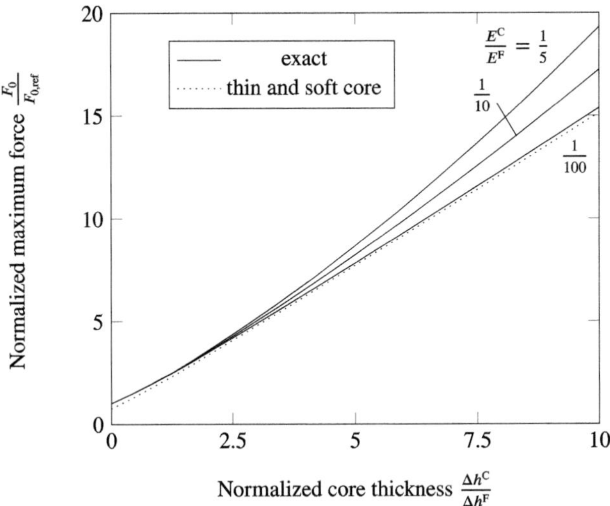

Fig. 5.12 Normalized maximum force for sandwich elements

$$\times \left[\frac{1}{4} + \frac{3}{4}\left(\frac{\Delta h^{\mathrm{C}}}{\Delta h^{\mathrm{F}}} + 1 \right)^2 + \frac{1}{8}\left(\frac{E^{\mathrm{C}}}{E^{\mathrm{F}}} \right)\left(\frac{\Delta h^{\mathrm{C}}}{\Delta h^{\mathrm{F}}} \right)^3 \right] \tag{5.63}$$

$$= \underbrace{\frac{2b\left(\Delta h^{\mathrm{F}}\right)^2 R^{\mathrm{F}}_{\mathrm{p0.2}}}{3L}}_{F_{0,\mathrm{ref}}} \times \frac{\Delta h^{\mathrm{F}}}{\frac{\Delta h^{\mathrm{C}}}{2} + \Delta h^{\mathrm{F}}} \times$$

$$\times \left[\frac{1}{4} + \frac{3}{4}\left(\frac{\Delta h^{\mathrm{C}}}{\Delta h^{\mathrm{F}}} + 1 \right)^2 + \frac{1}{8}\left(\frac{E^{\mathrm{C}}}{E^{\mathrm{F}}} \right)\left(\frac{\Delta h^{\mathrm{C}}}{\Delta h^{\mathrm{F}}} \right)^3 \right], \tag{5.64}$$

or in the normalized representation (see Fig. 5.12):

$$\frac{F_0}{F_{0,\mathrm{ref}}} = \frac{1}{1 + \frac{1}{2}\left(\frac{\Delta h^{\mathrm{C}}}{\Delta h^{\mathrm{F}}} \right)} \left[\frac{1}{4} + \frac{3}{4}\left(\frac{\Delta h^{\mathrm{C}}}{\Delta h^{\mathrm{F}}} + 1 \right)^2 + \frac{1}{8}\left(\frac{E^{\mathrm{C}}}{E^{\mathrm{F}}} \right)\left(\frac{\Delta h^{\mathrm{C}}}{\Delta h^{\mathrm{F}}} \right)^3 \right]. \tag{5.65}$$

For thin face sheets and soft cores, this relationship can be significantly simplified:

$$\frac{F_0}{F_{0,\mathrm{ref}}} \approx \frac{1}{1 + \frac{1}{2}\left(\frac{\Delta h^{\mathrm{C}}}{\Delta h^{\mathrm{F}}} \right)} \times \frac{3}{4}\left(\frac{\Delta h^{\mathrm{C}}}{\Delta h^{\mathrm{F}}} + 1 \right)^2. \tag{5.66}$$

Finally, the lightweight index of the sandwich based on Eqs. (5.65) and (5.60) results in:

$$
M = \frac{F_0}{mg} = \frac{\dfrac{2b\left(\Delta h^{\mathrm{F}}\right)^2 R^{\mathrm{F}}_{\mathrm{p}0.2}}{3L} \times \dfrac{\Delta h^{\mathrm{F}}}{\frac{\Delta h^{\mathrm{C}}}{2} + \Delta h^{\mathrm{F}}}}{g2\varrho^{\mathrm{F}} Lb\Delta h^{\mathrm{F}}\left(1 + \dfrac{1}{2}\dfrac{\varrho^{\mathrm{C}}}{\varrho^{\mathrm{F}}}\dfrac{\Delta h^{\mathrm{C}}}{\Delta h^{\mathrm{F}}}\right)} \times
$$

$$
\times \left[\frac{1}{4} + \frac{3}{4}\left(\frac{\Delta h^{\mathrm{C}}}{\Delta h^{\mathrm{F}}} + 1\right)^2 + \frac{1}{8}\left(\frac{E^{\mathrm{C}}}{E^{\mathrm{F}}}\right)\left(\frac{\Delta h^{\mathrm{C}}}{\Delta h^{\mathrm{F}}}\right)^3\right] \tag{5.67}
$$

$$
= \underbrace{\frac{R^{\mathrm{F}}_{\mathrm{p}0.2}}{3\varrho^{\mathrm{F}} g \dfrac{L^2}{\Delta h^{\mathrm{F}}}}}_{M_{\mathrm{ref}}} \times \frac{1}{1 + \frac{1}{2}\left(\frac{\Delta h^{\mathrm{C}}}{\Delta h^{\mathrm{F}}}\right)} \times
$$

$$
\times \frac{\left[\dfrac{1}{4} + \dfrac{3}{4}\left(\dfrac{\Delta h^{\mathrm{C}}}{\Delta h^{\mathrm{F}}} + 1\right)^2 + \dfrac{1}{8}\left(\dfrac{E^{\mathrm{C}}}{E^{\mathrm{F}}}\right)\left(\dfrac{\Delta h^{\mathrm{C}}}{\Delta h^{\mathrm{F}}}\right)^3\right]}{\left(1 + \dfrac{1}{2}\dfrac{\varrho^{\mathrm{C}}}{\varrho^{\mathrm{F}}}\dfrac{\Delta h^{\mathrm{C}}}{\Delta h^{\mathrm{F}}}\right)}, \tag{5.68}
$$

or in the normalized representation (see Fig. 5.13):

$$
\frac{M}{M_{\mathrm{ref}}} = \frac{\left[\dfrac{1}{4} + \dfrac{3}{4}\left(\dfrac{\Delta h^{\mathrm{C}}}{\Delta h^{\mathrm{F}}} + 1\right)^2 + \dfrac{1}{8}\left(\dfrac{E^{\mathrm{C}}}{E^{\mathrm{F}}}\right)\left(\dfrac{\Delta h^{\mathrm{C}}}{\Delta h^{\mathrm{F}}}\right)^3\right]}{\left(1 + \dfrac{1}{2}\left(\dfrac{\Delta h^{\mathrm{C}}}{\Delta h^{\mathrm{F}}}\right)\right) \times \left(1 + \dfrac{1}{2}\dfrac{\varrho^{\mathrm{C}}}{\varrho^{\mathrm{F}}}\dfrac{\Delta h^{\mathrm{C}}}{\Delta h^{\mathrm{F}}}\right)}. \tag{5.69}
$$

For thin face sheets and soft cores, this relationship can be significantly simplified:

$$
\frac{M}{M_{\mathrm{ref}}} = \frac{\dfrac{3}{4}\left(\dfrac{\Delta h^{\mathrm{C}}}{\Delta h^{\mathrm{F}}} + 1\right)^2}{\left(1 + \dfrac{1}{2}\left(\dfrac{\Delta h^{\mathrm{C}}}{\Delta h^{\mathrm{F}}}\right)\right) \times \left(1 + \dfrac{1}{2}\dfrac{\varrho^{\mathrm{C}}}{\varrho^{\mathrm{F}}}\dfrac{\Delta h^{\mathrm{C}}}{\Delta h^{\mathrm{F}}}\right)}. \tag{5.70}
$$

5.5.2   Tensile/Compressive Load

For this case, according to Eq. (5.11), the mean tensile stiffness for the configuration according to Fig. 5.7 can be set as follows:

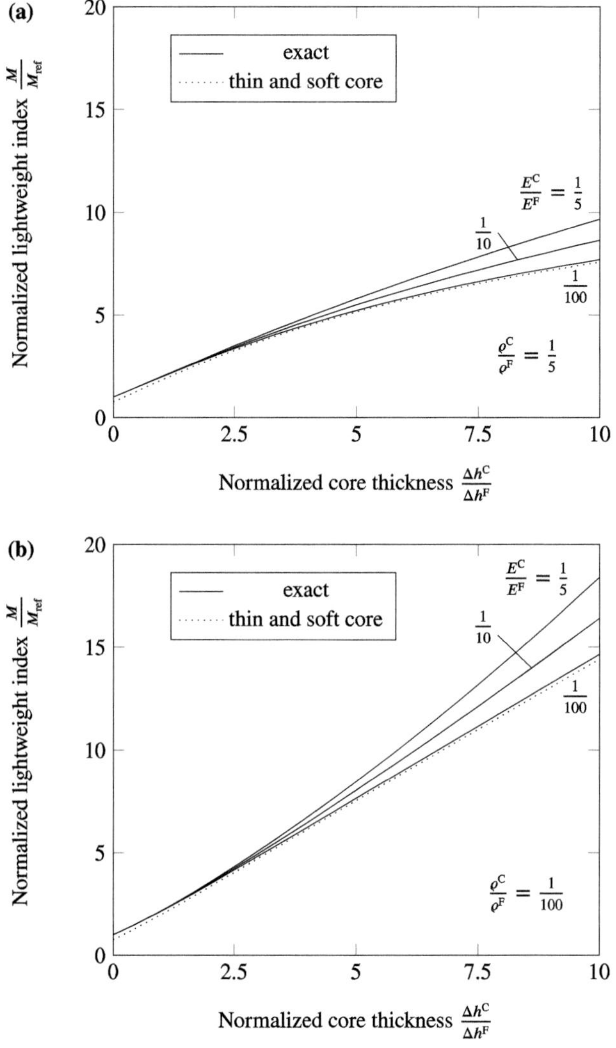

Fig. 5.13 Normalized lightweight index for sandwich elements under bending load: (a) density ratio $\frac{\varrho^{\mathrm{C}}}{\varrho^{\mathrm{F}}} = \frac{1}{5}$ and (b) density ratio $\frac{\varrho^{\mathrm{C}}}{\varrho^{\mathrm{F}}} = \frac{1}{100}$

$$\overline{EA} = \sum_{k=1}^{3} E^k \Delta h^k b = E^{\mathrm{F}} \Delta h^{\mathrm{F}} b + E^{\mathrm{C}} \Delta h^{\mathrm{C}} b + E^{\mathrm{F}} \Delta h^{\mathrm{F}} b$$

$$= \underbrace{2 E^{\mathrm{F}} \Delta h^{\mathrm{F}} b}_{\overline{EA}_{\mathrm{F}}} + \underbrace{E^{\mathrm{C}} \Delta h^{\mathrm{C}} b}_{\overline{EA}_{\mathrm{C}}} \, . \tag{5.71}$$

The mean tensile stiffness is therefore made up of two parts: The part $\overline{EA}_{\mathrm{F}}$, which describes the tensile stiffness of the face sheets, and the part $\overline{EA}_{\mathrm{C}}$, which describes the tensile stiffness of the core.

This results in the stress distribution in the core $\left(-\frac{\Delta h^{\mathrm{C}}}{2} \leq z \leq \frac{\Delta h^{\mathrm{C}}}{2}\right)$ according to Eq. (5.13)

$$\sigma_{x,\mathrm{C}} = \frac{F_0 E^{\mathrm{C}}}{2E^{\mathrm{F}}\Delta h^{\mathrm{F}}b + E^{\mathrm{C}}\Delta h^{\mathrm{C}}b}, \tag{5.72}$$

or in the upper face sheet $\left(\frac{\Delta h^{\mathrm{C}}}{2} \leq z \leq \frac{\Delta h^{\mathrm{C}}}{2} + \Delta h^{\mathrm{F}}\right)$ or in the lower face sheet $\left(-\frac{\Delta h^{\mathrm{C}}}{2} - \Delta h^{\mathrm{F}} \leq z \leq -\frac{\Delta h^{\mathrm{C}}}{2}\right)$:

$$\sigma_{x,\mathrm{F}} = \frac{F_0 E^{\mathrm{F}}}{2E^{\mathrm{F}}\Delta h^{\mathrm{F}}b + E^{\mathrm{C}}\Delta h^{\mathrm{C}}b}. \tag{5.73}$$

The graphical representation of these two courses is shown in Fig. 5.14a.

For soft cores, i.e. $E^{\mathrm{C}} \ll E^{\mathrm{F}}$, the tensile stiffness of the core can be neglected compared to the face sheets:

$$\overline{EA} \approx E^{\mathrm{F}}\left(2\Delta h^{\mathrm{F}}b + \underbrace{\frac{E^{\mathrm{C}}}{E^{\mathrm{F}}}}_{\ll 1}\Delta h^{\mathrm{C}}b\right) = 2E^{\mathrm{F}}\Delta h^{\mathrm{F}}b. \tag{5.74}$$

This results in the normal stress distribution (see Fig. 5.14b) in the core

$$\sigma_{x,\mathrm{C}} \approx \frac{F_0 E^{\mathrm{C}}}{2E^{\mathrm{F}}\Delta h^{\mathrm{F}}b} = \underbrace{\frac{E^{\mathrm{C}}}{E^{\mathrm{F}}}}_{\ll 1}\left(\frac{F_0}{2\Delta h^{\mathrm{F}}b}\right) \approx 0, \tag{5.75}$$

or in the upper face sheet $\left(\frac{\Delta h^{\mathrm{C}}}{2} \leq z \leq \frac{\Delta h^{\mathrm{C}}}{2} + \Delta h^{\mathrm{F}}\right)$ or in the lower face sheet $\left(-\frac{\Delta h^{\mathrm{C}}}{2} - \Delta h^{\mathrm{F}} \leq z \leq -\frac{\Delta h^{\mathrm{C}}}{2}\right)$:

$$\sigma_{x,\mathrm{F}} \approx \frac{F_0 E^{\mathrm{F}}}{2E^{\mathrm{F}}\Delta h^{\mathrm{F}}b} = \frac{F_0}{2\Delta h^{\mathrm{F}}b}. \tag{5.76}$$

The approximation formulas do not change if additional thin layers are taken into account.

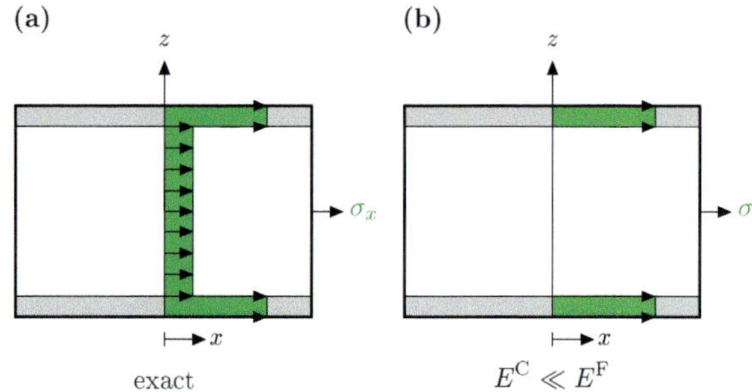

Fig. 5.14 Normal stress distribution under tensile load: (**a**) exact and (**b**) soft core

5.5.3   Shear Load

The exact shear stress distributions result from Eqs. (5.17) and (5.23) for the face sheet

$$
\tau_{zx,\mathrm{F}}(z) = \frac{E^{\mathrm{F}} Q_z(x)}{2\overline{EI_y}} \left[\left(\frac{\Delta h^{\mathrm{C}}}{2} + \Delta h^{\mathrm{F}} \right)^2 - z^2 \right]
\tag{5.77}
$$

$$
= \frac{E^{\mathrm{F}} Q_z(x) \left[\left(\dfrac{\Delta h^{\mathrm{C}}}{2} + \Delta h^{\mathrm{F}} \right)^2 - z^2 \right]}{2 \left(\dfrac{E^{\mathrm{F}} b (\Delta h^{\mathrm{F}})^3}{6} + \dfrac{E^{\mathrm{F}} b \Delta h^{\mathrm{F}} (h_c)^2}{2} + \dfrac{E^{\mathrm{C}} b (\Delta h^{\mathrm{C}})^3}{12} \right)},
\tag{5.78}
$$

or for the core layer

$$
\tau_{zx,\mathrm{C}}(z) = \frac{Q_z(x)}{2\overline{EI_y}} \left[E^{\mathrm{C}} \left(\left(\frac{\Delta h^{\mathrm{C}}}{2} \right)^2 - z^2 \right) + E^{\mathrm{F}} \left(\Delta h^{\mathrm{C}} + \Delta h^{\mathrm{F}} \right) \Delta h^{\mathrm{F}} \right]
\tag{5.79}
$$

$$
= \frac{Q_z(x) \left[E^{\mathrm{C}} \left(\left(\dfrac{\Delta h^{\mathrm{C}}}{2} \right)^2 - z^2 \right) + E^{\mathrm{F}} \left(\Delta h^{\mathrm{C}} + \Delta h^{\mathrm{F}} \right) \Delta h^{\mathrm{F}} \right]}{2 \left(\dfrac{E^{\mathrm{F}} b (\Delta h^{\mathrm{F}})^3}{6} + \dfrac{E^{\mathrm{F}} b \Delta h^{\mathrm{F}} (h_c)^2}{2} + \dfrac{E^{\mathrm{C}} b (\Delta h^{\mathrm{C}})^3}{12} \right)}.
\tag{5.80}
$$

These two parabolic curves arse shown in Fig. 5.15a.

For soft cores, i.e. $E^{\mathrm{C}} \ll E^{\mathrm{F}}$, the average bending stiffness can be approximated using Eq. (5.39) and according to the procedure in Sect. 5.4, the shear stress distribution in the face sheet results from integration over the normal stress gradient:

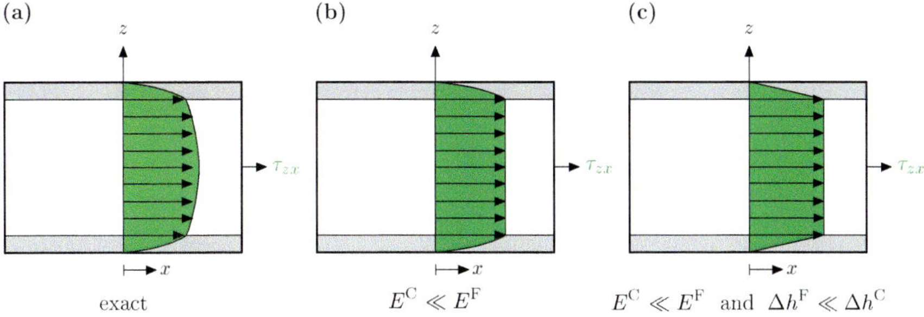

Fig. 5.15 Shear stress distribution under shear load: **(a)** exact, **(b)** soft core, **(c)** soft core and thin face sheets

$$\tau_{zx,\mathrm{F}}(z) = \int\limits_{z}^{\frac{\Delta h^C}{2}+\Delta h^F} \frac{\mathrm{d}\sigma_{x,\mathrm{F}}(x)}{\mathrm{d}x}\,\mathrm{d}z' + c_3 = \int\limits_{z}^{\frac{\Delta h^C}{2}+\Delta h^F} \frac{E^F Q_z(x)}{EI_y} z'\,\mathrm{d}z' + c_3$$

$$= \frac{E^F Q_z(x)}{2EI_y}\left[\left(\frac{\Delta h^C}{2}+\Delta h^F\right)^2 - z^2\right] + c_3 . \tag{5.81}$$

The constant of integration c_3 can also be determined here as $c_3 = 0$ using the condition that no shear stresses occur at the free edge. This finally gives the shear stress distribution in the face sheet:

$$\tau_{zx,\mathrm{F}}(z) = \frac{E^F Q_z(x)}{2EI_y}\left[\left(\frac{\Delta h^C}{2}+\Delta h^F\right)^2 - z^2\right]$$

$$= \frac{Q_z(x)}{\frac{b(\Delta h^F)^3}{3} + b(\Delta h^F)(\Delta h_c)^2}\left[\left(\frac{\Delta h^C}{2}+\Delta h^F\right)^2 - z^2\right]. \tag{5.82}$$

At the transition point between the core and the face sheet, the following value results from Eq. (5.82):

$$\tau_{zx,\mathrm{F}}\left(\frac{\Delta h^C}{2}\right) = \frac{Q_z(x)}{\frac{b(\Delta h^F)^3}{3} + b(\Delta h^F)(h_c)^2}\left[\left(\frac{\Delta h^C}{2}\right)^2 + \Delta h^C\Delta h^F + \left(\Delta h^F\right)^2 - \left(\frac{\Delta h^C}{2}\right)^2\right]$$

$$= \frac{Q_z(x)\left[\Delta h^C + \Delta h^F\right]}{b\left(\frac{(\Delta h^F)^2}{3} + (h_c)^2\right)} = \frac{Q_z(x)h_c}{b\left(\frac{(\Delta h^F)^2}{3} + (h_c)^2\right)} \tag{5.83}$$

$$= \frac{Q_z(x)}{2EI_y}E^F\Delta h^F h_c . \tag{5.84}$$

Accordingly, the shear stress for the core is:

$$\tau_{zx,\mathrm{C}}(z) = \underbrace{\int_{z}^{\frac{\Delta h^{\mathrm{C}}}{2}} \frac{\mathrm{d}\sigma_{x,\mathrm{C}}(x)}{\mathrm{d}x}\,\mathrm{d}z}_{=\,0,\ \text{see Eq.\,5.41}} + c_2 \,. \tag{5.85}$$

The integration constant c_2 results from the transition condition for the stress τ_{zx} between the core and the face sheet, i.e. identical stresses $\tau_{zx,\mathrm{C}}(z = \Delta h^{\mathrm{C}}/2) = \tau_{zx,\mathrm{F}}(z = \Delta h^{\mathrm{C}}/2)$:

$$c_2 = \frac{Q_z(x)}{2\overline{EI}_y}E^{\mathrm{F}}\Delta h^{\mathrm{F}}h_{\mathrm{c}} = \frac{Q_z(x)h_{\mathrm{c}}}{b\left(\frac{(\Delta h^{\mathrm{F}})^2}{3} + (h_{\mathrm{c}})^2\right)}\,, \tag{5.86}$$

and thus also the constant shear stress in the core

$$\tau_{zx,\mathrm{C}}(z) = \frac{Q_z(x)}{2\overline{EI}_y}E^{\mathrm{F}}\Delta h^{\mathrm{F}}h_{\mathrm{c}} = \frac{Q_z(x)h_{\mathrm{c}}}{b\left(\frac{(\Delta h^{\mathrm{F}})^2}{3} + (h_{\mathrm{c}})^2\right)}\,. \tag{5.87}$$

These two courses are shown in Fig. 5.15b.

For soft cores, i.e. $E^{\mathrm{C}} \ll E^{\mathrm{F}}$, and thin face sheets, i.e. $\Delta h^{\mathrm{F}} \ll \Delta h^{\mathrm{C}}$, the shear stress distribution in the face sheet results from integration over the normal stress gradient:

$$\tau_{zx,\mathrm{F}}(z) = \int_{z}^{\frac{\Delta h^{\mathrm{C}}}{2}+\Delta h^{\mathrm{F}}} \frac{\mathrm{d}\sigma_{x,\mathrm{F}}(x)}{\mathrm{d}x}\,\mathrm{d}z' + c_3 = \int_{z}^{\frac{\Delta h^{\mathrm{C}}}{2}+\Delta h^{\mathrm{F}}} \frac{Q_z(x)}{b\,\Delta h^{\mathrm{F}}\Delta h^{\mathrm{C}}}\,\mathrm{d}z' + c_3$$

$$= \frac{Q_z(x)}{b\,\Delta h^{\mathrm{F}}\Delta h^{\mathrm{C}}} \int_{z}^{\frac{\Delta h^{\mathrm{C}}}{2}+\Delta h^{\mathrm{F}}} \mathrm{d}z' + c_3 = \frac{Q_z(x)}{b\,\Delta h^{\mathrm{F}}\Delta h^{\mathrm{C}}}\left[\frac{\Delta h^{\mathrm{C}}}{2} + \Delta h^{\mathrm{F}} - z\right] + c_3 \,. \tag{5.88}$$

Here, too, the constant of integration c_3 can be determined as $c_3 = 0$ using the condition that no shear stresses occur at the free edge. This finally gives the shear stress distribution in the face sheet for soft cores and thin face sheets:

$$\tau_{zx,\mathrm{F}}(z) = \frac{Q_z(x)}{b\,\Delta h^{\mathrm{F}}\Delta h^{\mathrm{C}}}\left[\left(\frac{\Delta h^{\mathrm{C}}}{2} + \Delta h^{\mathrm{F}}\right) - z\right] \,.$$

At the transition point between core and face sheet, Eq. (5.89) results in the following value:

$$\tau_{zx,\mathrm{F}}\left(\tfrac{\Delta h^{\mathrm{C}}}{2}\right) = \frac{Q_z(x)}{b\Delta h^{\mathrm{F}}\Delta h^{\mathrm{C}}}\left[\left(\frac{\Delta h^{\mathrm{C}}}{2}+\Delta h^{\mathrm{F}}\right)-\frac{\Delta h^{\mathrm{C}}}{2}\right] = \frac{Q_z(x)}{b\Delta h^{\mathrm{C}}}. \qquad (5.89)$$

Alternatively, the last relation can also be expressed as follows:

$$\tau_{zx,\mathrm{F}}\left(\tfrac{\Delta h^{\mathrm{C}}}{2}\right) = \frac{Q_z(x)}{b\Delta h^{\mathrm{C}}} \approx \frac{Q_z(x)}{b\Delta h^{\mathrm{C}}\left(1+\frac{\Delta h^{\mathrm{F}}}{\Delta h^{\mathrm{C}}}\right)} = \frac{Q_z(x)}{b(\Delta h^{\mathrm{C}}+\Delta h^{\mathrm{F}})} = \frac{Q_z(x)}{bh_{\mathrm{c}}}. \qquad (5.90)$$

Accordingly, the shear stress for the core is:

$$\tau_{zx,\mathrm{C}}(z) = \underbrace{\int_{z}^{\frac{\Delta h^{\mathrm{C}}}{2}} \frac{\mathrm{d}\sigma_{x,\mathrm{C}}(x)}{\mathrm{d}x}\,\mathrm{d}z}_{=\,0,\ \text{see Eq. 5.45}} +c_2 = \frac{Q_z(x)}{b\Delta h^{\mathrm{C}}} = \frac{Q_z(x)}{bh_{\mathrm{c}}}. \qquad (5.91)$$

These two courses are shown in Fig. 5.15c.

Finally, all stress courses are shown comparatively in Tables 5.3, 5.4 and 5.5.

5.5.4 Bending Deformation of Sandwich Beams

A simple theory for determining the deflection of technical sandwich beams is presented in the following. Here it is assumed that the total deflection $u_z(x)$ is composed of a bending component $u_{z,\mathrm{b}}(x)$ and a shear component $u_{z,\mathrm{s}}(x)$ (see Fig. 5.16):

$$u_z(x) = u_{z,\mathrm{b}}(x) + u_{z,\mathrm{s}}(x). \qquad (5.92)$$

In this simple theory for sandwich beams with thin face sheets and a soft core—also known as the method of partial deflections in the literature, [4]—the bending deformation is calculated using the classical differential equations (see Eqs. (2.9)–(2.11)):

$$\frac{\mathrm{d}^2}{\mathrm{d}x^2}\left(EI_y\frac{\mathrm{d}^2 u_{z,\mathrm{b}}(x)}{\mathrm{d}x^2}\right) = q_z(x), \qquad (5.93)$$

$$\frac{\mathrm{d}}{\mathrm{d}x}\left(EI_y\frac{\mathrm{d}^2 u_{z,\mathrm{b}}(x)}{\mathrm{d}x^2}\right) = -Q_z(x), \qquad (5.94)$$

$$EI_y\frac{\mathrm{d}^2 u_{z,\mathrm{b}}(x)}{\mathrm{d}x^2} = -M_y(x), \qquad (5.95)$$

Table 5.3 Comparison of the normal stress distributions due to bending stress in the sandwich according to the Euler-Bernoulli beam theory. Approximations with soft core ($E^C \ll E^F$) and soft core with thin cover layers ($E^C \ll E^F$, $\Delta h^F \ll \Delta h^C$)

Approach	Face sheets $\sigma_{x,\mathrm{F}}(z)$	Core $\sigma_{x,\mathrm{C}}(z)$	Graphic
Exact	$\dfrac{M_y(x)\,E^F}{\dfrac{E^F b(\Delta h^F)^3}{6} + \dfrac{E^F b\Delta h^F(h_c)^2}{2} + \dfrac{E^C b(\Delta h^C)^3}{12}} \times z$	$\dfrac{M_y(x)\,E^C}{\dfrac{E^F b(\Delta h^F)^3}{6} + \dfrac{E^F b\Delta h^F(h_c)^2}{2} + \dfrac{E^C b(\Delta h^C)^3}{12}} \times z$	
$E^C \ll E^F$	$\dfrac{M_y(x)}{\dfrac{b(\Delta h^F)^3}{6} + \dfrac{b\Delta h^F(h_c)^2}{2}} \times z$	0	
$E^C \ll E^F,\ \Delta h^F \ll \Delta h^C$	$\dfrac{M_y(x)}{b\Delta h^F \Delta h^C}$	0	

Table 5.4 Comparison of the normal stress distributions due to tensile stress in the sandwich according to the bar theory. Soft core approximations ($E^C \ll E^F$)

Approach	Face sheets $\sigma_{x,F}(z)$	Core $\sigma_{x,C}(z)$	Graphic
Exact	$\dfrac{F_0 E^F}{2E^F \Delta h^F b + E^C \Delta h^C b}$	$\dfrac{F_0 E^C}{2E^F \Delta h^F b + E^C \Delta h^C b}$	
$E^C \ll E^F$	$\dfrac{F_0}{2\Delta h^F b}$	0	

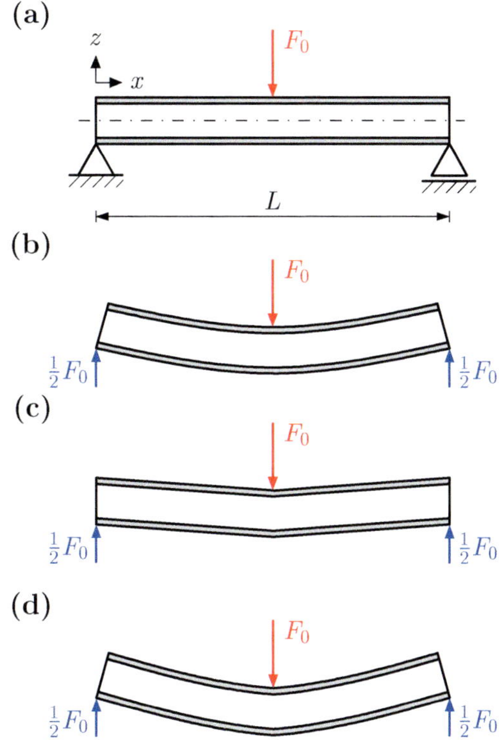

Fig. 5.16 Bending of sandwich beams: (**a**) undeformed, (**b**) bending deformation, (**c**) shear deformation, (**d**) total deformation

Table 5.5 Comparison of the shear stress distributions due to shear force loading in the sandwich according to the Euler-Bernoulli beam theory. Approximations with soft core ($E^C \ll E^F$) and soft core with thin cover layers ($E^C \ll E^F$, $\Delta h^F \ll \Delta h^C$)

Approach	Face sheets $\tau_{zx,F}(z)$	Core $\tau_{zx,C}(z)$	Graphic
Exact	$\dfrac{E^F Q_z(x)\left[\left(\frac{\Delta h^C}{2}+\Delta h^F\right)^2 - z^2\right]}{2\left(\frac{E^F b(\Delta h^F)^3}{6}+\frac{E^F b \Delta h^F (h_c)^2}{2}+\frac{E^C b(\Delta h^C)^3}{12}\right)}$	$\dfrac{Q_z(x)\left[E^C\left(\left(\frac{\Delta h^C}{2}\right)^2 - z^2\right)+E^F\left(\Delta h^C+\Delta h^F\right)\Delta h^F\right]}{2\left(\frac{E^F b(\Delta h^F)^3}{6}+\frac{E^F b \Delta h^F (h_c)^2}{2}+\frac{E^C b(\Delta h^C)^3}{12}\right)}$	
$E^C \ll E^F$	$\dfrac{Q_z(x)}{\frac{b(\Delta h^F)^3}{3}+b(\Delta h^F)(\Delta h_c)^2}\left[\left(\frac{\Delta h^C}{2}+\Delta h^F\right)^2 - z^2\right]$	$\dfrac{Q_z(x)h_c}{b\left(\frac{(\Delta h^F)^2}{3}+(h_c)^2\right)}$	
$E^C \ll E^F,$ $\Delta h^F \ll \Delta h^C$	$\dfrac{Q_z(x)}{b\Delta h^F \Delta h^C}\left[\left(\frac{\Delta h^C}{2}+\Delta h^F\right) - z\right]$	$\dfrac{Q_z(x)}{b\Delta h^C} \approx \dfrac{Q_z(x)}{bh_c}$	

where the mean bending stiffness according to Eq. (5.44) is to be used:

$$\overline{EI_y} \approx \underbrace{\frac{E^F b \Delta h^F (h_c)^2}{2}}_{\overline{EI}_{y,\mathrm{FSt}}}. \tag{5.96}$$

To determine the shear deformation, the shear stress[9] in the core for sandwich beams with thin face sheets and soft core is considered according to Eq. (5.91)

$$\tau_{zx,\mathrm{C}}(x) = \frac{Q_z(x)}{b\Delta h^C} = \frac{Q_z(x)}{bh_c}, \tag{5.97}$$

or the shear strain (angular distortion) is considered using Hooke's law ($\tau = G\gamma$) :

$$\gamma_{zx,\mathrm{C}}(x) = \frac{Q_z(x)}{Gb\Delta h^C} = \frac{Q_z(x)}{Gbh_c}. \tag{5.98}$$

For further derivation of the shear differential equation, consider a deformed sandwich element under pure shear deformation, see Fig. 5.17.

From the right-angled triangle $1'2'3'$, the following geometric relationship results:

$$\frac{\overline{1'2'}}{\overline{2'3'}} = \frac{\overline{1'2'}}{\Delta h^C} = |\tan(\gamma)| \approx |\gamma|. \tag{5.99}$$

Note that the shear strain γ is negative ($\gamma < 0$) in the way drawn in Fig. 5.17b. If one now looks at the right-angled triangle 123 (see Fig. 5.17b and the details in Fig. 5.18), a further geometric relationship results:

$$\frac{\overline{12}}{\overline{23}} = \frac{\overline{12}}{h_c} = |\tan(\alpha)| \approx |\alpha| \approx \frac{\left|\mathrm{d}u_{z,\,s}\right|}{\mathrm{d}x}. \tag{5.100}$$

Finally, the kinematic relationship, i.e., the relationship between the deformation and the shear strain, results using the geometric identity $\overline{1'2'} = \overline{12}$:

$$\frac{\mathrm{d}u_{z,\,s}}{\mathrm{d}x} = \frac{\Delta h^C}{h_c} \times \gamma. \tag{5.101}$$

If one also considers the relation for the shear strain according to Eq. (5.98), i.e. the combination of the equilibrium with the constitutive law, the shear differential equation results in:

---

[9] This relationship can also be seen as the basic equation of equilibrium, i.e. between the inner reactions (here: τ_{xy}) and the outer loads (here: Q_z, as a reaction to the outer shear forces).

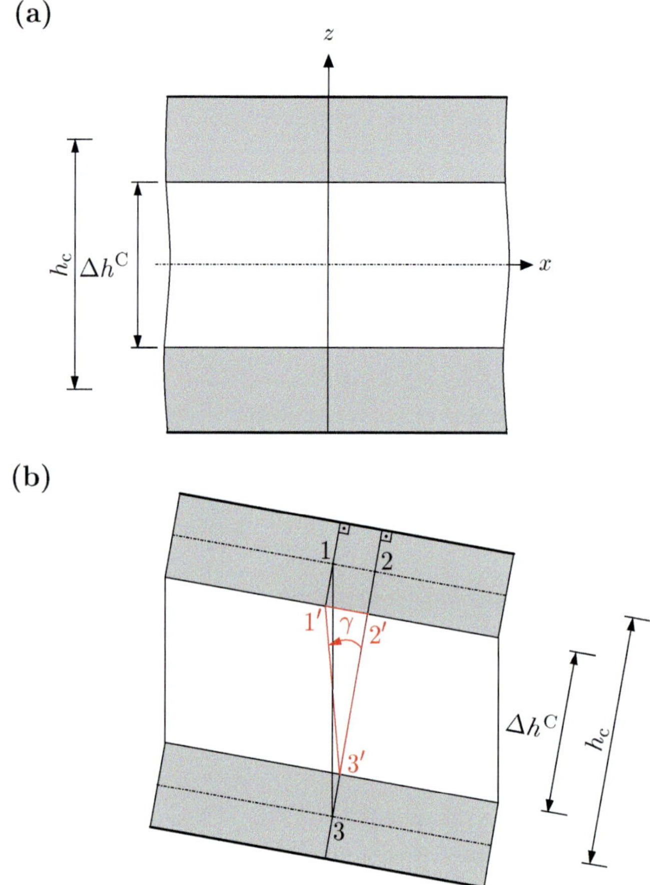

Fig. 5.17 Technical sandwich beam (the thickness of the cover layers is overdrawn): **(a)** undeformed initial state, **(b)** pure shear deformation (see Fig. 5.16c)

$$\frac{\mathrm{d}u_{z,\,\mathrm{s}}}{\mathrm{d}x} = \frac{Q_z(x)}{Gbh_{\mathrm{c}}} = \frac{Q_z(x)}{Gb\frac{h_{\mathrm{c}}^2}{\Delta h^{\mathrm{C}}}}.\tag{5.102}$$

By means of

$$A = \frac{bh_{\mathrm{c}}^2}{\Delta h^{\mathrm{C}}} \approx bh_{\mathrm{c}}\tag{5.103}$$

the following simplified expression results

$$\frac{\mathrm{d}u_{z,\,\mathrm{s}}}{\mathrm{d}x} = \frac{Q_z(x)}{AG},\tag{5.104}$$

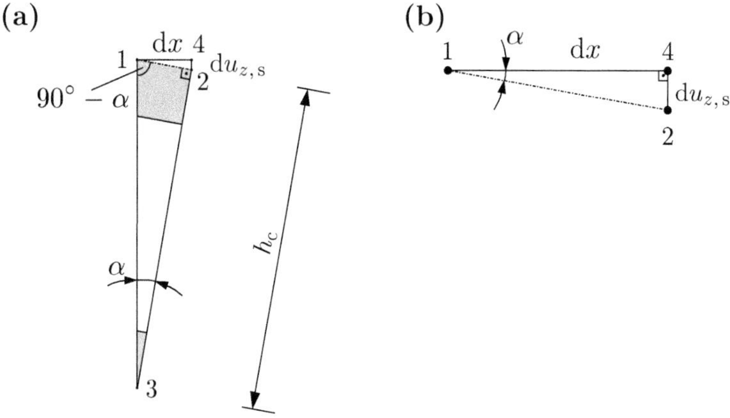

Fig. 5.18 Derivation of the shear differential equation

Table 5.6 Various differential equations for calculating the deflection for sandwich beams with thin face sheets and a soft core using the partial deflection method

Differential equation	Stiffness
Bending deformation	
$\dfrac{\mathrm{d}^2}{\mathrm{d}x^2}\left(EI_y\dfrac{\mathrm{d}^2u_{z,\mathrm{b}}(x)}{\mathrm{d}x^2}\right)=q_z(x)$	$\overline{EI_y}\approx\dfrac{E^{\mathrm{F}}b\Delta h^{\mathrm{F}}(h_{\mathrm{c}})^2}{2}$
$\dfrac{\mathrm{d}}{\mathrm{d}x}\left(EI_y\dfrac{\mathrm{d}^2u_{z,\mathrm{b}}(x)}{\mathrm{d}x^2}\right)=-Q_z(x)$	
$\overline{EI_y}\dfrac{\mathrm{d}^2u_{z,\mathrm{b}}(x)}{\mathrm{d}x^2}=-M_y(x)$	
Shear deformation	
$\dfrac{\mathrm{d}u_{z,\mathrm{s}}}{\mathrm{d}x}=\dfrac{Q_z(x)}{AG^{\mathrm{C}}}$	$AG^{\mathrm{C}}=\dfrac{bh_{\mathrm{c}}^2}{\Delta h^{\mathrm{C}}}G^{\mathrm{C}}\approx bh_{\mathrm{c}}G^{\mathrm{C}}$

where AG represents the shear stiffness of the sandwich with thin face sheets and a soft core.

The different differential equations for calculating the deflection are shown in Table 5.6 for comparison.

In the following, the deflection for a sandwich beam under 3-point bending with point load is calculated using the partial deflection method, see Fig. 5.19.

For this configuration, the internal shear force and bending moment distributions in the range $0 \leq x \leq L/2$ result in (see also Fig. 9.41):

$$Q_z(x) = -\frac{F_0}{2} \quad \text{and} \quad M_y(x) = -\frac{F_0 x}{x}. \tag{5.105}$$

The bending and shear deformations result in general—assuming thin face sheets and a soft core—by integrating the differential equations according to Table 5.6 to:

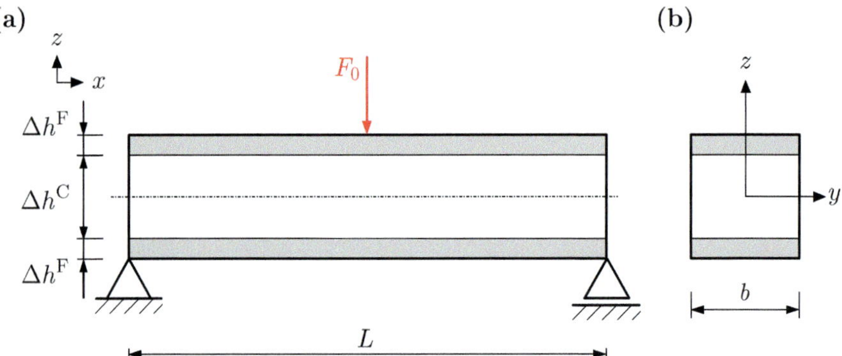

Fig. 5.19 Sandwich beam under 3-point bending with point load: **(a)** boundary conditions and external load; **(b)** beam cross-section

$$u_{z,\mathrm{b}}(x) = \frac{1}{\overline{EI_y}}\left(\frac{F_0 x^3}{12} + c_1 x + c_2\right),\tag{5.106}$$

$$u_{z,\mathrm{s}}(x) = -\frac{F_0 x}{2G^C A} + c_3.\tag{5.107}$$

Using the boundary conditions $u_{z,\mathrm{b}}(0) = u_{z,\mathrm{s}}(0) = 0$ and $\frac{\mathrm{d}u_{z,\mathrm{b}}(L/2)}{\mathrm{d}x} = 0$ the three constants of integration result to $c_1 = -\frac{F_0 L^2}{16}$, $c_2 = 0$ and $c_3 = 0$. Thus, the total deflection of the sandwich beam along the x axis is:

$$u_z(x) = u_{z,\mathrm{b}}(x) + u_{z,\mathrm{s}}(x)\tag{5.108}$$

$$= \frac{F_0 L^3}{48\overline{EI_y}} \times \left(4\left(\frac{x}{L}\right)^3 - 3\left(\frac{x}{L}\right)\right) - \frac{F_0 L}{4AG^C} \times 2\left(\frac{x}{L}\right),\tag{5.109}$$

or the maximum value of the deflection at $x = \frac{L}{2}$:

$$u_z\left(\frac{L}{2}\right) = -\frac{F_0 L^3}{48\overline{EI_y}} - \frac{F_0 L}{4AG^C}.\tag{5.110}$$

The ratio of the partial deflections at $x = \frac{L}{2}$ is:

$$\frac{u_{z,\mathrm{b}}}{u_{z,\mathrm{s}}} = \frac{\dfrac{F_0 L^3}{48\overline{EI_y}}}{\dfrac{F_0 L}{4AG^C}} = \frac{1}{6} \times \frac{G^C L^2}{E^F \Delta h^C \Delta h^F}.\tag{5.111}$$

Considering a concrete example of a sandwich beam with thin face sheets and a soft core, i.e., $L = 2000\,\mathrm{mm}$, $\Delta h^C = 150\,\mathrm{mm}$, $\Delta h^F = 5\,\mathrm{mm}$, $E^F = 74{,}000\,\mathrm{MPa}$ and

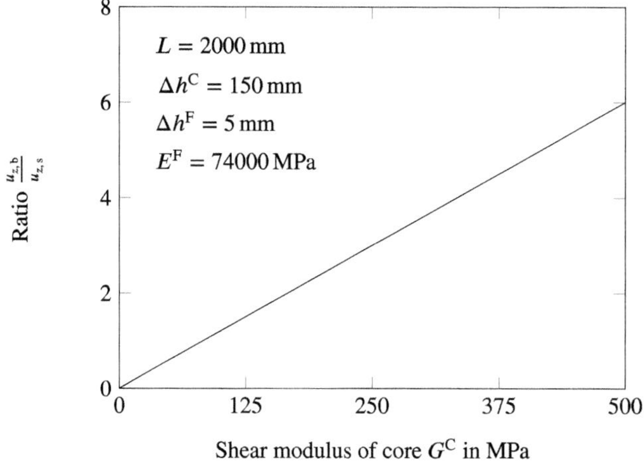

$$\text{Ratio } \frac{u_{z,\mathrm{b}}}{u_{z,\mathrm{s}}}$$

Shear modulus of core G^{C} in MPa

Fig. 5.20 Ratio of partial deflections as a function of core layer shear modulus: sandwich beam under 3-point bending with point load

$G^{\mathrm{C}} = 11\,\mathrm{MPa}$, the resulting ratio is $0.132 = 13.2\%$. The evaluation of Eq. (5.111) as a function of the shear modulus of the core layer is shown in Fig. 5.20.

It can be seen that for small values of the shear modulus of the core layer there are definitely clear contributions to the shear deformation. This is the case, although the beam is more than ten times longer ($L = 2000\,\mathrm{mm}$) than high ($10 \times 160 = 1600\,\mathrm{mm}$) and thus at a homogeneous beam, the shear component is usually negligible in a first approximation.

5.6 Exercises

5.6.1 Knowledge Questions

- Describe a sandwich element.
- Give typical properties in terms of geometry and materials for a *technical* sandwich element.
- Describe the function of the face sheets and the core in a technical sandwich.
- Name typical joining techniques between core and face sheets.
- Name typical types of loads that can occur in a sandwich element.
- Sketch the typical course of normal stress and normal strain versus the height for a symmetrical sandwich with three layers (core 'softer' than face sheets) under bending load.
- Sketch the typical course of normal stress and normal strain versus the height for a symmetrical sandwich with three layers (core 'softer' than face sheets) under tensile loading.
- Sketch the typical behavior of the shear stress and shear strain versus the height for a symmetrical sandwich with three layers (core 'softer' than face sheets) under shear loading.

- Sketch the course of the normal stress in a technical sandwich under bending load for: (a) exact consideration, (b) soft core and (c) soft core with thin face sheets.
- What is the thickness ratio of the core to the face sheet in terms of 'thick face sheets'?
- What is the thickness ratio of the core to the face sheet in terms of 'thin face sheets'?
- What is the ratio of the thickness of the core to the face sheet in terms of 'very thin face sheets'?
- Sketch the course of the normal stress in a technical sandwich under tensile loading for: (a) exact consideration and (b) soft core.
- Sketch the course of the shear stress in a technical sandwich under shear force loading for: (a) exact consideration, (b) soft core and (c) soft core with thin face sheets.

5.6.2 Calculation Problems
5.6.1 Average Bending Stiffness of a Sandwich

Determine the general average bending stiffness $\overline{EI}_y$ for the sandwich shown in Fig. 5.21, which consists of three layers and a cross-section as an I-profile. Afterwards, simplify the result for the special case $b^{\mathrm{C}} = b^{\mathrm{F}}/3$ and $\Delta h^{\mathrm{F}} = \Delta h^{\mathrm{C}}/4$.

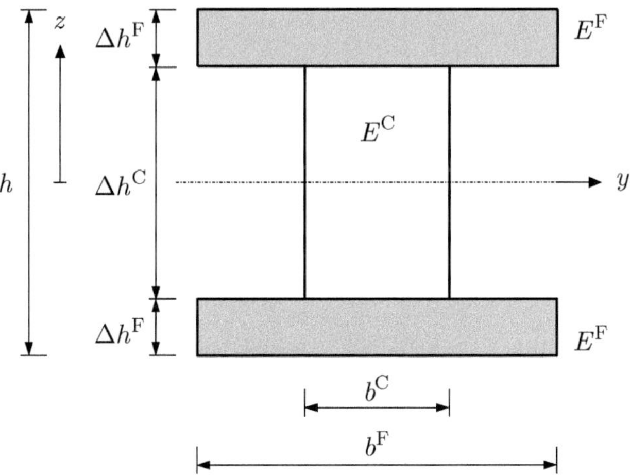

Fig. 5.21 Sandwich of three layers and cross section as I-profile

5.6.2 Simplification of the average bending stiffness for a homogeneous sandwich

Simplify the expression for the average bending stiffness $\overline{EI}_y$ according to Eq. (5.1), i.e.

$$\overline{EI}_y = \sum_{k=1}^{3} E^k \left[\tfrac{1}{12} b \left(\Delta h^k \right)^3 + b \Delta h^k \left(z_{\mathrm{c}}^k \right)^2 \right],\tag{5.112}$$

for the case of a homogeneous sandwich with three layers (see Fig. 5.22). Each of the three layers has the height $\tfrac{1}{3}h$ and the elastic modulus E.

5.6.3 Calculation of the Shear Stress Distribution for a Circular Cross-Section

A beam with a circular cross-section (radius R) is given. Calculate the shear stress distribution τ_{zx} over the cross-section under the influence of a shear force $Q_z(x)$. Consider the distribution in the center of the circular cross-section, i.e. for $y = 0$.

5.6.4 Limit Case of the Normalized Stiffness of a Sandwich

Simplify the expression for the normalized stiffness of a sandwich according to Eq. (5.57) for the limiting case $E^{\mathrm{F}} = E^{\mathrm{C}} = E$ and $\Delta h^{\mathrm{F}} = \Delta h^{\mathrm{C}} = \tfrac{1}{3}$. This is a homogeneous body of total height '1'. Clarify how the result to understand (why is the ratio not equal to '1'?).

5.6.5 Deflection of a Sandwich Beam with Distributed Load Using the Partial Deflection Method

For the sandwich beam shown in Fig. 5.23, the bending line $u_z(x)$ and the maximum value of the deflection must be determined. It should be assumed that it is a beam with thin face sheets and a soft core. Furthermore, calculate the general ratio of the partial deflections and also provide the numerical value for $L = 2000\,\mathrm{mm}$, $\Delta h^{\mathrm{C}} = 150\,\mathrm{mm}$, $\Delta h^{\mathrm{F}} = 5\,\mathrm{mm}$, $E^{\mathrm{F}} = 74{,}000\,\mathrm{MPa}$ and $G^{\mathrm{C}} = 11\,\mathrm{MPa}$.

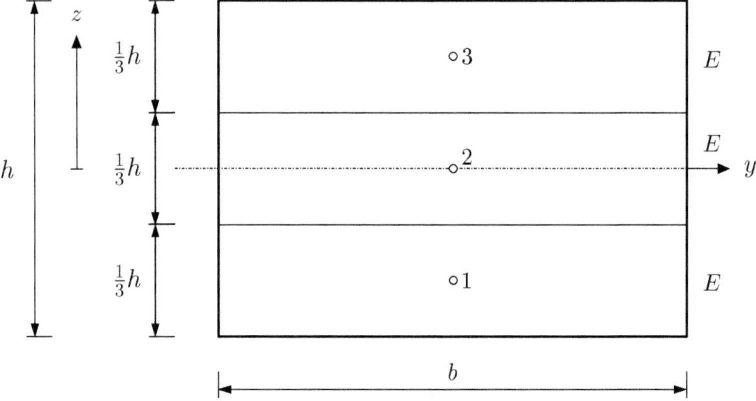

Fig. 5.22 Homogeneous sandwich made of three layers

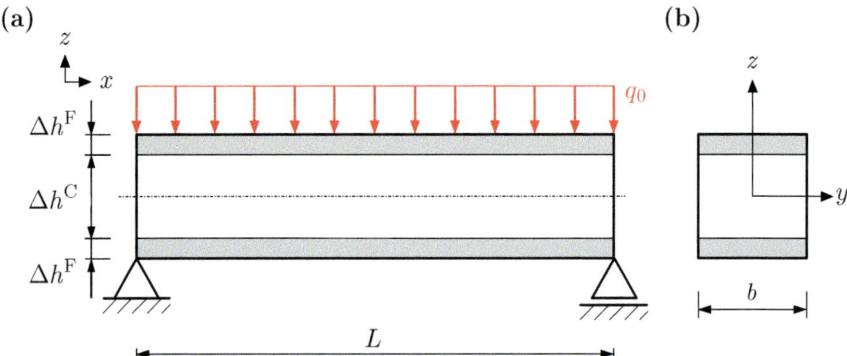

Fig. 5.23 Sandwich beam with constant distributed load: (**a**) boundary conditions and external load; (**b**) beam cross-section

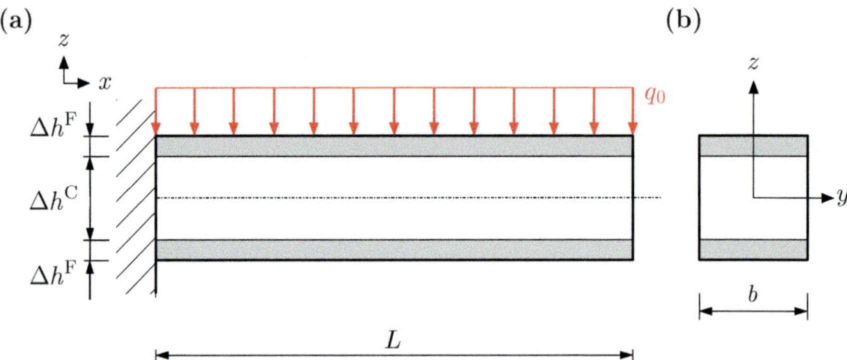

Fig. 5.24 Sandwich beam with constant distributed load: (**a**) boundary condition and external load; (**b**) beam cross-section

5.6.6 Deflection of a Sandwich Beam (Cantilever) with Distributed Load Using the Partial Deflection Method

For the sandwich beam shown in Fig. 5.24, the bending line $u_z(x)$ and the maximum value of the deflection must be determined. It should be assumed that it is a beam with thin face sheets and a soft core. Furthermore, calculate the general ratio of the partial deflections and also provide the numerical value for $L = 2000$ mm, $\Delta h^{\mathrm{C}} = 150$ mm, $\Delta h^{\mathrm{F}} = 5$ mm, $E^{\mathrm{F}} = 74{,}000$ MPa and $G^{\mathrm{C}} = 11$ MPa.

5.6.7 Deflection of a Sandwich Beam (Cantilever) with Shear Force Using the Partial Deflection Method

For the sandwich beam shown in Fig. 5.25, the bending line $u_z(x)$ and the maximum deflection value must be determined. It should be assumed that this is a beam with thin face sheets and a soft core. Furthermore, calculate the general ratio of the partial deflections and also provide the numerical value for $L = 2000$ mm, $\Delta h^{\mathrm{C}} = 150$ mm, $\Delta h^{\mathrm{F}} = 5$ mm, $E^{\mathrm{F}} = 74{,}000$ MPa and $G^{\mathrm{C}} = 11$ MPa.

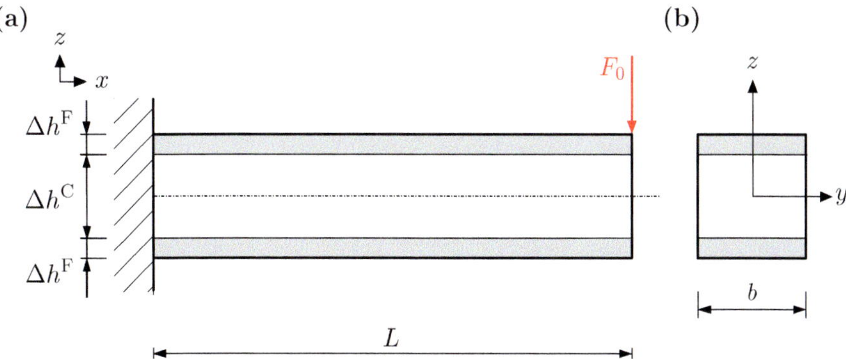

Fig. 5.25 Sandwich beam with shear force: **(a)** boundary condition and external load; **(b)** beam cross-section

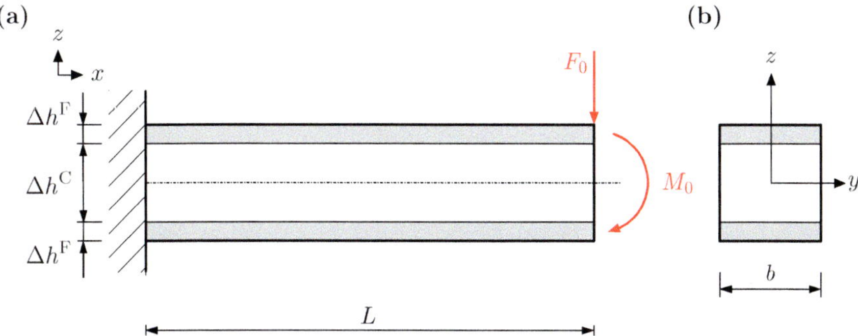

Fig. 5.26 Sandwich beam with shear force and single moment: **(a)** boundary condition and external loads; **(b)** beam cross-section

5.6.8 Deflection of a Sandwich Beam (Cantilever) with Shear Force and Bending Moment Using the Partial Deflection Method

For the sandwich beam shown in Fig. 5.26, the bending line $u_z(x)$ and the maximum deflection value must be determined. In addition to the single force F_0, a single moment of magnitude $M_0 = \frac{1}{2} F_0 L$ acts on the beam. It should be assumed that this is a beam with thin face sheets and a soft core. Furthermore, calculate the general ratio of the partial deflections and also provide the numerical value for $L = 2000$ mm, $\Delta h^{\mathrm{C}} = 150$ mm, $\Delta h^{\mathrm{F}} = 5$ mm, $E^{\mathrm{F}} = 74{,}000$ MPa and $G^{\mathrm{C}} = 11$ MPa.

5.6.9 Deflection of a Sandwich Beam with Linearly Distributed Load Using the Partial Deflection Method

For the sandwich beam shown in Fig. 5.27, the bending line $u_z(x)$ must be determined. It should be assumed that this is a beam with thin face sheets and a soft core. Furthermore, calculate the general ratio of the partial deflections and also provide the numerical value at $x = \frac{2L}{3}$ for $L = 2000$ mm, $\Delta h^{\mathrm{C}} = 150$ mm, $\Delta h^{\mathrm{F}} = 5$ mm, $E^{\mathrm{F}} = 74{,}000$ MPa and $G^{\mathrm{C}} = 11$ MPa.

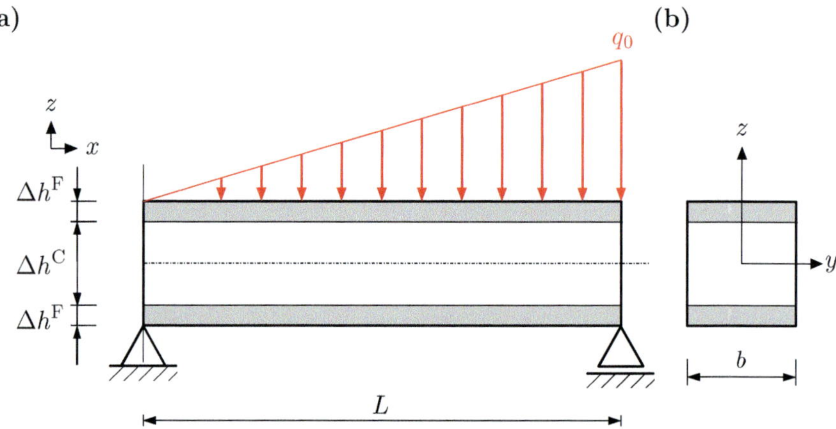

Fig. 5.27 Sandwich beam with linearly distributed load (maximum value: $|q_0|$): **(a)** boundary condition and external load; **(b)** beam cross-section

References

1. Allen, H.G.: Analysis and Design of Structural Sandwich Panels. Pergamon Press, Oxford (1969)
2. Aßmus, M., Bergmann, S., Eisenträger, J., Naumenko, K., Altenbach, H.: Consideration of non-uniform and non-orthogonal mechanical loads for structural analysis of photovoltaic composite structures. In Altenbach, H., Goldstein, R., Murashkin, E. (eds.) Mechanics for Materials and Technologies, pp. 73–122. Springer, Cham (2017)
3. Hertel, H.: Leichtbau: Bauelemente, Bemessungen und Konstruktionen von Flugzeugen und anderen Leichtbauwerken. Springer, Berlin (1960)
4. Klein, B.: Leichtbau-Konstruktion: Berechnungsgrundlagen und Gestaltung. Friedr. Vieweg & Sohn, Braunschweig (1989)
5. Klein, B.: Leichtbau-Konstruktion: Berechnungsgrundlagen und Gestaltung. Vieweg+Teubner, Wiesbaden (2009)
6. Öchsner, A.: Elasto-Plasticity of Frame Structure Elements: Modeling and Simulation of Rods and Beams. Springer, Berlin (2014)
7. Stamm, K., Witte, H.: Sandwichkonstruktionen. Springer-Verlag, Wien (1974)

Sandwich Structures: Limit Load

6

Abstract

In this chapter, the basics of lightweight design in regard to materials and shape are further deepened. In particular, the different failure mechanisms of sandwich beams in relation to face sheets, core and interlayer are analyzed. In addition to the classic failure by reaching the yield stress or strength, global and local instabilities are considered.

6.1 Failure Modes

The various failure modes for technical sandwich beams are shown in Table 6.1. Here, between failure mechanisms of the face sheets (yielding, instability and local deformation excess) and the core (yielding and instability) can be distinguished. The failure of the connecting layer between core and face sheet and the global instability failure must be considered as further failure mechanisms. The various forms of instability are shown in Table 6.2 with a distinction between global and local forms of failure.

6.1.1 Global Instability Failure

To derive the buckling formula[1] for a sandwich beam with thin face sheets and a soft core, consider a configuration hinged on both sides under a compressive force F, see Fig. 6.1a. The entire deformation is composed of the pure bending component and the shear component, see Fig. 6.1b.

---

[1] For a better understanding, it is recommended here that the reader first understands the buckling force derivation according to Euler for homogeneous and isotropic Euler-Bernoulli beams, see Appendix A.1.3.

A. Öchsner, *Lightweight Design*,
https://doi.org/10.1007/978-3-658-48162-9_6

Table 6.1 Failure modes in technical sandwich beams, cf. [8]

Failure mode	Load case (section)
Face sheets	
• Failure by reaching the yield stress or strength under tension or compression	Bending (5.5.1, 5.5.2)
• Exceeding the instability limit under compressive loading, i.e. local wrinkling (local instability) of the compressively loaded face sheet	Bending (6.1.3)
• Exceeding the instability limit under compressive loading, i.e. local symmetrical or asymmetrical wrinkling (local instability) of both face sheets	Compression (6.1.5, 6.1.4)
• Strong local deformations of the face sheets due to local force application	–
Core	
• Failure by reaching the shear yield stress or shear strength	Bending, shear force (5.5.3)
• Exceeding the instability limit under shear loading, e.g. buckling of the webs of a honeycomb core	Bending, shear force (—)
Other causes	
• Failure of the joining (e.g. adhesive layer) between the face sheet and core	Bending (6.1.2)
• Global instability failure (overall instability), i.e. buckling of the sandwich beam	Compression (6.1.1)

The equilibrium is now established for the first time on the deformed member[2], see Fig. 6.1b.

The equilibrium of moments at the cutting site x yields (see Fig. 6.2a):

$$\overset{\curvearrowright}{\sum} M_y = 0 \quad \Leftrightarrow \quad -F \times \left(u_{z,\mathrm{b}} + u_{z,\mathrm{s}}\right) + M_y(x) = 0 , \tag{6.1}$$

$$\Rightarrow \quad M_y(x) = +F \times \left(u_{z,\mathrm{b}} + u_{z,\mathrm{s}}\right) = +F u_z . \tag{6.2}$$

Thus, the following formulation for the bending deformation can be specified using the bending differential equation according to Eq. (5.95):

$$E I_y \frac{\mathrm{d}^2 u_{z,\mathrm{b}}}{\mathrm{d}x^2} = -M_y(x) = -F \times \left(u_{z,\mathrm{b}} + u_{z,\mathrm{s}}\right) , \tag{6.3}$$

---

[2] For the derivation of the classical Euler-Bernoulli differential equations, the equilibrium is established on the non-deformed member, see [5].

Table 6.2 Instabilities in technical sandwich beams: **(a)** buckling, **(b)** symmetrical wrinkling of the face sheets, **(c)** asymmetrical wrinkling of the face sheets, **(d)** wrinkling of the compressively loaded face sheet, cf. [8]

	Schematic representation	Load case (section)
	Global instability	
(a)		Compression (6.1.1)
	Local instability	
(b)		Compression (6.1.5)
(c)		Compression (6.1.4)
(d)		Bending (6.1.3)

or after a one-time differentiation with respect to the x coordinate:

$$\overline{EI_y}\frac{\mathrm{d}^3 u_{z,\mathrm{b}}}{\mathrm{d}x^3} = -\frac{\mathrm{d}M_y(x)}{\mathrm{d}x} = -F \times \left(\frac{\mathrm{d}u_{z,\mathrm{b}}}{\mathrm{d}x} + \frac{\mathrm{d}u_{z,\mathrm{s}}}{\mathrm{d}x}\right). \tag{6.4}$$

The following geometric relationship for small angles can be derived from the right-angled force triangle according to Fig. 6.2b:

$$\frac{Q_z}{F} = \sin\left(\frac{\mathrm{d}u_{z,\mathrm{b}}}{\mathrm{d}x} + \frac{\mathrm{d}u_{z,\mathrm{s}}}{\mathrm{d}x}\right) \approx \frac{\mathrm{d}u_{z,\mathrm{b}}}{\mathrm{d}x} + \frac{\mathrm{d}u_{z,\mathrm{s}}}{\mathrm{d}x} = \frac{\mathrm{d}u_z}{\mathrm{d}x}. \tag{6.5}$$

The last relation can be transformed immediately using the shear differential equation (5.104) after the first-order derivative of the shear deformation:

$$\frac{\mathrm{d}u_{z,\mathrm{s}}}{\mathrm{d}x} = \frac{F}{AG^{\mathrm{C}}\left(1 - \frac{F}{AG^{\mathrm{C}}}\right)} \times \frac{\mathrm{d}u_{z,\mathrm{b}}}{\mathrm{d}x}, \tag{6.6}$$

where the area according to Eq. (5.103) is given as $A = bh_{\mathrm{c}}^2/\Delta h^{\mathrm{C}} \approx bh_{\mathrm{c}}$.

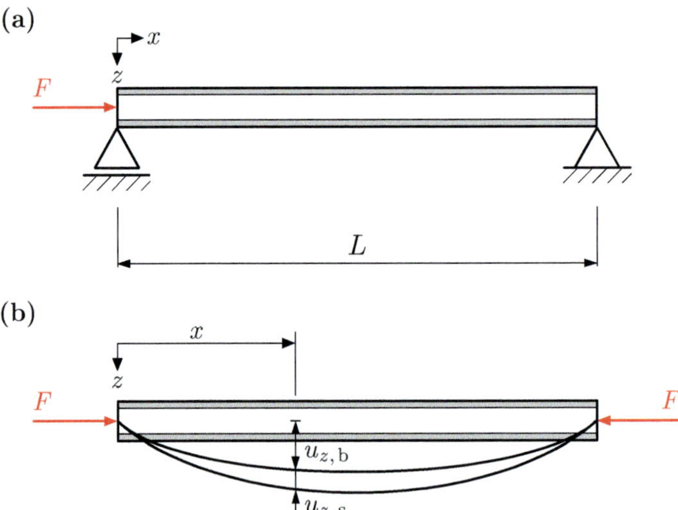

Fig. 6.1 Sandwich beam hinged on both sides under compressive load: **(a)** initial configuration and **(b)** deformation

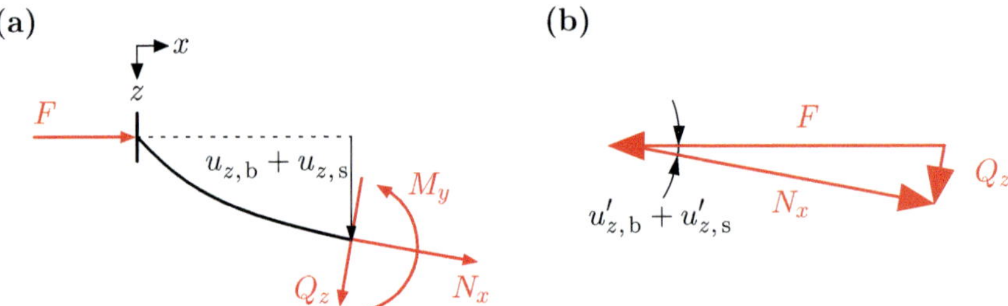

Fig. 6.2 Sandwich beam hinged on both sides under compressive load: **(a)** cutting free at location x; **(b)** force triangle (without consideration of the current signs)

If one now inserts Eq. (6.6) into the differential equation according to (6.4), the result is:

$$\overline{EI_y}\frac{\mathrm{d}^3 u_{z,\mathrm{b}}}{\mathrm{d}x^3} = -F \times \frac{\mathrm{d}u_{z,\mathrm{b}}}{\mathrm{d}x} - F \times \frac{F}{AG^{\mathrm{C}}\left(1 - \frac{F}{AG^{\mathrm{C}}}\right)} \times \frac{\mathrm{d}u_{z,\mathrm{b}}}{\mathrm{d}x}, \qquad (6.7)$$

or transformed to:

$$\frac{\mathrm{d}^3 u_{z,\mathrm{b}}}{\mathrm{d}x^3} + \underbrace{\frac{F}{EI_y\left(1 - \frac{F}{AG^{\mathrm{C}}}\right)}}_{\lambda^2} \frac{\mathrm{d}u_{z,\mathrm{b}}}{\mathrm{d}x} = 0, \qquad (6.8)$$

or as:

$$\frac{\mathrm{d}^3 u_{z,\mathrm{b}}}{\mathrm{d}x^3} + \lambda^2 \frac{\mathrm{d}u_{z,\mathrm{b}}}{\mathrm{d}x} = 0. \qquad (6.9)$$

The general solution for such a differential equation results in:

$$u_z(x) = c_1 \times \sin(\lambda x) + c_2 \times \cos(\lambda x) + c_3 \,. \tag{6.10}$$

The first- and second-order derivatives of this general function result in:

$$\frac{\mathrm{d}^1 u_{z,\mathrm{b}}}{\mathrm{d}x^1} = c_1 \alpha \cos(\lambda x) - c_2 \lambda \sin(\lambda x) \,, \tag{6.11}$$

$$\frac{\mathrm{d}^2 u_{z,\mathrm{b}}}{\mathrm{d}x^2} = -c_1 \lambda^2 \sin(\lambda x) - c_2 \lambda^2 \cos(\lambda x) \,. \tag{6.12}$$

If one inserts this second-order derivative into the differential equation according to (6.3), i.e. $M_y(x) = F\left(u_{z,\mathrm{b}} + u_{z,\mathrm{s}}\right) = -\overline{EI_y}\frac{\mathrm{d}^2 u_{z,\mathrm{b}}}{\mathrm{d}x^2}$, the total deflection is obtained as:

$$u_{z,\mathrm{b}} + u_{z,\mathrm{s}} = -\frac{\overline{EI_y}}{F}\left(c_1 \lambda^2 \sin(\lambda x) - c_2 \lambda^2 \cos(\lambda x)\right) \,, \tag{6.13}$$

or with $\lambda^2 = \dfrac{F}{\overline{EI_y}\left(1-\frac{F}{AG^{\mathrm{C}}}\right)}$ finally to:

$$u_{z,\mathrm{b}}(x) + u_{z,\mathrm{s}}(x) = u_z(x) = \frac{c_1 \sin(\lambda x) + c_2 \cos(\lambda x)}{1 - \frac{F}{AG^{\mathrm{C}}}} \,. \tag{6.14}$$

Using the boundary conditions $u_z(0) = 0$ and $u_z(L) = 0$, the unknown constants c_1 and c_2 can be approached.

The first boundary condition, i.e. $u_z(0) = 0$, results in:

$$0 = c_1 \times \sin(0) + c_2 \times \cos(0) \quad \Leftrightarrow \quad c_2 = 0 \,. \tag{6.15}$$

From the second boundary condition, i.e. $u_z(L) = 0$, the result is:

$$0 = c_1 \times \sin(\lambda \times L) \,. \tag{6.16}$$

If the product is to be zero, one of the two factors, i.e. c_1 or $\sin(\lambda \times L)$ must be zero. $c_1 = 0$ is a trivial solution (with $c_1 = c_2 = 0$ according to Eq. (6.14): $u_z = 0$, thus no deformation). Therefore $\sin(\lambda \times L) = 0$ has to be looked at more closely. The condition $\sin(\lambda \times L) = 0$ means that $\lambda \times L = k \times \pi$ with $k = 0, 1, 2, \cdots$ (see Fig. A.6).

The condition $\lambda \times L = 0$ would mean $F = 0$ (see definition of λ according to Eq. (6.8)). This results in the reasonable condition:

$$\lambda \times L = \pi \quad \Leftrightarrow \quad \lambda^2 = \frac{\pi^2}{L^2}.$$
(6.17)

If one inserts the last result into the definition equation of λ according to Eq. (6.8), the result is:

$$\frac{\pi^2}{L^2} = \frac{F}{EI_y\left(1 - \frac{F}{AG^C}\right)},$$
(6.18)

or after a short transformation:

$$F_{cr} = \frac{\frac{\pi^2 \overline{EI_y}}{L^2}}{1 + \frac{\pi^2 \overline{EI_y}}{L^2 \times AG^C}} = \frac{F_{cr}^E}{1 + \frac{F_{cr}^E}{AG^C}},$$
(6.19)

where $F_{cr}^E = \pi^2 \overline{EI_y}/L^2$ is the Euler buckling force for homogeneous beams, see Eq. (A.23). Furthermore, it should be noted that the area according to Eq. (5.103) is given as $A = bh_c^2/\Delta h^C \approx bh_c$. For other support conditions, the length L in the equation for the buckling force according to Euler can be replaced by the so-called buckling length L_{cr} (see Table A.2). Thus, a total of four support cases can be calculated. Equation (6.19) is also often represented in the following form (see [2]):

$$\frac{1}{F_{cr}} = \frac{1}{F_{cr}^E} + \frac{1}{AG^C},$$
(6.20)

where the following special cases can be distinguished:

- $G \to \infty \quad \Rightarrow \quad F_{cr} = F_{cr}^E.$
- G finite $\quad \Rightarrow \quad F_{cr} < F_{cr}^E.$
- G small $\quad \Rightarrow \quad F_{cr} \to AG^C.$

An alternative way to derive the buckling formula is shown in [6]. For this purpose, the bending and shear differential equations (see (5.95) or (5.104)) are rearranged for the second-order derivatives:

$$\frac{d^2 u_{zb}}{dx^2} = -\frac{M_y(x)}{EI_y},$$
(6.21)

$$\frac{d^2 u_{z,s}}{dx^2} = \frac{1}{AG^C}\frac{dQ_z(x)}{dx}.$$
(6.22)

Both formulations can be composed additively to the total curvature[3]:

$$\frac{\mathrm{d}^2 u_{z,}}{\mathrm{d}x^2} = \frac{\mathrm{d}^2 u_{z,\mathrm{b}}}{\mathrm{d}x^2} + \frac{\mathrm{d}^2 u_{z,\mathrm{s}}}{\mathrm{d}x^2} = -\frac{M_y(x)}{EI_y} + \frac{1}{AG^C}\frac{\mathrm{d}Q_z(x)}{\mathrm{d}x}. \tag{6.23}$$

If one uses in the last expression of the differential equation the bending moment according to Eq. (6.2), i.e. $M_y(x) = +F u_z(x)$, and the shear force according to Eq. (6.5), i.e. $Q_z(x) = F\frac{\mathrm{d}u_z(x)}{\mathrm{d}x}$, the following differential equation results after transformation:

$$\frac{\mathrm{d}^2 u_z(x)}{\mathrm{d}x^2} + \underbrace{\frac{F}{EI_y\left(1 - \frac{F}{AG^C}\right)}}_{\lambda^2} u_z(x) = 0 \quad \Leftrightarrow \quad \frac{\mathrm{d}^2 u_z}{\mathrm{d}x^2} + \lambda^2 u_z(x) = 0. \tag{6.24}$$

The solution is as shown in the appendix A.1.3 and one finally gets the result according to Eq. (6.19).

6.1.2 Shear failure of the Connecting Layer

To assess whether the connecting layer (e.g. an adhesive layer) fails, the shear stress at the point $\pm \Delta h^C/2$ must be evaluated.

According to the exact theory (see Eq. (5.80)), the shear stress is:

$$\tau_{zx}\left(\frac{\Delta h^C}{2}\right) = \frac{Q_z(x)\left[E^F\left(\Delta h^C + \Delta h^F\right)\Delta h^F\right]}{2\left(\frac{E^F b(\Delta h^F)^3}{6} + \frac{E^F b \Delta h^F (h_c)^2}{2} + \frac{E^C b(\Delta h^C)^3}{12}\right)}. \tag{6.25}$$

For soft cores, i.e. $E^C \ll E^F$, the result is (see Eq. (5.84)):

$$\tau_{zx}\left(\frac{\Delta h^C}{2}\right) = \frac{Q_z(x)\left[\Delta h^C + \Delta h^F\right]}{b\left(\frac{(\Delta h^F)^2}{3} + (h_c)^2\right)} = \frac{Q_z(x)h_c}{b\left(\frac{(\Delta h^F)^2}{3} + (h_c)^2\right)}$$

$$= \frac{Q_z(x)}{2EI_y}E^F \Delta h^F h_c. \tag{6.26}$$

---

[3] To be precise, the following applies to the curvature: $\kappa(x) = -\frac{\mathrm{d}^2 u_z(x)}{\mathrm{d}x^2}$.

Table 6.3 Mechanical properties of some adhesives: G: shear modulus; τ_p: shear yield limit; τ_aB: shear strength; γ_aB: shear strain at failure. Based on [7]

Adhesive	Manufacturer	G in MPa	τ_p in MPa	τ_aB in MPa	γ_aB in %
Epoxide					
Araldite AV138	Huntsman	1559	25.0	30.2	5.50
Hysol EA 9394	Loctite	1140	25.0	40.4	8.36
Hysol EA 9321	Loctite	1030	20.0	33.0	6.35
Supreme 10HT	Master Bond	1460	37.1	37.1	16.1
Araldite AV 119	Huntsman	1260	47.0	47.0	50.7
Hysol EA 9359.3	Loctite	660.0	35.3	35.3	63.0
Araldite 2015	Huntsman	560.0	14.0	20.0	40.3
Polyurethane					
Sikaflex 256	Sika	1.351	8.26	8.26	330
Bismaleimide					
Redux 326	Hexcel comp.	1615	37.9	37.9	3.70
Modified acrylate					
DP-8005	3M	178.6	5.3	8.40	180

For soft cores, i.e. $E^\text{C} \ll E^\text{F}$, and thin face sheets, i.e. $\Delta h^\text{F} \ll \Delta h^\text{C}$, the shear stress results in (see Eq. (5.89)):

$$\tau_{zx}\left(\frac{\Delta h^\text{C}}{2}\right) = \frac{Q_z(x)}{b\,\Delta h^\text{C}} = \frac{Q_z(x)}{bh_\text{c}}. \tag{6.27}$$

Typical material properties of adhesive layers are summarized in Table 6.3.

6.1.3   Local Wrinkling of the Compressive Face Sheet (Bending Load)

Local wrinkling of the face sheets in the bending load case can occur in the face sheet loaded in compression[4]. The mechanical mechanical replacement model model shown in Fig. 6.3 was proposed in the literature as a modeling approach, see [2].

---

[4] In the bending load case, one face sheet is loaded in tension and the other in compression.

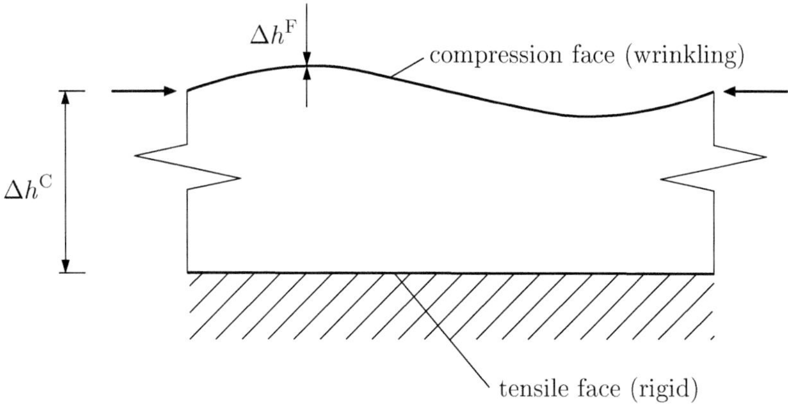

Fig. 6.3 Simplified model for wrinkling of the compression-loaded top layer (bending load case). Adapted from [2]

Here it is assumed that the tensile face sheet remains completely planar since it is modeled as a rigid layer. Furthermore, the core is modeled as a homogeneous and isotropic continuum[5].

The critical wrinkling stress is generally found to be

$$\sigma_{\mathrm{cr}} = B_1 \left(E^{\mathrm{F}}\right)^{\frac{1}{3}} \left(E^{\mathrm{C}}\right)^{\frac{2}{3}} , \tag{6.28}$$

where the factor B_1 depends on the geometry and material parameters. Using the relationship between the elastic constants for isotropic materials, i.e.

$$G^{\mathrm{C}} = \frac{E^{\mathrm{C}}}{2(1 + \nu^{\mathrm{C}})}, \tag{6.29}$$

this results in the alternative relationship for the wrinkling stress:

$$\sigma_{\mathrm{cr}} = \underbrace{B_1 \left[2(1 + \nu^{\mathrm{C}})\right]^{\frac{1}{3}}}_{B_1'} \times \left(E^{\mathrm{F}} E^{\mathrm{C}} G^{\mathrm{C}}\right)^{\frac{1}{3}} = B_1' \times \left(E^{\mathrm{F}} E^{\mathrm{C}} G^{\mathrm{C}}\right)^{\frac{1}{3}} . \tag{6.30}$$

The following conservative approximation for the factor B_1' is proposed by [1,4]:

$$\sigma_{\mathrm{cr}} \approx \frac{1}{2} \times \left(E^{\mathrm{F}} E^{\mathrm{C}} G^{\mathrm{C}}\right)^{\frac{1}{3}} . \tag{6.31}$$

---

[5] In other modeling approaches, the core layer was assumed to be the elastic spring bedding of the face sheet. However, such approaches neglect the shear stiffness in the xz plane and are therefore inadequate.

For sandwich structures with thin face sheets and a soft core, under the condition $k = \frac{\Delta h^{\mathrm{F}}}{\Delta h^{\mathrm{C}}} \left(\frac{E^{\mathrm{F}}}{E^{\mathrm{C}}}\right)^{1/3} < 0.25$, the following values for B_1 were determined in [2,3]:

$$B_1(\nu^{\mathrm{C}} = 0.25) = 0.575\,, \tag{6.32}$$

$$B_1(\nu^{\mathrm{C}} = 0.50) = 0.543\,, \tag{6.33}$$

which results in the following values for B_1':

$$B_1'(\nu^{\mathrm{C}} = 0.25) = 0.575 \left[2(1 + \tfrac{1}{4})\right]^{\frac{1}{3}} = 0.7804\,, \tag{6.34}$$

$$B_1'(\nu^{\mathrm{C}} = 0.50) = 0.543 \left[2(1 + \tfrac{1}{2})\right]^{\frac{1}{3}} = 0.7831\,. \tag{6.35}$$

For other material and geometry combinations, the following calculation must be carried out: For specific values of ν^{C} and $k = \frac{\Delta h^{\mathrm{F}}}{\Delta h^{\mathrm{C}}} \left(\frac{E^{\mathrm{F}}}{E^{\mathrm{C}}}\right)^{1/3}$, i.e.

$$
\begin{aligned}
B_1 &= \frac{k^2 \Theta^2}{12} + \frac{f(\Theta)}{k} \tag{6.36}\\
&= \frac{k^2 \Theta^2}{12} + \frac{1}{k}\left(\frac{2}{\Theta} \times \frac{(3 - \nu^{\mathrm{C}}) \sinh(\Theta) \cosh(\Theta) + (1 + \nu^{\mathrm{C}})\Theta}{(1 + \nu^{\mathrm{C}})(3 - \nu^{\mathrm{C}})^2 \sinh^2(\Theta) - (1 + \nu^{\mathrm{C}})^3 \Theta^2}\right), \tag{6.37}
\end{aligned}
$$

whereby the argument as $\Theta = \frac{2\pi \Delta h^{\mathrm{C}}}{\lambda}$ is given and λ represents the wavelength of the wrinkle wave of the compressive face sheet (see Fig. 6.4 and 6.5).

One can see from Fig. 6.4 that an enlargement of the k value with constant ν^{C} leads to a shift of the minimum to smaller Θ values. From Fig. 6.5 it can be seen that the influence of the Poisson's ratio is not so dominant.

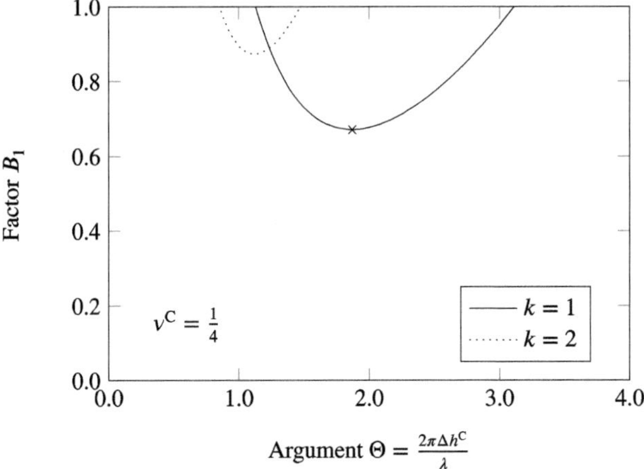

Fig. 6.4 Position of the local minimum of the function $B_1 = B_1(\Theta)$: influence of the parameter $k = \frac{\Delta h^{\mathrm{F}}}{\Delta h^{\mathrm{C}}} \left(\frac{E^{\mathrm{F}}}{E^{\mathrm{C}}}\right)^{1/3}$

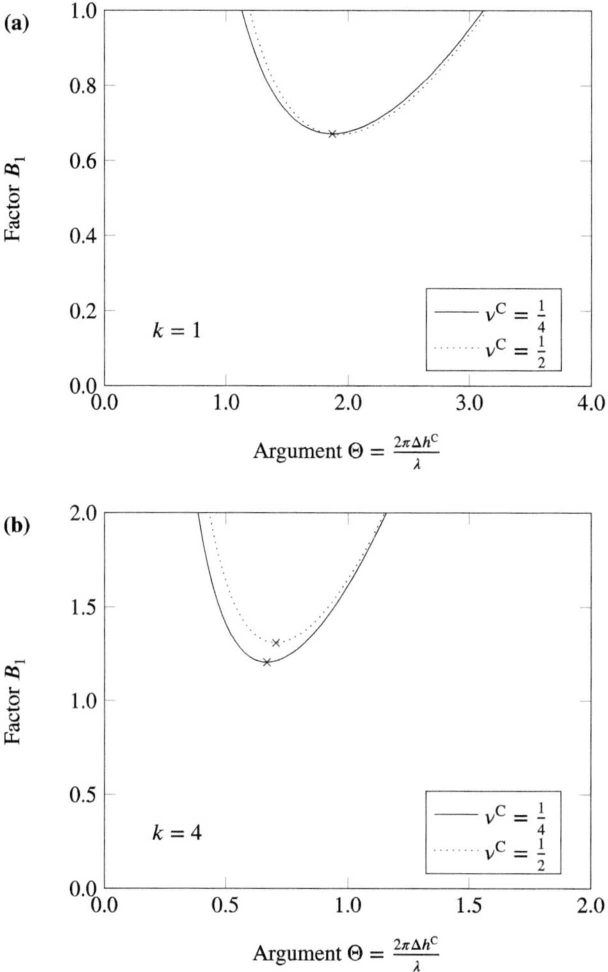

Fig. 6.5 Position of the local minimum of the function $B_1 = B_1(\Theta)$: influence of the Poisson's ratio ν^C of the core for **(a)** $k = 1$ und **(b)** $k = 4$

The local minimum of the function $B_1 = B_1(\Theta)$ results from the condition

$$\frac{\partial B_1(\Theta)}{\partial \Theta} = \frac{k^2 \Theta}{6} + \frac{1}{k} \times \frac{\partial f(\Theta)}{\partial \Theta} \stackrel{!}{=} 0, \tag{6.38}$$

where the Newton's method can be used for the numerical determination of the root[6]. If one repeats the procedure of determining the minimum for a value range of the

---

[6] A Python3 program is provided in the appendix A.2 to automatically evaluate Eq. (6.38) for given value ranges of k and ν^C.

geometry and material factor k with a constant Poisson's ratio ν^C, determination diagrams, as in Figs. 6.6– 6.8 are generated. These diagrams allow the critical factor B_1 to be read off easily without the need for numerical iteration to determine the minimum. The disadvantage, however, is that each of these diagrams is only valid for a specific Poisson's ratio.

From the diagrams in Figs. 6.6–6.8 it can be seen that the B_1 value approaches a constant value for $k \leq 0.25$. These values are summarized in Fig. 6.9 as a function of the Poisson's ratio. One can see a slightly falling trend of the B_1 value. Furthermore, Figs. 6.6–6.8 show that the B_1 value and the normalized wavelength are monotonically increasing functions.

6.1.4   Local Antisymmetric Wrinkling of Both Face Sheets (Compressive Load)

Local antisymmetric wrinkling can occur in compression-loaded sandwich beams and is characterized by the face sheets deforming antisymmetrically to the center line in the transverse direction, see Fig. 6.10.

In the case of thin face sheets and a soft core, the critical wrinkling stress σ_{cr} can be approximated according to the basic equation from Sect. 6.1.3, i.e.

$$\sigma_{cr} = B_1 \left(E^F\right)^{\frac{1}{3}} \left(E^C\right)^{\frac{2}{3}} . \tag{6.39}$$

However, another function $f(\Theta)$ must be used to calculate the factor B_1:

$$B_1 = \frac{k^2\Theta^2}{12} + \frac{f(\Theta)}{k} \tag{6.40}$$

$$= \frac{k^2\Theta^2}{12} + \frac{1}{k}\left(\frac{2}{\Theta} \times \frac{\cosh(\Theta) - 1}{(1 + \nu^C)(3 - \nu^C)\sinh(\Theta) + (1 + \nu^C)^2\Theta}\right) . \tag{6.41}$$

Here, too, the determination of the local minimum of the function $B_1(\Theta)$ allows the factor B_1 for Eq. (6.39) to be determined.

When applying Newton's method for small values of k, it should be noted that the gradient of the function $B_1(\Theta)$ has local extreme values close to $\Theta = 0$, see Fig. 6.11 b. It is therefore useful for convergence if the start value of the Newton's iteration $(\Theta_{i=0})$ is to the right-hand side of the root.

Determination diagrams for the factor B_1 as a function of the geometry and material factor k are shown in Figs. 6.12–6.14. These diagrams allow the critical factor B_1 to be read off easily without the need for numerical iteration to determine the minimum. The disadvantage here, however, is that each of these diagrams is only valid for one specific Poisson's ratio ν^C[7].

---

[7] A Python3 program is provided in the appendix A.1.5 to automatically evaluate Eq. (6.38) for given value ranges of k and ν^C.

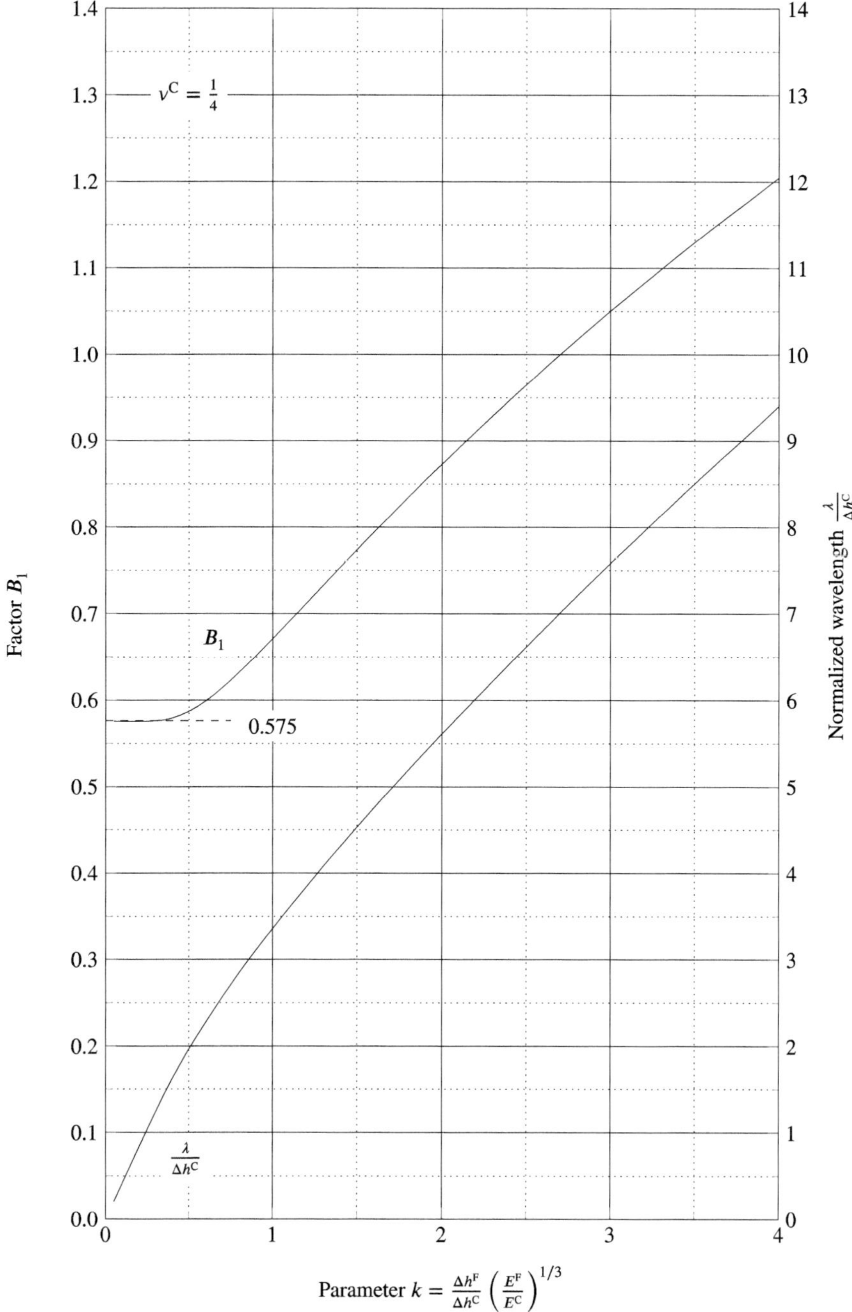

Fig. 6.6 Determination diagram of the factor B_1 and the wavelength λ for the case $v^C = \frac{1}{4}$. Based on [2]

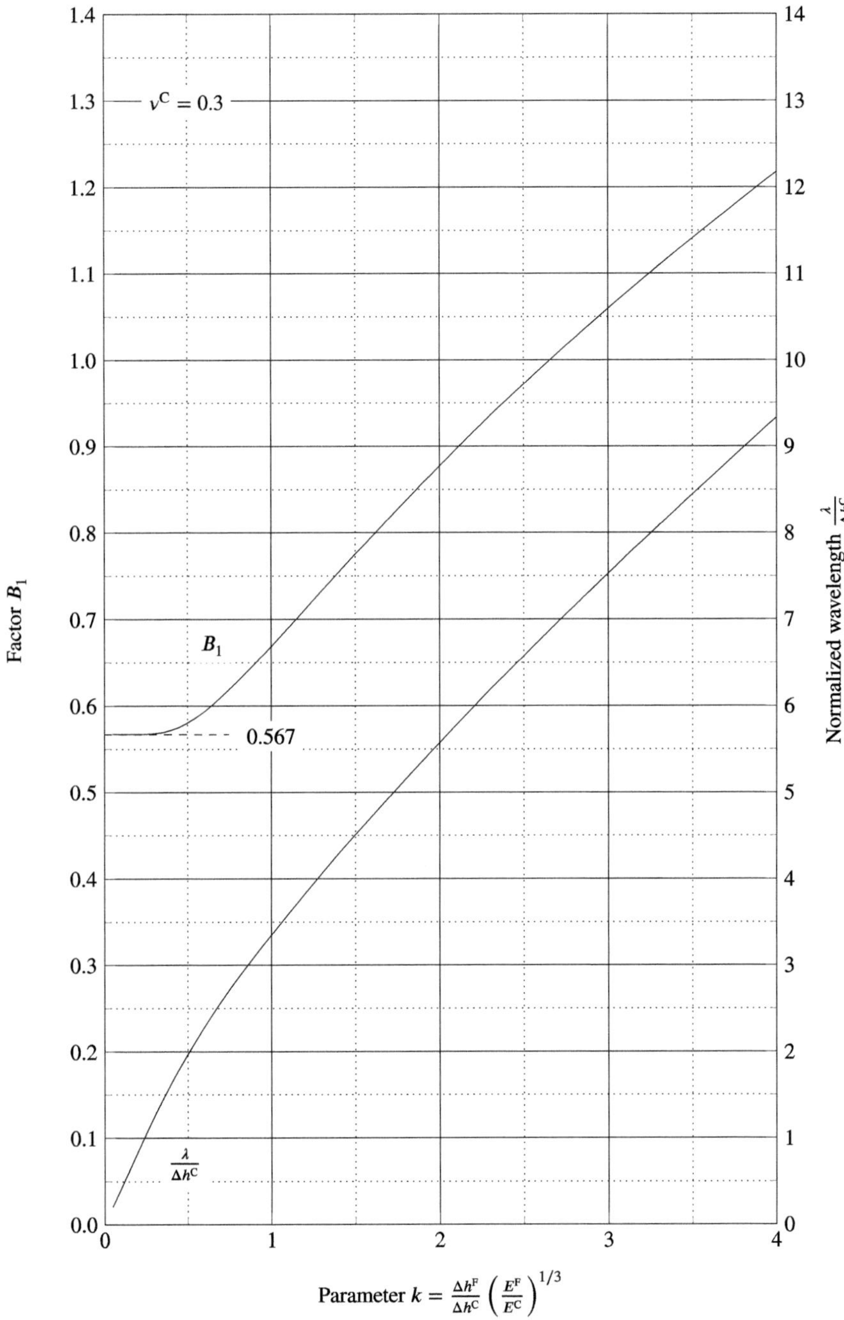

Fig. 6.7 Determination diagram of the factor B_1 and the wavelength λ for the case $v^C = 0.3$

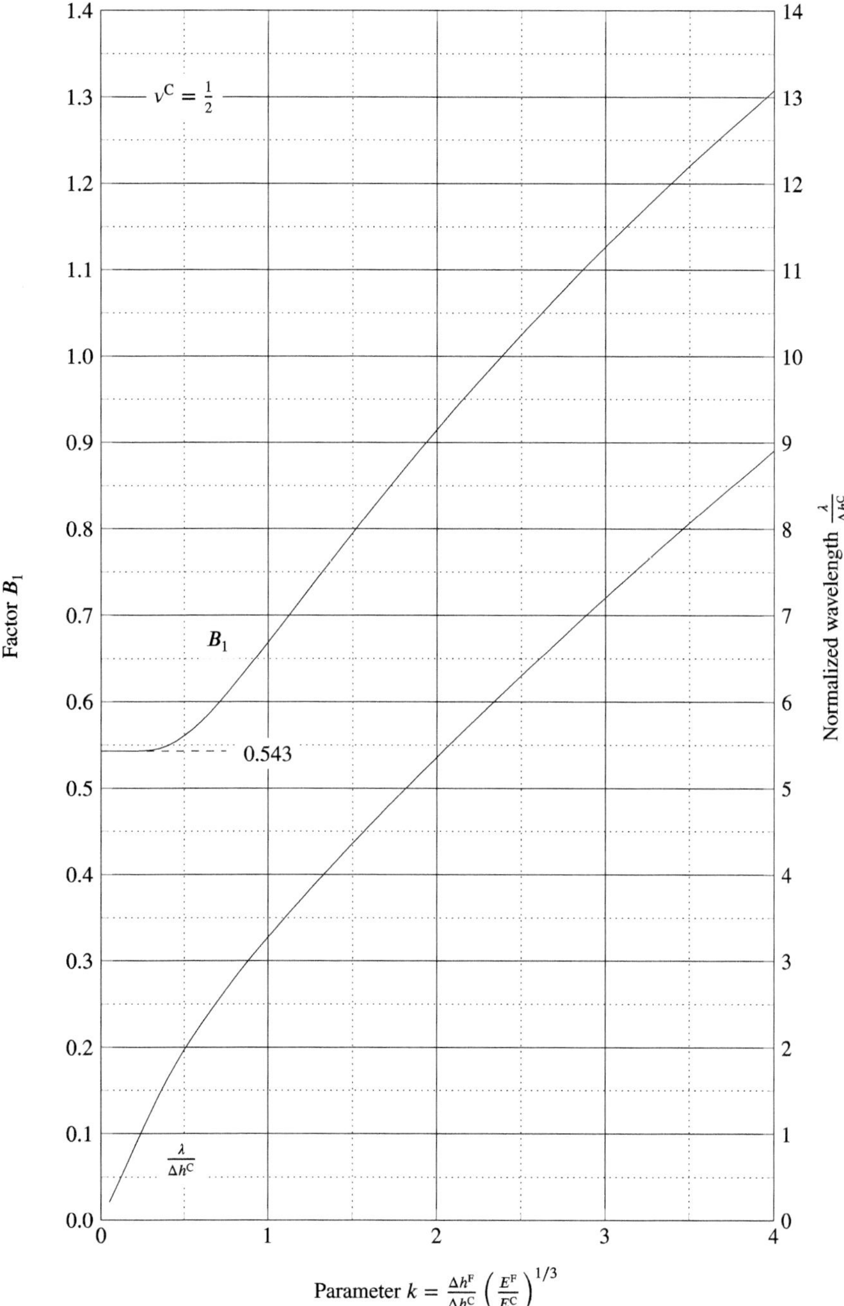

Fig. 6.8 Determination diagram of the factor B_1 and the wavelength λ for the case $\nu^C = \frac{1}{2}$. Based on [2]

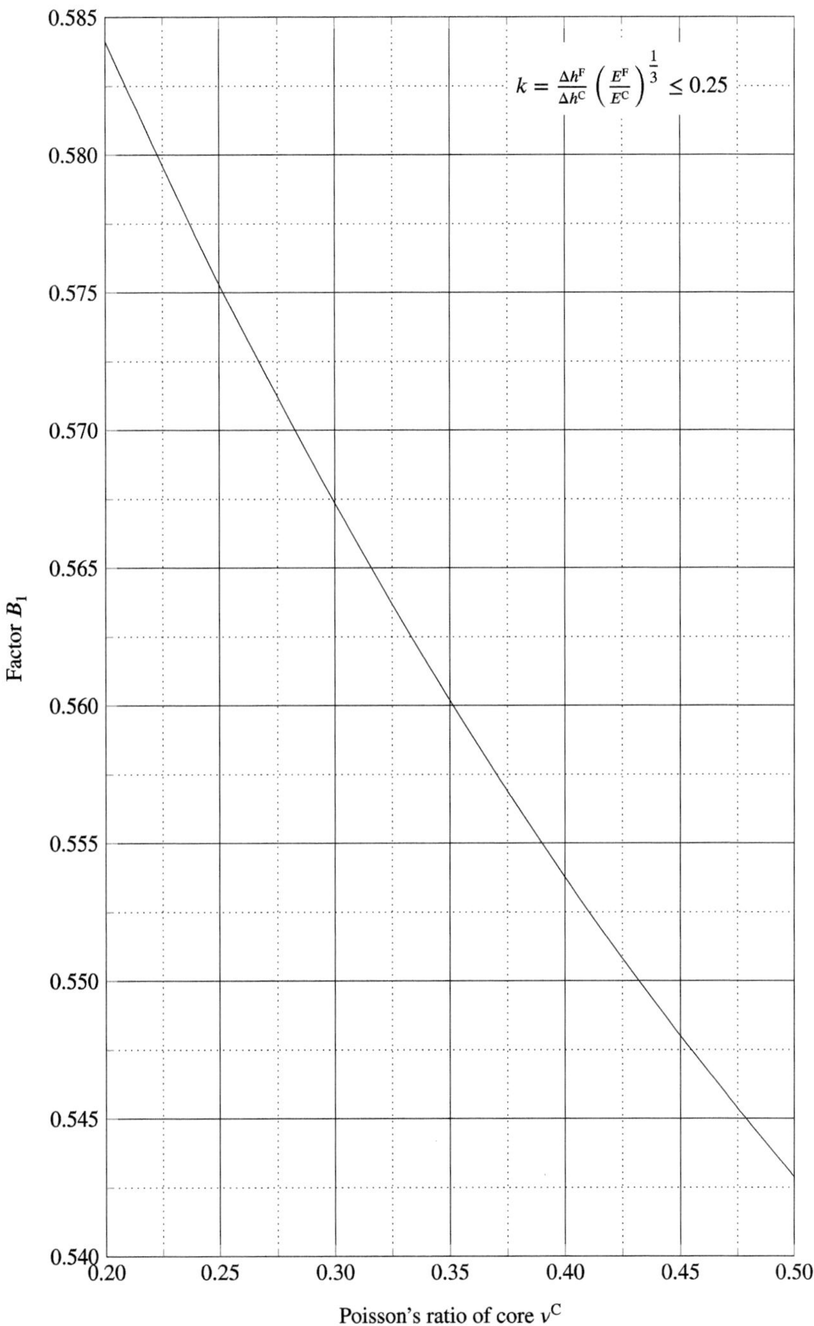

Fig. 6.9 Determination diagram of the factor B_1 for the case $k \leq 0.25$

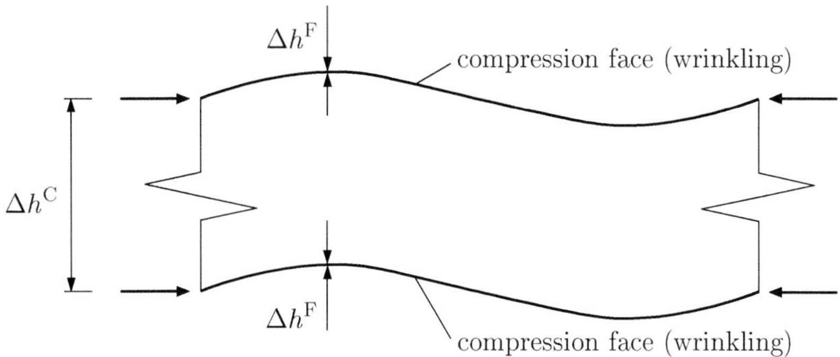

Fig. 6.10 Simplified representation of the antisymmetric wrinkling of the face sheets (compression load case). Based on [2]

Furthermore, one can see from Figs. 6.12–6.14 that the curves for B_1 and for the normalized wavelength $\frac{\lambda}{\Delta h^{\mathrm{C}}}$ only go up to a certain abscissa value k, which is determined by the following relationship:

$$\left(\frac{1 - \nu^{\mathrm{C}}}{8(1 + \nu^{\mathrm{C}})}\right)^{\frac{1}{3}} . \tag{6.42}$$

From the diagrams in Figs. 6.12–6.14 it can be seen that the B_1 value approaches a constant value for $k \leq 0.20$. These values are summarized in Fig. 6.15 as a function of the Poisson's ratio. One can see a slightly falling trend of the B_1 value. Furthermore, Figs. 6.12–6.14 show that the B_1 value is a monotonically decreasing function and the normalized wavelength is a monotonically increasing function.

6.1.5 Local Symmetric Wrinkling of Both Face Sheets (Compressive Load)

Local symmetric wrinkling can occur in compression-loaded sandwich beams and is characterized by the face sheets deforming symmetrically to the centerline in the transverse direction, see Fig. 6.16.

In the case of thin face sheets and a soft core, the critical wrinkling stress σ_{cr} can also be approximated according to the basic equation from Sect. 6.1.3, i.e.

$$\sigma_{\mathrm{cr}} = B_1 \left(E^{\mathrm{F}}\right)^{\frac{1}{3}} \left(E^{\mathrm{C}}\right)^{\frac{2}{3}} . \tag{6.43}$$

(a)

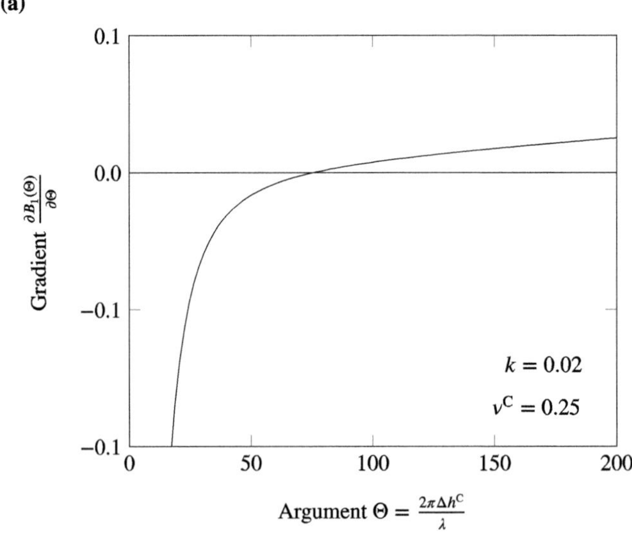

(b)

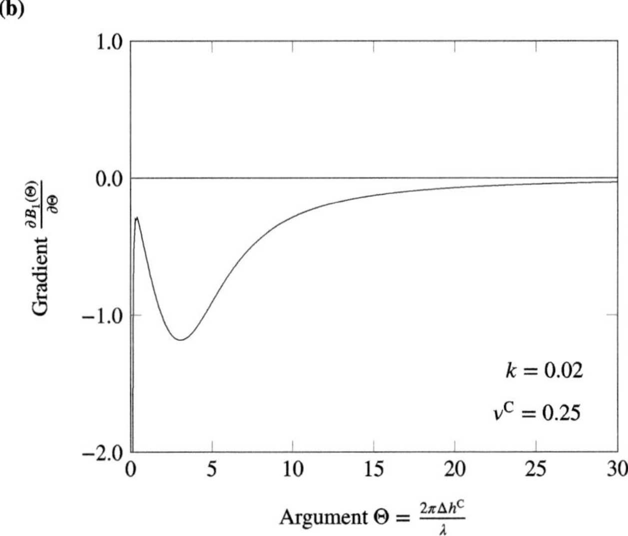

Fig. 6.11 Determination of the root of the derivative of the function $B_1 = B_1(\Theta)$: **(a)** large range of values with root at $\Theta = 75.85$ and (**b**) local extrema for $\Theta \to 0$

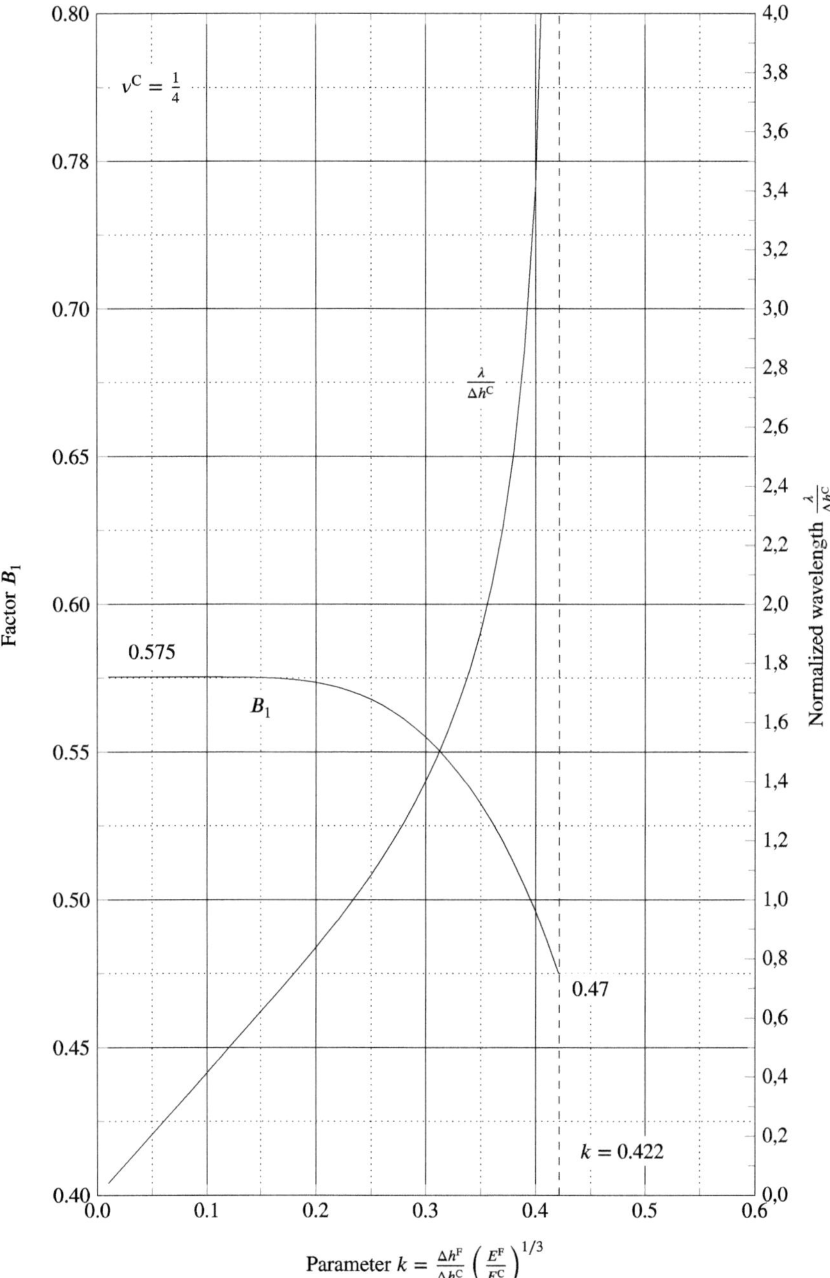

Fig. 6.12 Determination diagram of the factor B_1 and the wavelength λ for the case $\nu^{C} = \frac{1}{4}$ in antisymmetric wrinkling. Based on [2]

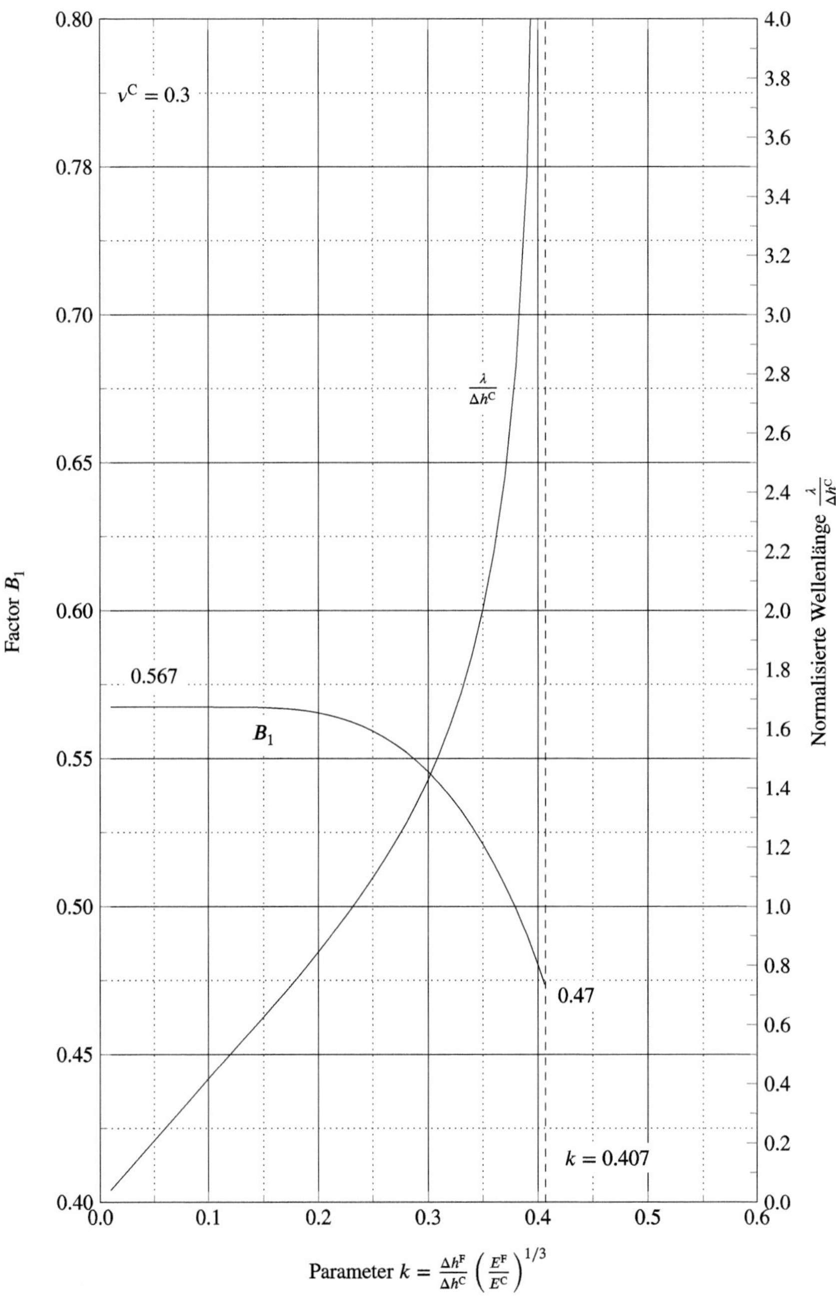

Fig. 6.13 Determination diagram of the factor B_1 and the wavelength λ for the case $v^C = 0.3$ in antisymmetric wrinkling

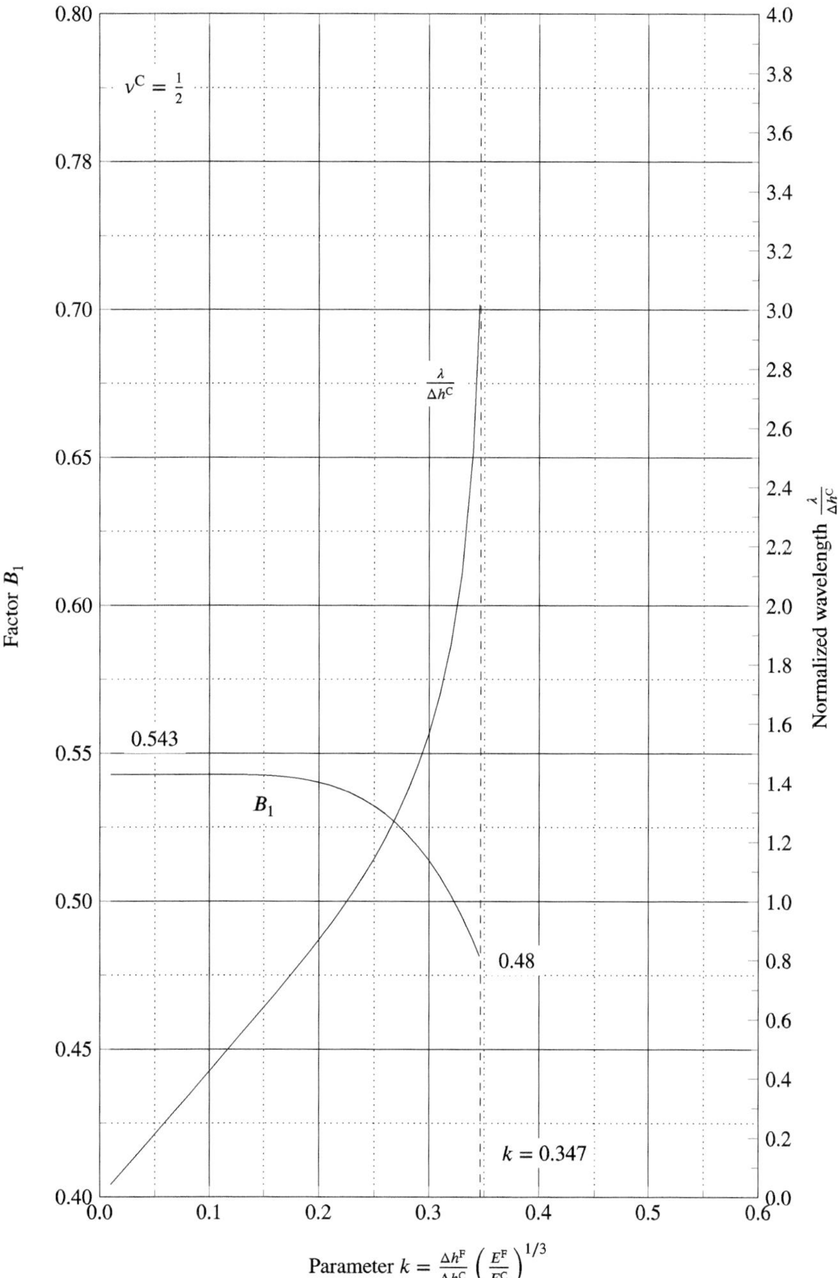

Fig. 6.14 Determination diagram of the factor B_1 and the wavelength λ for the case $\nu^C = 0.5$ in antisymmetric wrinkling. Based on [2]

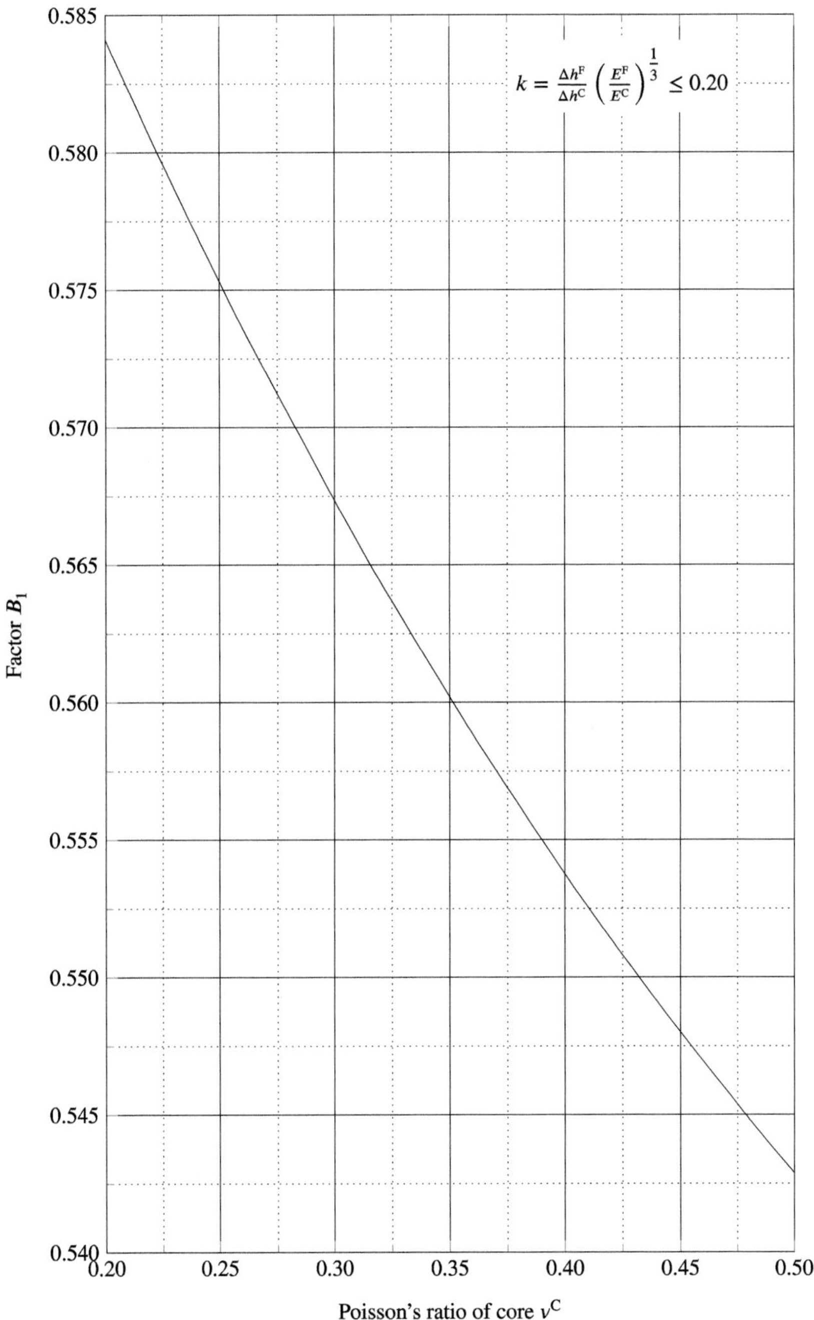

Fig. 6.15 Determination diagram of the factor B_1 for the case $k \leq 0.20$ in antisymmetric wrinkling

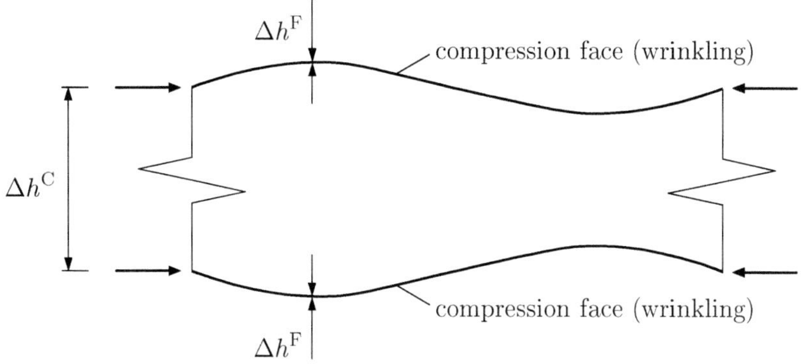

Fig. 6.16 Simplified representation of the symmetric wrinkling of the face sheets (compression load case). Based on [2]

However, another function $f(\Theta)$ should be used to calculate the factor B_1:

$$B_1 = \frac{k^2\Theta^2}{12} + \frac{f(\Theta)}{k} \tag{6.44}$$

$$= \frac{k^2\Theta^2}{12} + \frac{1}{k}\left(\frac{2}{\Theta} \times \frac{\cosh(\Theta)+1}{3\sinh(\Theta)-\Theta}\right). \tag{6.45}$$

Here, too, the determination of the local minimum of the function $B_1(\Theta)$ allows the factor B_1 for Eq. (6.43) to be determined. As in the other cases, the smaller value of B_1 occurs with larger Poisson's ratios ν^{C}. Therefore, Fig. 6.17 provides a determination diagram for the factor B_1 as a function of the geometry and material factor k. Here, too, small values of k, i.e. $k \leq 0.25$, result in a constant value of 0.630 for $\nu^{\mathrm{C}} = 0.0$. Furthermore, Fig. 6.17 shows that the B_1 value and the normalized wavelength are monotonically increasing functions.

The following conclusions can be drawn from the diagrams for antisymmetric (see Figs. 6.12–6.14) and symmetric wrinkling (see Fig. 6.17), see [2]:

- For very small values of k, i.e. $k < 0.2$, there are similar values for B_1. Thus, antisymmetric and symmetric wrinkling occur with more or less equal probability.

- In the range $0.2 < k < \left(\frac{1-\nu^{\mathrm{C}}}{8(1+\nu^{\mathrm{C}})}\right)^{\frac{1}{3}}$ antisymmetric wrinkling occurs at lower stresses since $B_1(\Theta)$ is a monotonically decreasing function for antisymmetric wrinkling, but monotonically increasing for symmetric wrinkling.

- For $k > \left(\frac{1-\nu^{\mathrm{C}}}{8(1+\nu^{\mathrm{C}})}\right)^{\frac{1}{3}}$ no antisymmetric wrinkling occurs and only symmetric wrinkling needs to be considered.

- For $k = \left(\frac{1-\nu^{\mathrm{C}}}{8(1+\nu^{\mathrm{C}})}\right)^{\frac{1}{3}}$ the B_1 value of antisymmetric wrinkling is the most conservative of all three wrinkling cases.

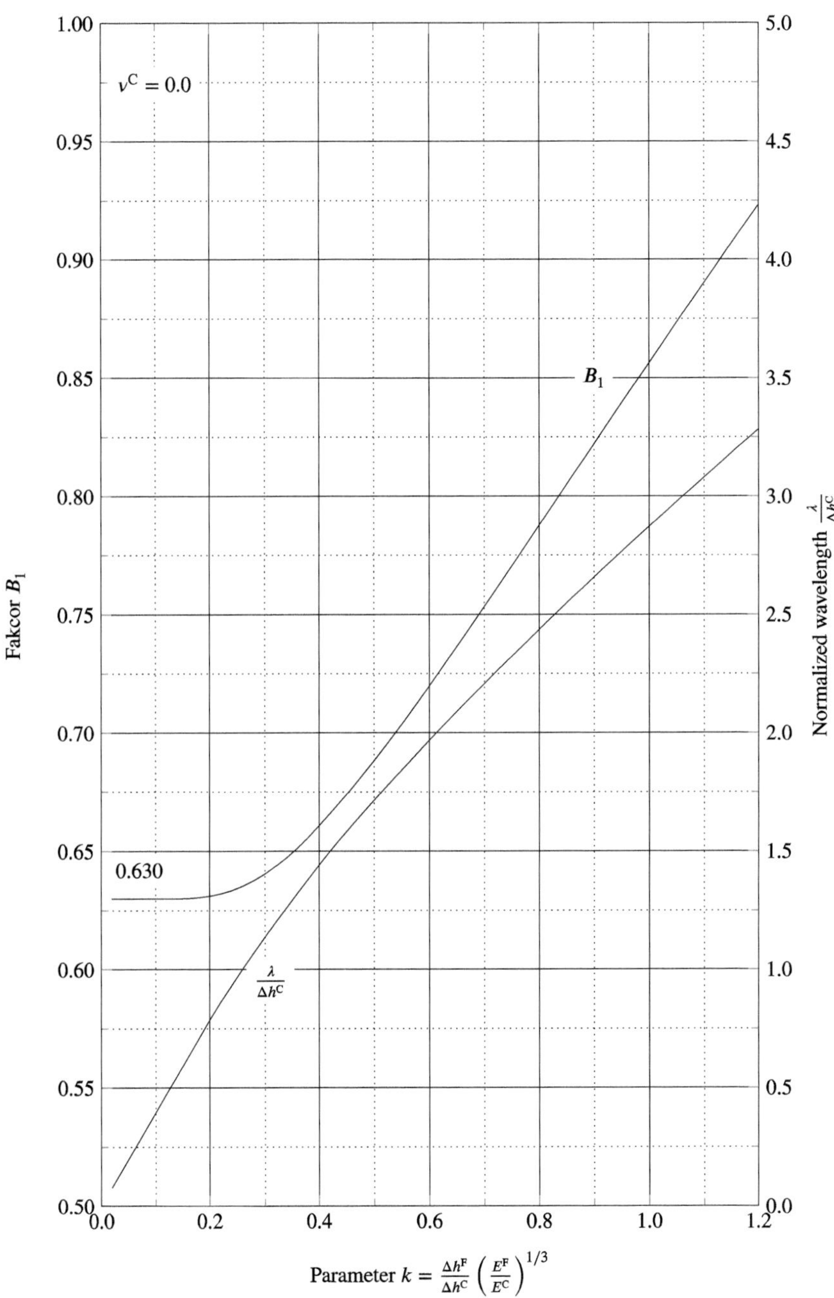

Fig. 6.17 Determination diagram of the factor B_1 and the wavelength λ for the case $v^C = 0.0$ in symmetric wrinkling. Based on [2]

Table 6.4 Determination of the factor $B_1 = \frac{k^2\Theta^2}{12} + \frac{f(\Theta)}{k}$ for local wrinkling. Based on [2]

Case (section)	Function $f(\Theta)$
Bending (6.1.3)	$\dfrac{2}{\Theta} \times \dfrac{(3 - \nu^C)\sinh(\Theta)\cosh(\Theta) + (1 + \nu^C)\Theta}{(1 + \nu^C)(3 - \nu^C)^2 \sinh^2(\Theta) - (1 + \nu^C)^3\Theta^2}$
Compression, asymmetric (6.1.4)	$\dfrac{2}{\Theta} \times \dfrac{\cosh(\Theta) - 1}{(1 + \nu^C)(3 - \nu^C)\sinh(\Theta) + (1 + \nu^C)^2\Theta}$
Compression, symmetric (6.1.5)	$\dfrac{2}{\Theta} \times \dfrac{\cosh(\Theta) + 1}{3\sinh(\Theta) - \Theta}\quad (\nu^C = 0)$

- For sufficiently long beams, global instability failure (see subchapter 6.1.1) occurs before local buckling.

At the end of this subchapter, the various formulations for calculating the factor B_1 are summarized and compared in Table 6.4.

6.2 Exercises

6.2.1 Knowledge Questions

- Describe typical failure modes of the face sheets of a sandwich beam.
- Describe typical failure modes of the core of a sandwich beam.
- Which failure modes can occur in a sandwich apart from the failure of the face sheets and the core?
- Under which types of loading can local wrinkling occur in a sandwich?
- Name the different types of instability that can occur in a sandwich beam.
- Name typical core materials for sandwich beams.
- What is the magnitude of the shear strength of classic adhesives?
- Which conservative approximation formula can be used to estimate the local wrinkling of the pressure face sheet (bending load case)?
- Give the simple formula for the buckling force of a sandwich considering Euler's buckling force.
- Sketch the bending moment distribution of a sandwich beam under symmetric 3-point and 4-point bending.

6.2.2 Calculation Problems

6.2.1 Failure Analysis of a Sandwich Beam with Distributed Load

Carry out the corresponding proofs of strength for the sandwich beam shown in Fig. 6.18. The sandwich consists of a hard foam core material bonded to two aluminum face sheets. The thickness of the adhesive layer can be neglected in the calculation. Furthermore, it should be assumed that the value of the distributed load q_0 has already been multiplied by a sufficient safety factor.

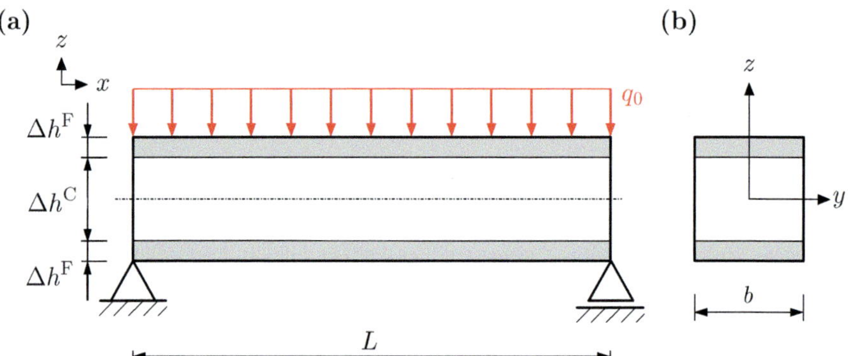

Fig. 6.18 Sandwich beam with constant distributed load: **(a)** boundary conditions and external load; **(b)** beam cross-section

Given are:

- Geometric dimensions: $L = 2000\,\text{mm}$, $b = 200\,\text{mm}$, $\Delta h^{\text{C}} = 150\,\text{mm}$, $\Delta h^{\text{F}} = 5\,\text{mm}$.
- Material properties of the hard foam core: $E^{\text{C}} = 30\,\text{MPa}$, $\nu^{\text{C}} = 0.364$, $\tau^{\text{C}}_{\text{aB}} = 0.5\,\text{MPa}$, $R^{\text{C}}_{\text{m}} = 0.90\,\text{MPa}$, $\sigma^{\text{C}}_{\text{dB}} = 0.38\,\text{MPa}$.
- Material properties of the face sheets: $E^{\text{F}} = 74{,}000\,\text{MPa}$, $R^{\text{F}}_{\text{p0.2}} = 364\,\text{MPa}$.
- Material property of the adhesive: $\tau_{\text{aB}} = 37.1\,\text{MPa}$.
- External load: $q_0 = 2.5\,\frac{\text{N}}{\text{mm}}$.

6.2.2 Failure Analysis of a Sandwich Beam (Cantilever) with Distributed Load

Carry out the corresponding proofs of strength for the sandwich beam shown in Fig. 6.19. The sandwich consists of a hard foam core material bonded to two aluminum face sheets. The thickness of the adhesive layer can be neglected in the calculation. Furthermore, it should be assumed that the value of the distributed load q_0 has already been multiplied by a sufficient safety factor.

Given are:

- Geometric dimensions: $L = 2000\,\text{mm}$, $b = 200\,\text{mm}$, $\Delta h^{\text{C}} = 150\,\text{mm}$, $\Delta h^{\text{F}} = 5\,\text{mm}$.
- Material properties of the hard foam core: $E^{\text{C}} = 30\,\text{MPa}$, $\nu^{\text{C}} = 0.364$, $\tau^{\text{C}}_{\text{aB}} = 0.5\,\text{MPa}$, $R^{\text{C}}_{\text{m}} = 0.90\,\text{MPa}$, $\sigma^{\text{C}}_{\text{dB}} = 0.38\,\text{MPa}$.
- Material properties of the face sheets: $E^{\text{F}} = 74{,}000\,\text{MPa}$, $R^{\text{F}}_{\text{p0.2}} = 364\,\text{MPa}$.
- Material property of the adhesive: $\tau_{\text{aB}} = 37.1\,\text{MPa}$.
- External load: $q_0 = 2.5\,\frac{\text{N}}{\text{mm}}$.

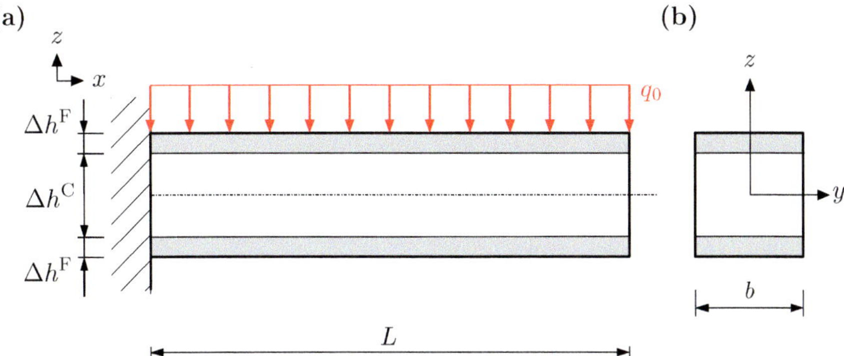

Fig. 6.19 Sandwich beam (cantilever) with constant distributed load: **(a)** boundary conditions and external load; **(b)** beam cross-section

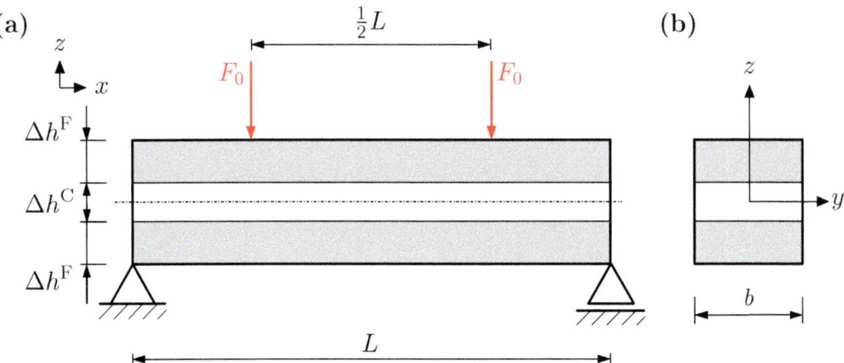

Fig. 6.20 Sandwich beam under 4-point bending: **(a)** boundary conditions and external loads; **(b)** beam cross-section

6.2.3 Failure Analysis of a Sandwich Beam Under 4-Point Bending

Carry out the corresponding proofs of strength for the sandwich beam shown in Fig. 6.20. The sandwich is constructed from a synthetic core material bonded to two aluminum face sheets. The thickness of the adhesive layer can be neglected in the calculation. Furthermore, it should be assumed that the value of the force F_0 has already been multiplied by a sufficient safety factor.

Given are:

- Geometric dimensions: $L = 2000\,\text{mm}$, $b = 200\,\text{mm}$, $\Delta h^{\text{C}} = 50\,\text{mm}$, $\Delta h^{\text{F}} = 50\,\text{mm}$.
- Material properties of the synthetic core: $E^{\text{C}} = 25000\,\text{MPa}$, $\nu^{\text{C}} = 0.4$, $\tau^{\text{C}}_{\text{aB}} = 40\,\text{MPa}$, $R^{\text{C}}_{\text{m}} = 80\,\text{MPa}$, $\sigma^{\text{C}}_{\text{dB}} = 60\,\text{MPa}$.
- Material properties of the face sheets: $E^{\text{F}} = 74{,}000\,\text{MPa}$, $R^{\text{F}}_{\text{p0.2}} = 364\,\text{MPa}$.
- Material property of the adhesive: $\tau_{\text{aB}} = 37.1\,\text{MPa}$.
- External load: $F_0 = 2500\,\text{N}$.

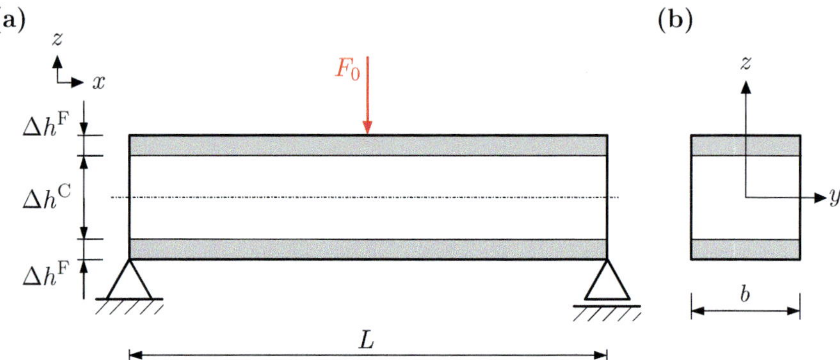

Fig. 6.21 Sandwich beam under 3-point bending: **(a)** boundary conditions and external load; **(b)** beam cross-section

6.2.4 Failure Analysis of a Sandwich Beam Under 3-Point Bending with Point Load

Carry out the corresponding proofs of strength for the sandwich beam shown in Fig. 6.21. The sandwich consists of a plastic core bonded to two aluminum face sheets. The thickness of the adhesive layer can be neglected in the calculation. Furthermore, it should be assumed that the value of the force F_0 has already been multiplied by a sufficient safety factor.

Given are:

- Geometric dimensions: $L = 2000\,\text{mm}$, $b = 200\,\text{mm}$, $\Delta h^{\mathrm{C}} = 100\,\text{mm}$, $\Delta h^{\mathrm{F}} = 25\,\text{mm}$.
- Material properties of the plastic core: $E^{\mathrm{C}} = 1500\,\text{MPa}$, $\nu^{\mathrm{C}} = 0.4$, $\tau^{\mathrm{C}}_{\mathrm{aB}} = 40\,\text{MPa}$, $R^{\mathrm{C}}_{\mathrm{m}} = 80\,\text{MPa}$, $\sigma^{\mathrm{C}}_{\mathrm{dB}} = 60\,\text{MPa}$.
- Material properties of the face sheets: $E^{\mathrm{F}} = 74{,}000\,\text{MPa}$, $R^{\mathrm{F}}_{\mathrm{p0.2}} = 364\,\text{MPa}$.
- Material property of the adhesive: $\tau_{\mathrm{aB}} = 37.1\,\text{MPa}$.
- External load: $F_0 = 5000\,\text{N}$.

6.2.5 Failure Analysis of a Sandwich Beam Under 3-Point Bending with Distributed Load

Carry out the corresponding proofs of strength for the sandwich beam shown in Fig. 6.22. The sandwich consists of a plastic core bonded to two aluminum face sheets. The thickness of the adhesive layer can be neglected in the calculation. Furthermore, it should be assumed that the value of the distributed load q_0 has already been multiplied by a sufficient safety factor.

Given are:

- Geometric dimensions: $L = 2000\,\text{mm}$, $b = 200\,\text{mm}$, $\Delta h^{\mathrm{C}} = 100\,\text{mm}$, $\Delta h^{\mathrm{F}} = 25\,\text{mm}$.
- Material properties of the plastic core: $E^{\mathrm{C}} = 1500\,\text{MPa}$, $\nu^{\mathrm{C}} = 0.4$, $\tau^{\mathrm{C}}_{\mathrm{aB}} = 40\,\text{MPa}$, $R^{\mathrm{C}}_{\mathrm{m}} = 80\,\text{MPa}$, $\sigma^{\mathrm{C}}_{\mathrm{dB}} = 60\,\text{MPa}$.

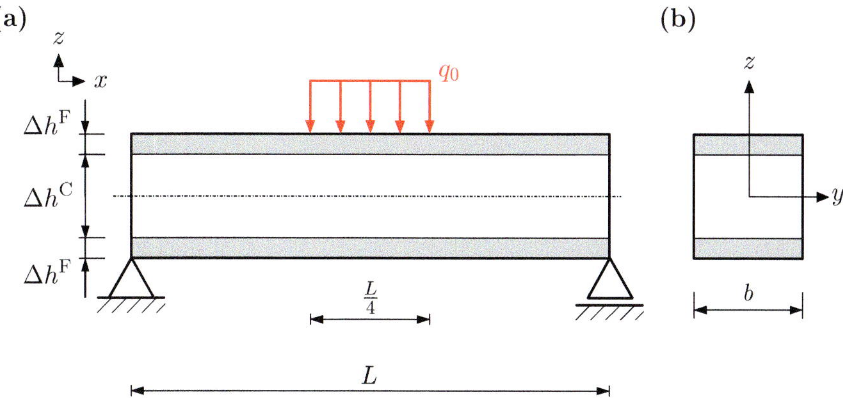

Fig. 6.22 Sandwich beam under 3-point bending with distributed load: **(a)** boundary conditions and external load; **(b)** beam cross-section

- Material properties of the face sheets: $E^{\mathrm{F}} = 74{,}000\,\mathrm{MPa}$, $R^{\mathrm{F}}_{\mathrm{p}0.2} = 364\,\mathrm{MPa}$.
- Material property of the adhesive: $\tau_{\mathrm{aB}} = 37.1\,\mathrm{MPa}$.
- External load: $q_0 = 10\,\mathrm{N/mm}$.

6.2.6 Failure Analysis of a Sandwich Beam Under 5-Point Bending with Point Loads

Carry out the corresponding proofs of strength for the sandwich beam shown in Fig. 6.23. The sandwich consists of a plastic core bonded to two aluminum face sheets. The thickness of the adhesive layer can be neglected in the calculation. Fur-

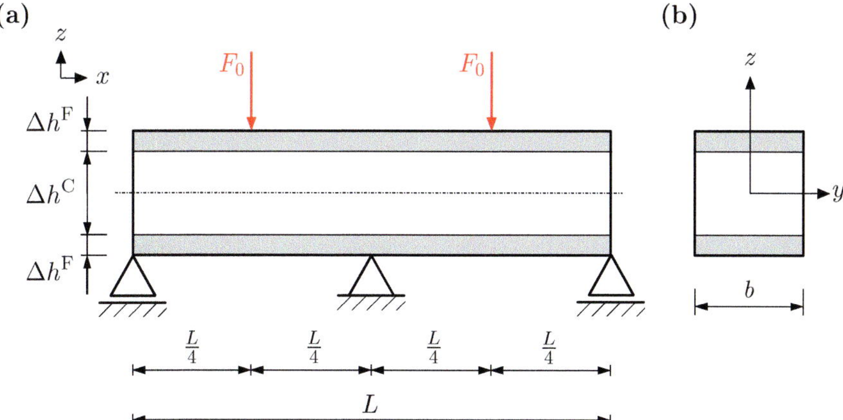

Fig. 6.23 Sandwich beam under 5-point bending with point loads: **(a)** boundary conditions and external loads; **(b)** beam cross-section

thermore, it should be assumed that the value of the forces F_0 is already multiplied by a sufficient safety factor.

Given are:

- Geometric dimensions: $L = 2000$ mm, $b = 200$ mm, $\Delta h^C = 100$ mm, $\Delta h^F = 25$ mm.
- Material properties of the plastic core: $E^C = 1500$ MPa, $\nu^C = 0.4$, $\tau_{aB}^C = 40$ MPa, $R_m^C = 80$ MPa, $\sigma_{dB}^C = 60$ MPa.
- Material properties of the face sheets: $E^F = 74{,}000$ MPa, $R_{p0.2}^F = 364$ MPa.
- Material property of the adhesive: $\tau_{aB} = 37.1$ MPa.
- External load: $F_0 = 2500$ N.

6.2.7 Failure Analysis of a Sandwich Beam Under 5-Point Bending with Distributed Load

Carry out the corresponding proofs of strength for the sandwich beam shown in Fig. 6.24. The sandwich consists of a plastic core bonded to two aluminum face sheets. The thickness of the adhesive layer can be neglected in the calculation. Furthermore, it should be assumed that the value of the distributed load q_0 has already been multiplied by a sufficient safety factor.

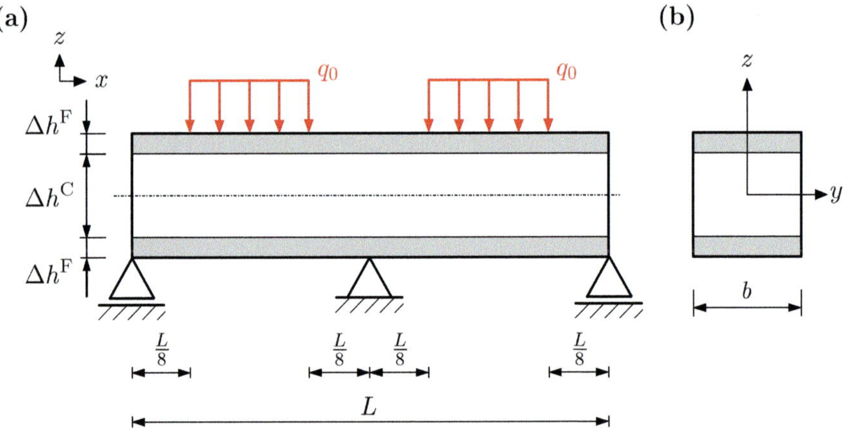

Fig. 6.24 Sandwich beam under 5-point bending with distributed loads: **(a)** boundary conditions and external loads; **(b)** beam cross-section

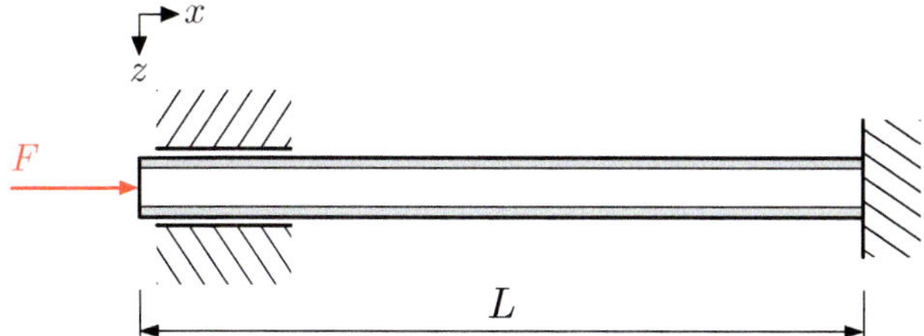

Fig. 6.25 Global instability failure of a sandwich beam clamped on both sides under compressive loading

Given are:

- Geometric dimensions: $L = 2000\,\text{mm}$, $b = 200\,\text{mm}$, $\Delta h^{\text{C}} = 100\,\text{mm}$, $\Delta h^{\text{F}} = 25\,\text{mm}$.
- Material properties of the plastic core: $E^{\text{C}} = 1500\,\text{MPa}$, $\nu^{\text{C}} = 0.4$, $\tau^{\text{C}}_{\text{aB}} = 40\,\text{MPa}$, $R^{\text{C}}_{\text{m}} = 80\,\text{MPa}$, $\sigma^{\text{C}}_{\text{dB}} = 60\,\text{MPa}$.
- Material properties of the face sheets: $E^{\text{F}} = 74{,}000\,\text{MPa}$, $R^{\text{F}}_{\text{p0.2}} = 364\,\text{MPa}$.
- Material property of the adhesive: $\tau_{\text{aB}} = 37.1\,\text{MPa}$.
- External load: $q_0 = 5\,\text{N/mm}$.

6.2.8 Global Instability Failure of a Sandwich Beam Clamped on Both Sides Under Compressive Loading

Derive the buckling force for the sandwich beam shown in Fig. 6.25 assuming thin face sheets and a soft core. Compare the result with the classic solution according to Euler for the 4th support condition.

6.2.9 Instability Failure of a Sandwich Beam Hinged on Both Ends Under Compressive Loading

For the sandwich beam shown in Fig. 6.26, calculate the critical stresses for global and local instability failure assuming thin face sheets and a soft core. Furthermore, sketch the course of the global buckling stress as a function of the beam length L.

Given are:

- Geometric dimensions: $L = 2000\,\text{mm}$, $b = 200\,\text{mm}$, $\Delta h^{\text{C}} = 100\,\text{mm}$, $\Delta h^{\text{F}} = 5\,\text{mm}$.
- Material properties of the isotropic core: $E^{\text{C}} = 200\,\text{MPa}$, $\nu^{\text{C}} = 0.4$.
- Material properties of the face sheets: $E^{\text{F}} = 74{,}000\,\text{MPa}$.

(a)

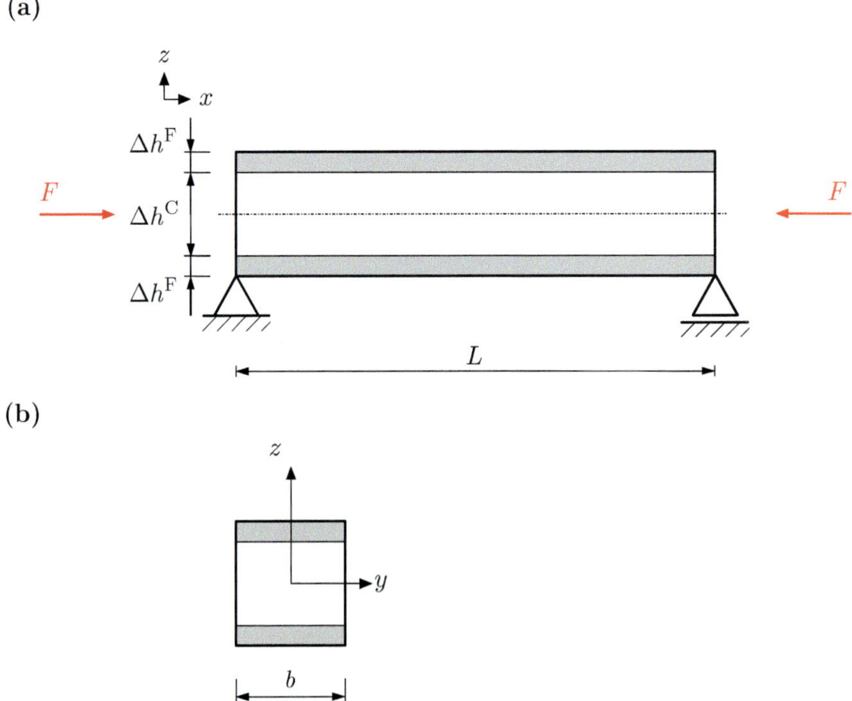

(b)

Fig. 6.26 (a) Sandwich beam hinged on both sides under compression; (b) beam cross-section

References

1. Allen, H.G.: Optimum design of sandwich struts and beams. In: Plastics in Building Structures, Proceedings of a Conference Held in London, 14-16 June 1965. Pergamon Press, Oxford (1966)
2. Allen, H.G.: Analysis and Design of Structural Sandwich Panels. Pergamon Press, Oxford (1969)
3. Gough, G.S., Elam, C.F., Tipper, G.H., De Bruyne, N.A.: The stabilisation of a thin sheet by a continuous supporting medium. J. R. Aeronaut. Soc. 44(349), 12–43 (1940). https://doi.org/10.1017/S036839310010495X
4. Hoff, N.J., Mautner, S.E.: The buckling of sandwich-type panels. J. Aeronaut. Sci. 12(3), 285–297 (1945). https://doi.org/10.2514/8.11246
5. Öchsner, A.: Elasto-Plasticity of Frame Structure Elements: Modeling and Simulation of Rods and Beams. Springer, Berlin (2014)
6. Plantema, F.J.: Sandwich Construction: the Bending and Buckling of Sandwich Beams, Plates, and Shells. John Wiley & Sons, New York (1966)
7. da Silva, L.F.M., Öchsner, A., Adams, R. (eds.): Handbook of Adhesion Technology. Springer, Cham (2018)
8. Stamm, K., Witte, H.: Sandwichkonstruktionen. Springer-Verlag, Wien (1974)

Abstract

In this chapter, the basics of lightweight design in regard to materials and shape are further deepened. In particular, approaches to the optimal dimensioning of sandwich beams are discussed. In the optimization strategy presented, a distinction is made between tensile and compressive loads or bending.

7.1 Optimal Dimensioning of Sandwich Beams

7.1.1 Tensile or Compressive Load

The following derivations for the optimization of a sandwich beam under tensile or compressive loading are limited to sandwich beams with thin face sheets and a soft core. A detailed description can be found in [1]. The general configuration with the geometric dimensions used can be seen in Fig. 7.1.

In the case of pure tensile loading for a technical sandwich beam, there acts only tensile stress in both face sheets (see Fig. 5.14). This means that the acting force F

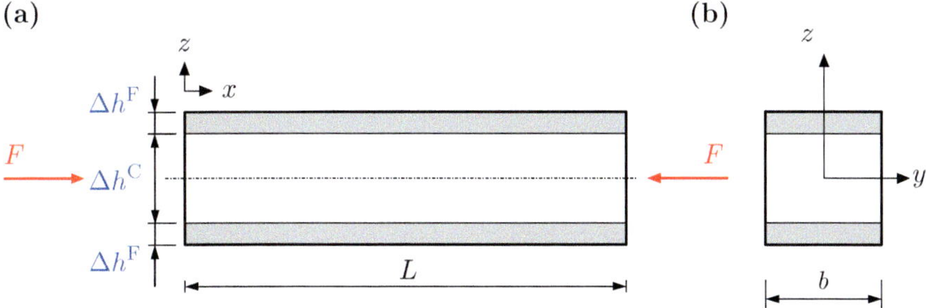

Fig. 7.1 Sandwich beam under compression: (**a**) general configuration and (**b**) beam cross-section

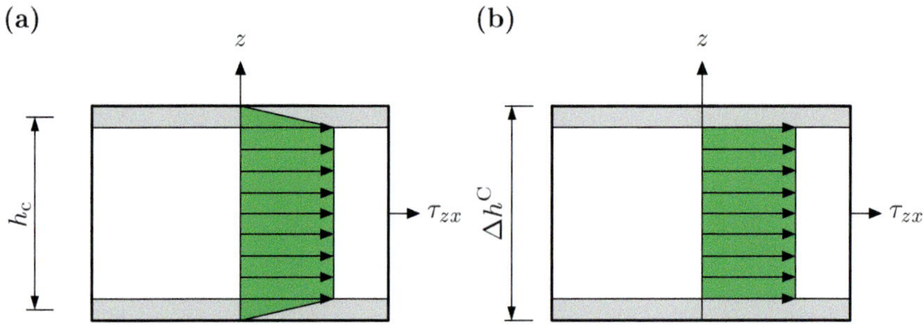

Fig. 7.2 Approximation of the shear area: **(a)** $A \approx b\Delta h_{\mathrm{c}}$ and **(b)** $A \approx b\Delta h^{\mathrm{C}}$

must follow the following condition:

$$F \leq 2b\Delta h^{\mathrm{F}} R_{\mathrm{p0.2}} . \tag{7.1}$$

The case of compressive loading is much more complex, since global buckling (see Sect. 6.1.1), local wrinkling (see Sects. 6.1.4 and 6.1.5) and yield failure (see Sect. 5.5.2) can occur.

According to Eq. (6.19), global buckling under compression occurs at the following critical force:

$$F = \frac{F_{\mathrm{cr}}^{\mathrm{E}}}{1 + \frac{F_{\mathrm{cr}}^{\mathrm{E}}}{AG^{\mathrm{C}}}} , \tag{7.2}$$

where $F_{\mathrm{cr}}^{\mathrm{E}} = \frac{\pi^2 \overline{EI_y}}{L^2}$ with $\overline{EI_y} \approx \frac{E^{\mathrm{F}} b\Delta h^{\mathrm{F}} (h_{\mathrm{c}})^2}{2}$ is the Euler buckling force for homogeneous beams and the area is given according to Eq. (5.103) as $A = \frac{bh_{\mathrm{c}}^2}{\Delta h^{\mathrm{C}}} \approx bh_{\mathrm{c}}$. For the further derivations, this approximation is further simplified to $A \approx b\Delta h^{\mathrm{C}}$[1] (see Fig. 7.2).

The failure of the face sheets occurs at the following force:

$$F = 2b\Delta h^{\mathrm{F}} \sigma_{\mathrm{cr}} , \tag{7.3}$$

where for the critical stress σ_{cr} the lower value of the 0.2% initial yield stress ($R_{\mathrm{p0.2}}$) or the wrinkling stress is to be used. To simplify the following derivations, the wrinkling stress according to Eq. (6.31) is approximated using $\sigma_{\mathrm{cr}} \approx \frac{1}{2} \times \left(E^{\mathrm{F}} E^{\mathrm{C}} G^{\mathrm{C}}\right)^{1/3}$.

To optimize a sandwich structure, the weight is usually reduced to a minimum. The total mass is composed proportionally of the face sheets and the core:

$$m = \varrho V \tag{7.4}$$

$$= \varrho^{\mathrm{C}} V^{\mathrm{C}} + \varrho^{\mathrm{F}} V^{\mathrm{F}} = \varrho^{\mathrm{C}} \Delta h^{\mathrm{C}} bL + \varrho^{\mathrm{F}} 2\Delta h^{\mathrm{F}} bL \tag{7.5}$$

$$= b\left(\varrho^{\mathrm{C}} \Delta h^{\mathrm{C}} L + \varrho^{\mathrm{F}} 2\Delta h^{\mathrm{F}} L\right) , \tag{7.6}$$

---

[1] This means that the entire shear stress acts only in the core.

or as length-related mass:

$$m^{\mathrm{n}} = \frac{m}{L} = b \left(\varrho^{\mathrm{C}} \Delta h^{\mathrm{C}} + \varrho^{\mathrm{F}} 2 \Delta h^{\mathrm{F}} \right) . \tag{7.7}$$

If one replaces the volume-related mass in Eq. (7.7), i.e. the density, through the volume-related costs of the core and face sheets, m^{n} can be interpreted as the length-related costs of the sandwich beam.

If one uses the length-specific normalizations for the face sheet, i.e. $\Delta h^{\mathrm{F,n}} = \Delta h^{\mathrm{F}}/L$, and the core, i.e. $\Delta h^{\mathrm{C,n}} = \Delta h^{\mathrm{C}}/L$, Eqs. (7.2), (7.3) and (7.7) can be formulated as follows:

$$\frac{F_{\mathrm{cr}}^{\mathrm{E}}}{1 + \dfrac{F_{\mathrm{cr}}^{\mathrm{E}}}{AG^{\mathrm{C}}}} \geq F \tag{7.8}$$

$$\frac{\dfrac{\Delta h^{\mathrm{F}}(\Delta h^{\mathrm{C}})^2}{L^2 \times L} \times bL}{\dfrac{2}{\pi^2 E^{\mathrm{F}}} + \dfrac{\Delta h^{\mathrm{F}} \Delta h^{\mathrm{C}}}{L^2} \times \dfrac{1}{G^{\mathrm{C}}}} \geq F \tag{7.9}$$

$$g_1 \left(\Delta h^{\mathrm{F,n}}, \Delta h^{\mathrm{C,n}} \right) = \frac{\Delta h^{\mathrm{F,n}} \left(\Delta h^{\mathrm{C,n}} \right)^2}{\dfrac{2}{\pi^2 E^{\mathrm{F}}} + \dfrac{\Delta h^{\mathrm{F,n}} \Delta h^{\mathrm{C,n}}}{G^{\mathrm{C}}}} \geq \frac{F}{bL}, \tag{7.10}$$

or Eq. (7.3)

$$g_2 \left(\Delta h^{\mathrm{F,n}}, \Delta h^{\mathrm{C,n}} \right) = 2 \Delta h^{\mathrm{F,n}} \sigma_{\mathrm{cr}} \geq \frac{F}{bL}, \tag{7.11}$$

or Eq. (7.7)

$$f \left(\Delta h^{\mathrm{F,n}}, \Delta h^{\mathrm{C,n}} \right) = \frac{m}{bL} = \frac{m^{\mathrm{n}}}{b} = \left(\varrho^{\mathrm{C}} \Delta h^{\mathrm{C,n}} + 2 \varrho^{\mathrm{F}} \Delta h^{\mathrm{F,n}} \right) . \tag{7.12}$$

Equation (7.12), i.e. $f \left(\Delta h^{\mathrm{F,n}}, \Delta h^{\mathrm{C,n}} \right)$, can be regarded as an objective function which is to be minimized under the constraints $g_1 \left(\Delta h^{\mathrm{F,n}}, \Delta h^{\mathrm{C,n}} \right)$ and $g_2 \left(\Delta h^{\mathrm{F,n}}, \Delta h^{\mathrm{C,n}} \right)$. A graphical representation of the objective function f in Fig. 7.3 shows that it is an inclined plane ($OABC$) through the origin.

The constraints g_i according to Eqs. (7.10) and (7.11) can first be illustrated in a $\Delta h^{\mathrm{C,n}}$-$\Delta h^{\mathrm{F,n}}$ coordinate system. To do this, both equations are solved for $\Delta h^{\mathrm{F,n}}$:

$$\Delta h^{\mathrm{F,n}} \geq \frac{\dfrac{F}{bL} \times \dfrac{2}{\pi^2 E^{\mathrm{F}}}}{\Delta h^{\mathrm{C,n}} \left(\Delta h^{\mathrm{C,n}} - \dfrac{F}{bL} \times \dfrac{1}{G^{\mathrm{C}}} \right)}, \tag{7.13}$$

$$\Delta h^{\mathrm{F,n}} \geq \frac{F}{2bL\sigma_{\mathrm{cr}}}. \tag{7.14}$$

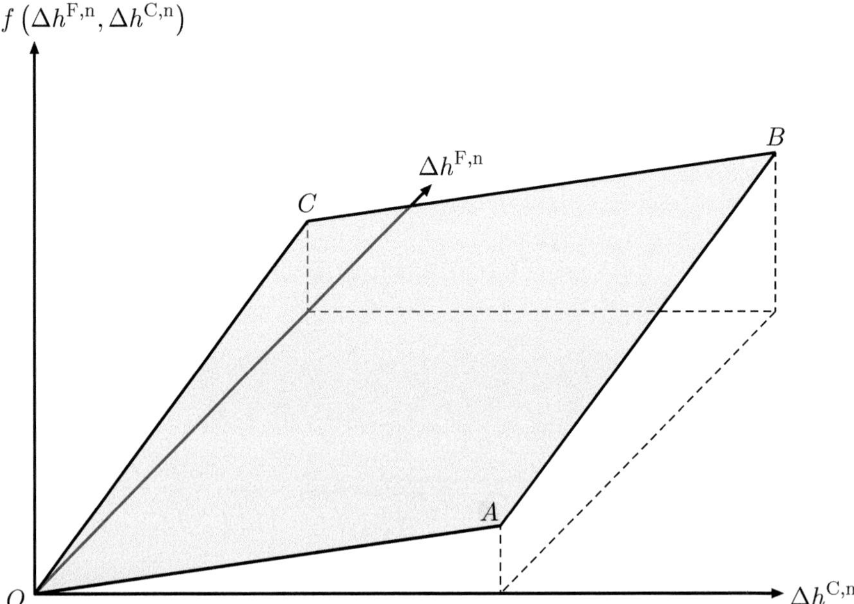

Fig. 7.3 Schematic representation of the objective function $f\left(\Delta h^{\mathrm{F,n}},\Delta h^{\mathrm{C,n}}\right)$. Adapted from [1]

Figure 7.4 shows the course of the two limit curves g_1 and g_2 in the $\Delta h^{\mathrm{C,n}}$-$\Delta h^{\mathrm{F,n}}$ plane. The gray area is the common permissible area.

If one transfers the limit curves from Fig. 7.4 to the three-dimensional representation of Fig. 7.3 and projects both curves onto the inclined plane f, Fig. 7.5 results.

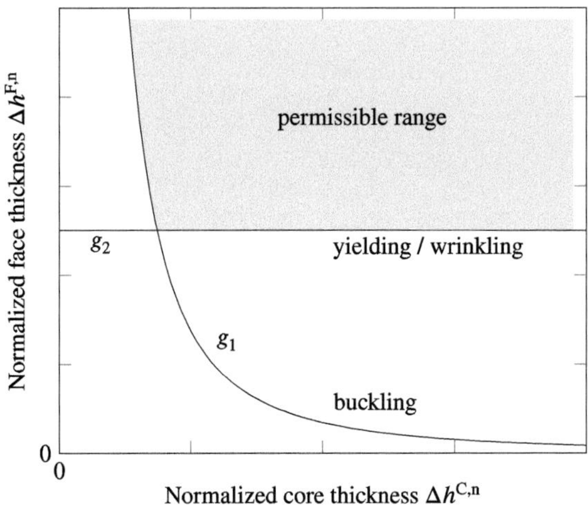

Fig. 7.4 Normalized face thickness as a function of the normalized core thickness based on the functions g_1 and g_2 according to Eqs. (7.10) and (7.11). Only areas above the two constraints are allowed

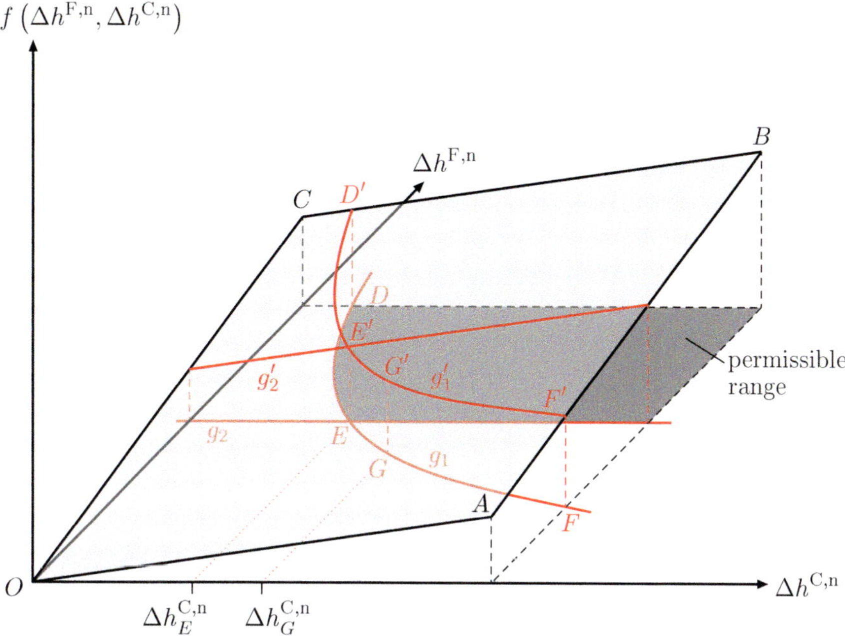

Fig. 7.5 Schematic representation of the objective function $f\left(\Delta h^{\mathrm{F,n}}, \Delta h^{\mathrm{C,n}}\right)$ and the constraints $g_1\left(\Delta h^{\mathrm{F,n}}, \Delta h^{\mathrm{C,n}}\right)$ and $g_2\left(\Delta h^{\mathrm{F,n}}, \Delta h^{\mathrm{C,n}}\right)$. Adapted from [1]

The points lying along the curve $D'E'F'$ on the inclined plane can be expressed as follows: From the first constraint in the formulation of Eq. (7.13) follows

$$\Delta h^{\mathrm{F,n}} \geq \frac{\frac{2}{\pi^2 E^{\mathrm{F}}}}{\Delta h^{\mathrm{C,n}}\left(\Delta h^{\mathrm{C,n}} \times \frac{bL}{F} - \frac{1}{G^{\mathrm{C}}}\right)}. \tag{7.15}$$

Inserting this relationship into the objective function f according to Eq. (7.12) results in

$$f\left(\Delta h^{\mathrm{C,n}}\right) = 2\varrho^{\mathrm{F}} \frac{\frac{2}{\pi^2 E^{\mathrm{F}}}}{\Delta h^{\mathrm{C,n}}\left(\Delta h^{\mathrm{C,n}} \times \frac{bL}{F} - \frac{1}{G^{\mathrm{C}}}\right)} + \varrho^{\mathrm{C}} \Delta h^{\mathrm{C,n}} \tag{7.16}$$

$$= \frac{\frac{4\varrho^{\mathrm{F}}}{\pi^2 E^{\mathrm{F}}}}{\Delta h^{\mathrm{C,n}}\left(\Delta h^{\mathrm{C,n}} \times \frac{bL}{F} - \frac{1}{G^{\mathrm{C}}}\right)} + \varrho^{\mathrm{C}} \Delta h^{\mathrm{C,n}}. \tag{7.17}$$

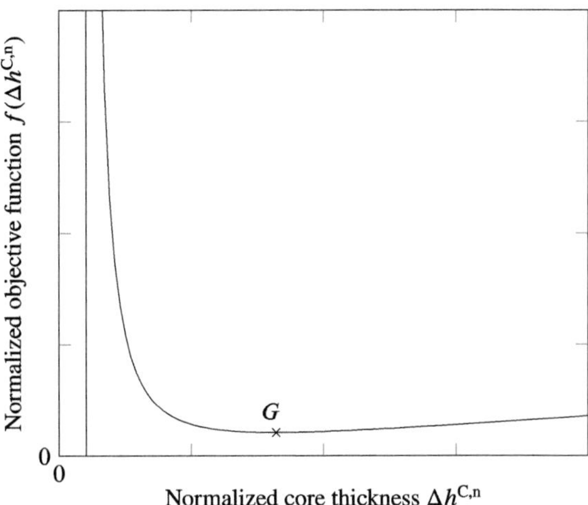

Fig. 7.6 Normalized objective function as a function of the normalized core thickness

The schematic course of the function $f\left(\Delta h^{C,n}\right)$ is shown in Fig. 7.6. The aim of the optimization is now to determine the minimum at the point $G(f_G, \Delta h_G^{C,n})$ in order to minimize the weight or the costs. This minimum can be obtained using the condition[2]

$$\frac{\partial f\left(\Delta h^{C,n}\right)}{\partial \Delta h^{C,n}} \stackrel{!}{=} 0 \,. \tag{7.18}$$

However, it should be noted that the minimum G must be located within the permissible range, see Fig. 7.5.

To assess the admissibility of point G, two cases must be distinguished. In the case of $\Delta h_G^{C,n} \leq \Delta h_E^{C,n}$, i.e. the point G lies on the monotonously falling curve section $D'E'$, the optimum point has been found, since both constraints g_1 and g_2 are fulfilled. For the second case with $\Delta h_G^{C,n} > \Delta h_E^{C,n}$, the point lies on the monotonically decreasing curve section $F'E'$. However, only the constraint g_1 is fulfilled here. The next point with minimum function value of f is E, i.e. the intersection of the two constraints. Therefore, in this case, point E is the optimal point with minimum weight. These two facts are shown again in the $\Delta h^{C,n}$-$\Delta h^{F,n}$ coordinate system in Fig. 7.7.

In order to be able to distinguish between cases with regard to the point E, its coordinates in a general representation are helpful. The point of intersection E of the constraints g_1 and g_2 in the $\Delta h^{C,n}$-$\Delta h^{F,n}$ coordinate system results from equating

---

[2] In the Appendix A.2, a Python3 program is provided to automatically carry out the evaluation of Eq. (7.18).

Fig. 7.7 Determination of the optimum point in the $\Delta h^{\mathrm{C,n}}$-$\Delta h^{\mathrm{F,n}}$ coordinate system: **(a)** point G for $\Delta h_G^{\mathrm{C,n}} \leq \Delta h_E^{\mathrm{C,n}}$, **(b)** point E for $\Delta h_G^{\mathrm{C,n}} > \Delta h_E^{\mathrm{C,n}}$

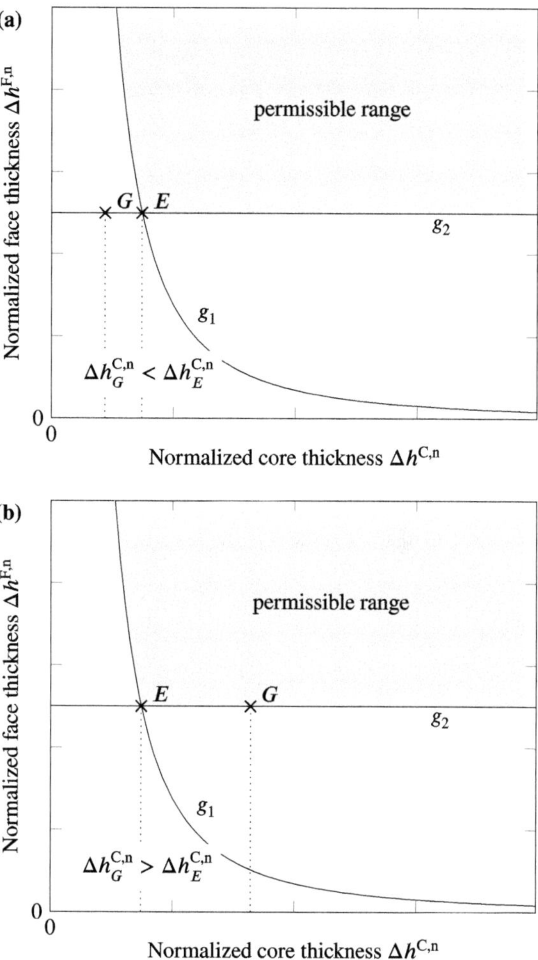

Eqs. (7.13) and (7.14), i.e.

$$\frac{\frac{F}{bL} \times \frac{2}{\pi^2 E^{\mathrm{F}}}}{\Delta h_E^{\mathrm{C,n}}\left(\Delta h_E^{\mathrm{C,n}} - \frac{F}{bL} \times \frac{1}{G^{\mathrm{C}}}\right)} = \frac{F}{2bL\sigma_{\mathrm{cr}}} \tag{7.19}$$

$$\Leftrightarrow \left(\Delta h_E^{\mathrm{C,n}} - \frac{F}{2bLG^{\mathrm{C}}}\right)^2 = \left(\frac{F}{2bLG^{\mathrm{C}}}\right)^2 + \frac{4\sigma_{\mathrm{cr}}}{\pi^2 E^{\mathrm{F}}} \tag{7.20}$$

$$\Rightarrow \Delta h_E^{\mathrm{C,n}} = \sqrt{\left(\frac{F}{2bLG^{\mathrm{C}}}\right)^2 + \frac{4\sigma_{\mathrm{cr}}}{\pi^2 E^{\mathrm{F}}}} + \frac{F}{2bLG^{\mathrm{C}}}, \tag{7.21}$$

or as complete coordinates of point E:

$$E\left(\Delta h_E^{C,n}, \Delta h_E^{F,n}\right) = \left(\sqrt{\left(\frac{F}{2bLG^C}\right)^2 + \frac{4\sigma_{cr}}{\pi^2 E^F}} + \frac{F}{2bLG^C}, \frac{F}{2bL\sigma_{cr}}\right). \tag{7.22}$$

7.1.2   Bending Load

The following derivations for optimizing a sandwich beam under bending loading are again limited to sandwich beams with thin face sheets and a soft core. A detailed description can be found in [1]. The general configuration with the geometric dimensions used can be taken from Fig. 7.8. It should be noted here that the external loads (e.g. single forces or distributed loads) were not drawn in since they are case-specific.

In the case of a bending load, local wrinkling (see Sect. 6.1.3) can occur in the pressure-loaded face sheet and/or plastic yield failure can occur in the tensile or compressive region (see Sect. 5.5.1).

To ensure that the face sheets do not fail, the following relationship (see Eq. (5.49)) must be fulfilled:

$$\sigma_{x,F} \approx \frac{M_{y,max}}{b\Delta h^F \Delta h^C} < \sigma_{cr}, \tag{7.23}$$

where for the critical stress σ_{cr} the lower value of the 0.2% initial yield stress ($R_{p0.2}$) or the wrinkling stress is to be used. To simplify the following derivations, the wrinkling stress is approximated according to Eq. (6.31) using $\sigma_{cr} \approx \frac{1}{2} \times \left(E^F E^C G^C\right)^{1/3}$.

The following relationship (see Eq. (5.91)) must be fulfilled so that no shear failure of the core or the connecting layer between core and cover layer occurs

$$\tau_{zx,C} \approx \frac{Q_{z,max}}{b\Delta h^C} < \tau_p, \tag{7.24}$$

where τ_p represents the shear yield stress of the core or the shear yield stress of the interlayer (see Table 6.3 for adhesives).

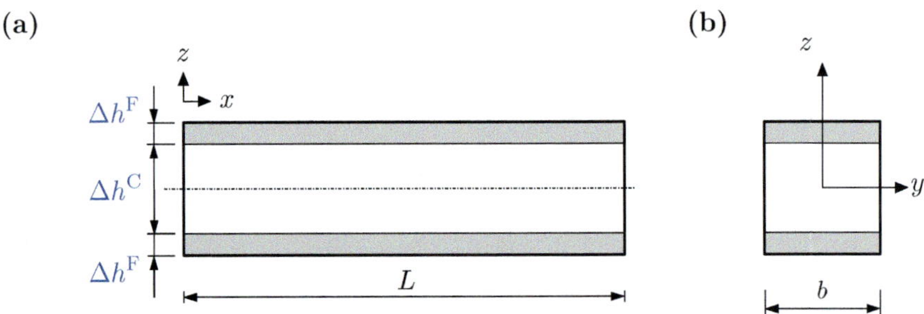

(a)

(b)

Fig. 7.8 andwich beam for optimization under bending loads: **(a)** general configuration and **(b)** beam cross-section

A maximum deflection is often specified as a boundary condition. According to the partial deflection method, two differential equations (see Table 5.6) have to be solved and the expression for the deflection depends on the boundary and loading conditions. For the special case of a sandwich beam under 3-point bending with a concentrated load in the middle (see Fig. 5.19), the maximum deflection results from Eq. (5.110). If the limit value is specified as a fraction of the beam length as $r_1 L$, the following additional condition results:[3]

$$\frac{FL^3}{48\overline{EI_y}} + \frac{FL}{4AG^C} < r_1 L \,, \tag{7.25}$$

or with $\overline{EI_y} \approx E^F b \Delta h^F (h_c)^2 / 2$ and $A = bh_c^2 / \Delta h^C \approx bh_c$ or with the additional simplifications $h_c \approx \Delta h^C$, i.e. $\overline{EI_y} \approx E^F b \Delta h^F (\Delta h^C)^2 / 2$ and $A \approx b \Delta h^C$:

$$\frac{2FL^3}{48E^F b \Delta h^F (\Delta h^C)^2} + \frac{FL}{4b \Delta h^C G^C} < r_1 L \,. \tag{7.26}$$

In order to optimize a sandwich structure, the weight is usually reduced to a minimum here as well. The total mass is composed proportionally of the face sheets and the core, see Eq. (7.4):

$$m = \varrho V = b \left(\varrho^C \Delta h^C L + \varrho^F 2 \Delta h^F L \right) \,, \tag{7.27}$$

or as length-related mass:

$$m^n = \frac{m}{L} = b \left(\varrho^C \Delta h^C + \varrho^F 2 \Delta h^F \right) \,. \tag{7.28}$$

If one substitutes the volume-related mass in Eq. (7.28), i.e. the density, through the volume-related costs of the core and face sheets, m^n can be interpreted as the length-related costs of the sandwich beam.

If one uses the length-specific normalizations $\Delta h^{F,n} = \Delta h^F / L$ and $\Delta h^{C,n} = \Delta h^C / L$, Eqs. (7.23), (7.24), (7.26) and (7.28) can be formulated as follows:

- Constraints g_i:

$$g_1(\Delta h^{C,n}, \Delta h^{F,n}) = \frac{M_{y,\max}}{\Delta h^{C,n} \Delta h^{F,n} b L^2} < \sigma_{\mathrm{cr}} \,, \tag{7.29}$$

$$g_2(\Delta h^{C,n}, \Delta h^{F,n}) = \frac{Q_{z,\max}}{\Delta h^{C,n} b L} < \tau_{\mathrm{p}} \,, \tag{7.30}$$

$$g_3(\Delta h^{C,n}, \Delta h^{F,n}) = \frac{2F}{48E^F b L \Delta h^{F,n} (\Delta h^{C,n})^2} + \frac{F}{4b L G^C \Delta h^{C,n}} < r_1 \,. \tag{7.31}$$

---

[3] The force F was assumed to be in the positive z-direction in order to avoid the minus sign in Eq. (5.110).

- Objective function f:

$$f\left(\Delta h^{\mathrm{F,n}}, \Delta h^{\mathrm{C,n}}\right) = \frac{m}{bL} = \frac{m^{\mathrm{n}}}{b} = \varrho^{\mathrm{C}}\Delta h^{\mathrm{C,n}} + 2\varrho^{\mathrm{F}}\Delta h^{\mathrm{F,n}} . \tag{7.32}$$

The constraints g_i according to Eqs. (7.29)–(7.31) can be illustrated again in a $\Delta h^{\mathrm{C,n}}$-$\Delta h^{\mathrm{F,n}}$ coordinate system. To do this, the three equations are solved for $\Delta h^{\mathrm{F,n}}$:

$$g_1: \qquad \Delta h^{\mathrm{F,n}}_{g_1} > \frac{M_{y,\max}}{bL^2\sigma_{\mathrm{cr}}\Delta h^{\mathrm{C,n}}} , \tag{7.33}$$

$$g_2: \qquad \Delta h^{\mathrm{C,n}}_{g_2} > \frac{Q_{z,\max}}{bL\tau_{\mathrm{p}}} , \tag{7.34}$$

$$g_3: \qquad \Delta h^{\mathrm{F,n}}_{g_3} > \frac{\frac{2F}{48E^{\mathrm{F}}bL(\Delta h^{\mathrm{C,n}})^2}}{r_1 - \frac{F}{4bLG^{\mathrm{C}}\Delta h^{\mathrm{C,n}}}} . \tag{7.35}$$

Figure 7.9 shows the three limit curves g_1, g_2 and g_3 in the $\Delta h^{\mathrm{C,n}}$-$\Delta h^{\mathrm{F,n}}$ plane. The gray area is the common permissible area. The poles of the limit curve g_3 are $\Delta h^{\mathrm{C,n}} = 0$ and

$$\Delta h^{\mathrm{C,n}} = \frac{F}{4bLG^{\mathrm{C}}r_1} . \tag{7.36}$$

It should be noted here that for g_3 the case of a sandwich beam under 3-point bending with a concentrated load in the middle was assumed[4] and thus $|M_{y,\max}| = FL/4$ and $|Q_{z,\max}| = F/2$ results.

The intersection points of the constraint curves (see Fig. 7.9) can be determined as follows:

The point E results from the intersection of the limit curves g_1 and g_2:

$$E = \left(\Delta h^{\mathrm{C,n}}_E \;\middle|\; \Delta h^{\mathrm{F,n}}_E\right) = \left(\frac{Q_{x,\max}}{bL\tau_{\mathrm{p}}} \;\middle|\; \frac{M_{y,\max}\tau_{\mathrm{p}}}{Q_{z,\max}L\sigma_{\mathrm{cr}}}\right) . \tag{7.37}$$

The point A results from the intersection of the limit curves g_1 and g_3:

$$A = \left(\Delta h^{\mathrm{C,n}}_A \;\middle|\; \Delta h^{\mathrm{F,n}}_A\right)$$

$$= \left(\frac{F}{M_{y,\max}r_1}\left(\frac{2L\sigma_{\mathrm{cr}}}{48E^{\mathrm{F}}} + \frac{M_{y,\max}}{4bLG^{\mathrm{C}}}\right) \;\middle|\; \frac{M^2_{y,\max}r_1}{FbL^2\sigma_{\mathrm{cr}}\left(\frac{2L\sigma_{\mathrm{cr}}}{48E^{\mathrm{F}}} + \frac{M_{y,\max}}{4bLG^{\mathrm{C}}}\right)}\right) . \tag{7.38}$$

---

[4] At this point it is pointed out again that Eqs. (7.26) and (7.35) are subject to this assumption and for other cases, i.e. supports and loads, have to be adjusted.

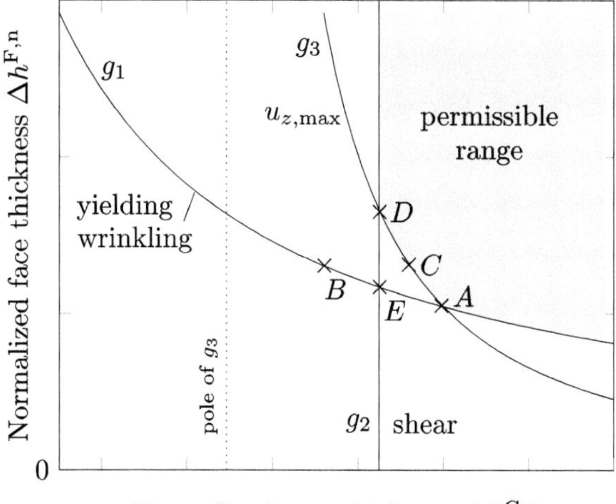

Fig. 7.9 Normalized face thickness as a function of the normalized core thickness based on the functions g_1, g_2 and g_3 according to Eqs. (7.33)–(7.35). Only the areas above the two constraints g_1 and g_3 and to the right of g_2 are allowed

The point D results from the intersection of the limit curves g_2 and g_3:

$$D = \left(\Delta h_D^{C,n} \;\middle|\; \Delta h_D^{F,n}\right) = \left(\frac{Q_{x,\max}}{bL\tau_p} \;\middle|\; \frac{2bLF\tau_p^2}{48E^F Q_{z,\max}^2 \left(r_1 - \frac{F\tau_p}{4G^C Q_{z,\max}}\right)}\right). \qquad (7.39)$$

The minimum C of the objective function f along the constraints g_3 for $u_{z,\max}$ results from inserting $\Delta h_{g_3}^{F,n}$ according to Eq. (7.35) into the objective function (7.32), i.e.

$$f\left(\Delta h^{C,n}\right) = \varrho^C \Delta h^{C,n} + 2\varrho^F \Delta h^{F,n}$$

$$= \varrho^C \Delta h^{C,n} + 2\varrho^F \frac{\dfrac{2F}{48E^F bL(\Delta h^{C,n})^2}}{r_1 - \dfrac{F}{4bLG^C \Delta h^{C,n}}}, \qquad (7.40)$$

and subsequent differentiation according to the variable $\Delta h^{C,n}$:

$$\frac{\partial f(\Delta h^{C,n})}{\partial \Delta h^{C,n}} \overset{!}{=} 0 \quad \Rightarrow \quad \Delta h_C^{C,n}. \qquad (7.41)$$

The root can be determined numerically, for example using Newton's method. The coordinates of the point sought are then obtained using Eq. (7.35): $C = \left(\Delta h_C^{C,n} \;\middle|\; \Delta h_C^{F,n}\right)$.

The minimum B of the objective function f along the constraints g_1 for yielding/wrinkling results from inserting $\Delta h_{g_1}^{F,n}$ according to Eq. (7.33) into the objective function (7.32), i.e.

$$f\left(\Delta h^{C,n}\right) = \varrho^{C}\Delta h^{C,n} + 2\varrho^{F}\Delta h^{F,n}$$

$$= \varrho^{C}\Delta h^{C,n} + 2\varrho^{F}\frac{M_{y,\max}}{bL^{2}\sigma_{\mathrm{cr}}\Delta h^{C,n}}, \tag{7.42}$$

and subsequent differentiation according to the variable $\Delta h^{C,n}$:

$$\frac{\partial f(\Delta h^{C,n})}{\partial \Delta h^{C,n}} = \varrho^{C} - 2\varrho^{F}\frac{M_{y,\max}}{bL^{2}\sigma_{\mathrm{cr}}\left(\Delta h^{C,n}\right)^{2}} \overset{!}{=} 0. \tag{7.43}$$

The last equation can be written in terms of the quantity sought, i.e. $\Delta h^{C,n}$, and the point we are looking for finally results in:

$$B = \left(\Delta h_{B}^{C,n} \mid \Delta h_{B}^{F,n}\right) = \left(\sqrt{2 \times \frac{\varrho^{F}}{\varrho^{C}} \times \frac{M_{y,\max}}{bL^{2}\sigma_{\mathrm{cr}}}} \,\middle|\, \sqrt{\frac{1}{2} \times \frac{\varrho^{C}}{\varrho^{F}} \times \frac{M_{y,\max}}{bL^{2}\sigma_{\mathrm{cr}}}}\right). \tag{7.44}$$

Depending on the position of these points relative to each other[5], different cases can be distinguished, see Fig. 7.10. For cases 1–3 according to Fig. 7.10a, the second pole of the limit curve g_3 (see Eq. (7.36)) lies to the right of the limit curve g_2 (see Eq. (7.34)):

$$\frac{F}{4bLG^{C}r_{1}} > \frac{Q_{z,\max}}{bL\tau_{\mathrm{p}}}. \tag{7.45}$$

Here, the following cases can be distinguished:

Case 1: $\Delta h_{C}^{C,n} < \Delta h_{A}^{C,n}$ applies, i.e. the minimum exists on the limit curve g_3 for the deflection.

Case 2: $\Delta h_{B}^{C,n} > \Delta h_{A}^{C,n}$ applies, i.e. the minimum exists on the limit curve g_1 for yielding/wrinkling.

Case 3: $\Delta h_{C}^{C,n} \not< \Delta h_{A}^{C,n}$ and $\Delta h_{B}^{C,n} \not> \Delta h_{A}^{C,n}$ applies, i.e. no minimum exists on the limit curves. The optimum point is represented by A (simultaneous yielding/wrinkling and deflection limit).

For cases 4–7 according to Fig. 7.10b, the second pole of the limit curve g_3 (see Eq. (7.36)) is to the left of the limit curve g_2 (see Eq. (7.34)) and furthermore both straight lines lie to the left of point A (see Eq. (7.38)):

$$\frac{F}{4bLG^{C}r_{1}} < \frac{Q_{z,\max}}{bL\tau_{\mathrm{p}}} < \Delta h_{A}^{C,n}. \tag{7.46}$$

Here, the following cases can be distinguished:

---

[5] A Python3 program is provided in the Appendix A.2 to automatically evaluate the points A–E.

Fig. 7.10 Optimization ranges for sandwich under bending load: (**a**) case 1–3, (**b**) case 4–7, (**c**) case 8–9

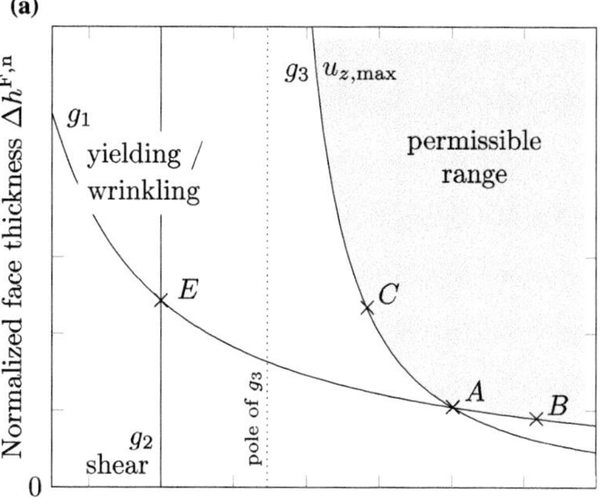

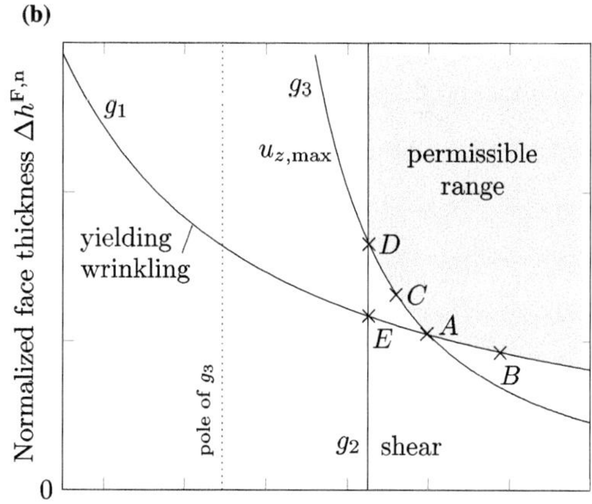

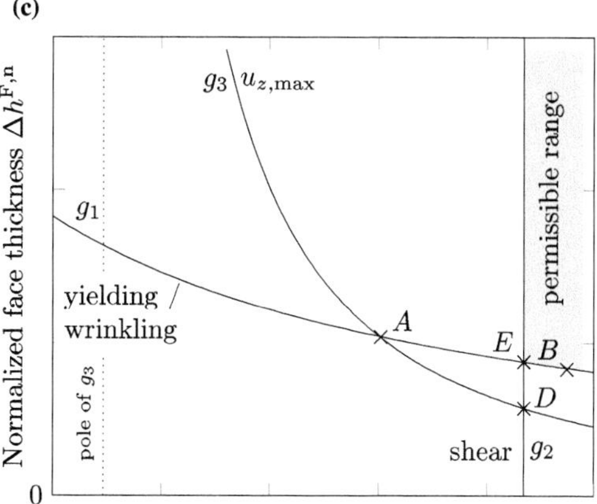

Case 4: $\Delta h_D^{C,n} < \Delta h_C^{C,n} < \Delta h_A^{C,n}$ applies, i.e. the minimum exists on the deflection
limit curve g_3.

Case 5: $\Delta h_B^{C,n} > \Delta h_A^{C,n}$ applies, i.e. the minimum exists on the yielding/wrinkling
limit curve g_1.

If neither case 4 nor case 5 is applicable, either A or D represents the optimal point:

Case 6: Point A represents the optimum (simultaneous yielding/wrinkling and
deflection limit).

Case 7: Point D represents the optimum (simultaneous shear failure and deflection
limit).

For cases 8–9 according to Fig. 7.10c, the limit curve g_2 (see Eq. (7.34)) lies to the
right of point A (see Eq. (7.38)):

$$\frac{Q_{z,\max}}{bL\tau_p} > \Delta h_A^{C,n}. \tag{7.47}$$

Here, the following cases can be distinguished:

Case 8: $\Delta h_B^{C,n} > \Delta h_E^{C,n}$ applies, i.e. the minimum exists on the yielding/wrinkling
limit curve g_1.

Case 9: $\Delta h_B^{C,n} < \Delta h_E^{C,n}$ applies, i.e. no minimum exists on the limit curves. The
optimum point is represented by E (simultaneous yielding/wrinkling and
shear failure).

It should be noted here that cases 1 and 2 or 4 and 5 could also occur simultaneously.

7.2 Exercises

7.2.1 Knowledge Questions

- Which different failure modes have to be considered when optimizing a technical
  sandwich beam under (a) tensile and (b) compressive loading?
- Which different failure modes have to be considered when optimizing a technical
  sandwich beam under bending load?

7.2.2 Calculation Problems

7.2.1 Optimization of a Sandwich Beam Under Compressive Loading
For the sandwich beam shown in Fig. 7.11, optimize the core and face sheet thick-
nesses assuming thin face sheets and a soft core.
Given are:

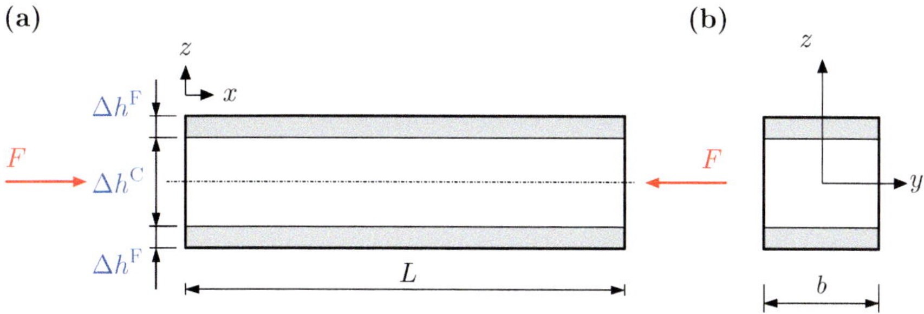

Fig. 7.11 Optimization of a sandwich beam under compressive loading: **(a)** general configuration and **(b)** beam cross-section

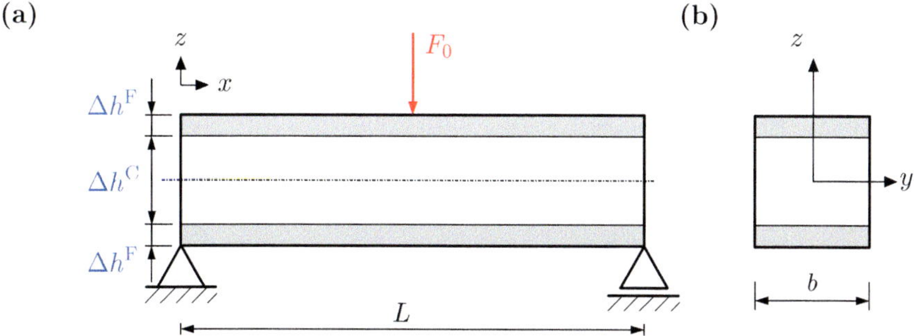

Fig. 7.12 Optimization of a sandwich beam under bending load due to a single force: **(a)** general configuration and **(b)** beam cross-section

- Geometric dimensions: $L = 2540\,\text{mm}$, $b = 305\,\text{mm}$.
- Material properties of the core: $E^{\text{C}} = 6.8948\,\text{MPa}$, $G^{\text{C}} = 3.4474\,\text{MPa}$, $\varrho^{\text{C}} = 240\,\text{kg/m}^3$.
- Material properties of the face sheets: $E^{\text{F}} = 68{,}948\,\text{MPa}$, $\varrho^{\text{F}} = 2691\,\text{kg/m}^3$, $R^{\text{F}}_{\text{p0.2}} = 247\,\text{MPa}$.
- External load: case (a): $F = 2670\,\text{N}$, case (b) $10 \times F = 26{,}700\,\text{N}$.

7.2.2 Optimization of a Sandwich Beam Under Bending Load Due to a Single Force

For the sandwich beam shown in Fig. 7.12, optimize the core and face sheet thicknesses assuming thin face sheets and a soft core.

Given are:

- Geometric dimensions: $L = 2540\,\text{mm}$, $b = 305\,\text{mm}$.
- Material properties of the core: $E^{\text{C}} = 6.8948\,\text{MPa}$, $G^{\text{C}} = 3.4474\,\text{MPa}$, $\varrho^{\text{C}} = 240\,\text{kg/m}^3$, $\tau^{\text{C}}_{\text{p}} = E^{\text{C}}/50$.

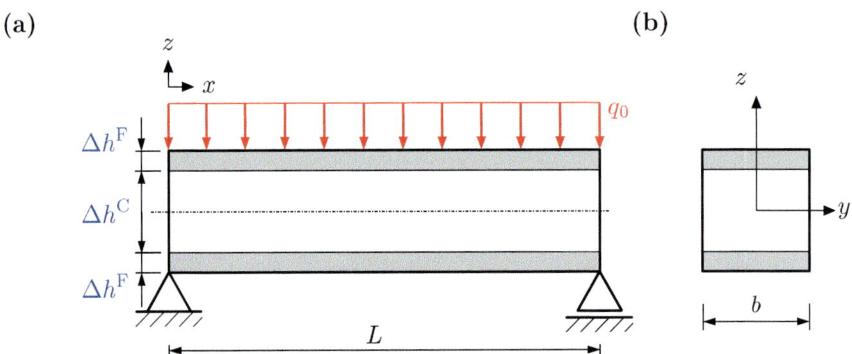

Fig. 7.13 Optimization of a sandwich beam under bending load due to a distributed load: **(a)** general configuration and **(b)** beam cross-section

- Material properties of the face sheets: $E^F = 68{,}948\,\text{MPa}$, $\varrho^F = 2691\,\text{kg/m}^3$, $R^F_{\text{p0.2}} = 247\,\text{MPa}$.
- External load: $F_0 = 2667\,\text{N}$.

Furthermore, $r_1 L$ with $r_1 = 0.003$ can be assumed for the maximum deflection.

7.2.3 Optimization of a Sandwich Beam Under Bending Load Due to a Distributed Load

For the sandwich beam shown in Fig. 7.13, optimize the core and face sheet thicknesses assuming thin face sheets and a soft core.
Given are:

- Geometric dimensions: $L = 2540\,\text{mm}$, $b = 305\,\text{mm}$.
- Material properties of the core: $E^C = 6.8948\,\text{MPa}$, $G^C = 3.4474\,\text{MPa}$, $\varrho^C = 240\,\text{kg/m}^3$, $\tau^C_{\text{p}} = E^C/50$.
- Material properties of the face sheets: $E^F = 68{,}948\,\text{MPa}$, $\varrho^F = 2691\,\text{kg/m}^3$, $R^F_{\text{p0.2}} = 247\,\text{MPa}$.
- External load: $q_0 = 1.05\,\text{N/mm}$.

Furthermore, $r_1 L$ with $r_1 = 0.003$ can be assumed for the maximum deflection.

7.2.4 Optimization of a Homogeneous Beam Under Bending Load Due to a Single Force

For the homogeneous beam shown in Fig. 7.14, optimize the dimensions b and h of the cross-section for minimum weight. The maximum normal and shear stress, the maximum deflection, and the maximum height-to-width ratio ($h \leq 20b$) in order to avoid instabilities, are to be taken into account as constraints.

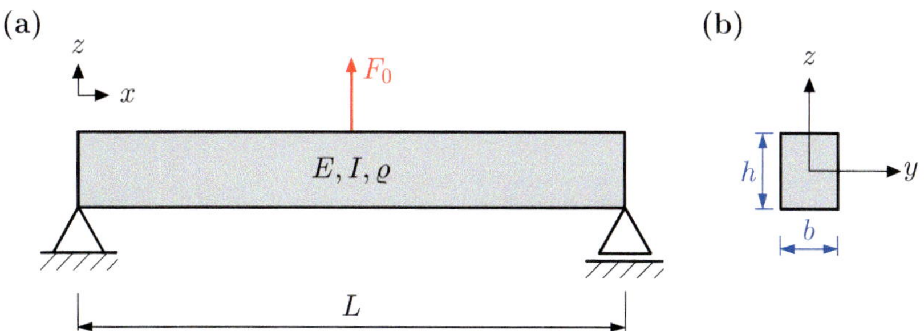

Fig. 7.14 Optimization of a homogeneous beam under bending load due to a single force: **(a)** general configuration and **(b)** beam cross-section

Given are:

- Geometric dimensions: $L = 2540\,\text{mm}$.
- Material properties of the beam: $E = 68{,}948\,\text{MPa}$, $\varrho = 2691\,\text{kg/m}^3$, $R_{p0.2} = 247\,\text{MPa}$, $\tau_p = R_{p0.2}/2$.
- External load: $F_0 = 2667\,\text{N}$.

Furthermore, $r_1 L$ with $r_1 = 0.03$ or $r_1 = 0.003$ can be assumed for the maximum deflection.

References

1. Allen, H.G.: Optimum design of sandwich struts and beams. In: Plastics in Building Structures, Proceedings of a Conference Held in London, 14–16 June 1965. Pergamon Press, Oxford (1966)

Further Lightweight Design Concepts and Calculation Methods

8

Abstract

This chapter briefly discusses lightweight design concepts in regard to conditions, manufacturing, and concepts. Finally, the importance of numerical methods in the analysis of complex structures is highlighted.

8.1 Further Lightweight Design Concepts

In the following, further lightweight design concepts are briefly discussed, although a detailed presentation and treatment is not intended here.

- *Lightweight design: conditions*: Consideration of external influencing factors, framework and boundary conditions (e.g. reduction in service life), legislation, costs, environmental factors, design measures (e.g. shortening the lever arm), safety-relevant aspects.

Figure 8.1 uses the example of a cantilever beam to explain two ways of increasing the lightweight potential within the framework of conditional lightweight design. In configuration 8.1b, a reduction in the length of the beam leads to a reduction in weight, whereas in configuration 8.1c, the cross-section is reduced and the function is retained by a support.

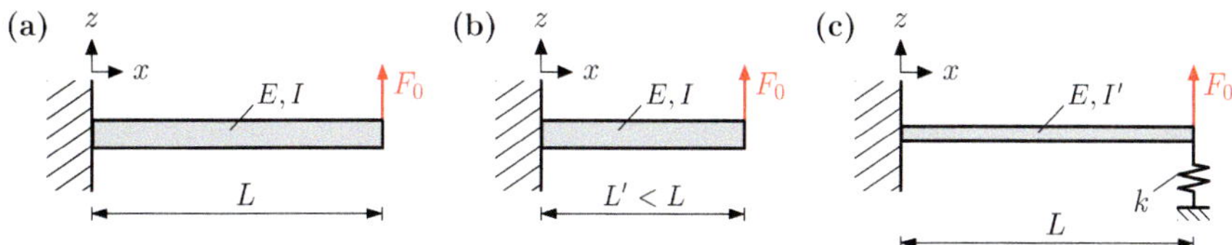

Fig. 8.1 Conditional lightweight design: (**a**) original configuration; (**b**) weight reduction by means of a shortened lever arm; (**c**) support

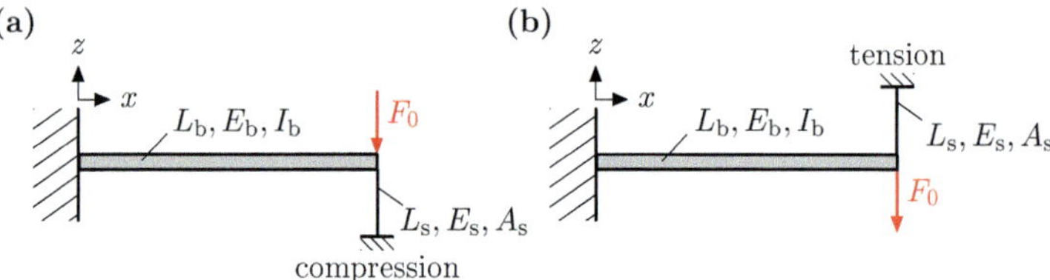

Fig. 8.2 Example of conditional lightweight design: Beam supported by a rod: (**a**) compression rod; (**b**) tension rod

When constructing the column according to Fig. 8.1c, however, it must be taken into account whether the structural element is subjected to tension and/or compression (see Fig. 8.2). In the case of compression, possible instability (buckling) must also be taken into account.

Using the special boundary condition

$$Q_z(x = L) = -\frac{E_s A_s u_z(x)}{L_s} - F_0 \,, \tag{8.1}$$

the bending line of the problem according to Fig. 8.2 can be determined by adjusting the equations (2.12) to (2.15):

$$u_z(x) = \frac{F_0 L_b^3}{3 E_b I_b} \times \frac{\frac{1}{2}\left[\frac{x}{L_b}\right]^3 - \frac{3}{2}\left[\frac{x}{L_b}\right]^2}{1 - \frac{E_s A_s}{L_s} \times \frac{L_b^3}{3 E_b I_b} \times \left(\frac{1}{2}\left[\frac{x}{L_b}\right]^3 - \frac{3}{2}\left[\frac{x}{L_b}\right]^2\right)} \,. \tag{8.2}$$

The maximum deflection for $x = L_b$ is:

$$u_z(x = L_b) = -\frac{F_0 L_b^3}{3 E_b I_b + \frac{E_s A_s}{L_s} \times L_b^3} \,. \tag{8.3}$$

A beam without support can be used as a reference configuration ('ref') (see Fig. 8.1a). For such a cantilever beam, the deflection—assuming a rectangular cross-section with side length a—at the load application point is

$$u_z(L_b) = -\frac{F_0 L_b^3}{3 E_b I_b} = -\frac{4 F_0 L_b^3}{E_b a^4} \,. \tag{8.4}$$

By using the support ('s') at the end of the beam, the original beam cross-section can be reduced to αa with $\alpha < 1$. If one requires that both configurations have the same deflection at $x = L_b$, the following condition arises:

$$-\frac{F_0 L_b^3}{3 E_b I_b + \frac{E_s A_s}{L_s} \times L_b^3} \overset{!}{=} -\frac{F_0 L_b^3}{3 E_b I_b} \,. \tag{8.5}$$

Table 8.1 Material costs of various materials. Based on [2]

	in $ per ton	c^*
Stainless steel (austenitic)	600	1.5
Al alloys	400	1.0
Ti alloys	10,000	25

If we also assume a bar with a square cross-section with side length $b = \frac{a}{40}$ and a bar length of $L_s = \frac{L_b}{3} = \frac{10a}{3}$, we obtain a factor for the cross-sectional reduction of $\alpha = 1/\sqrt{2}$ for the same material parameters $E_s = E_b$. If we compare the masses of both configurations, we obtain the following for the same materials:

$$\frac{m_{bs}}{m_{ref}} = \frac{(\alpha a)^2 L_b + \left(\frac{a}{40}\right)^2 \frac{L_b}{3}}{a^2 L_b} \approx 0.5 \,. \tag{8.6}$$

Another factor that is taken into account in the context of conditional lightweight design relates to the material costs. Table 8.1 gives some reference value for the metallic materials considered here. In the last column, the absolute costs were normalized with the most economic value and this ratio was abbreviated by c^*. The following lightweight index can therefore be introduced, which takes into account both a mechanical material value and the material costs:

$$M = \frac{F_0}{c^* F_G} \,. \tag{8.7}$$

If we take the example in Fig. 3.7 as a reference, a very different recommendation results when the material costs are taken into account, see Fig. 8.3. The Ti alloy performs the worst and the Al alloy achieves the highest lightweight index.

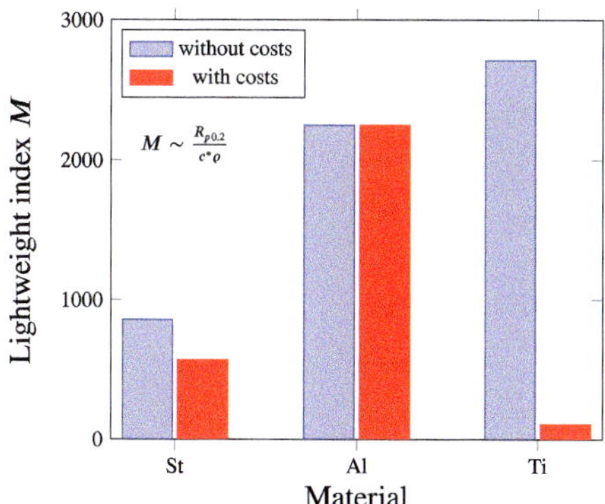

Fig. 8.3 Lightweight index for cantilever beams made of different materials with constant geometry ($L = 100$ mm, $h = b = 10$ mm) with stress criterion taking into account the material costs (see Table 8.1)

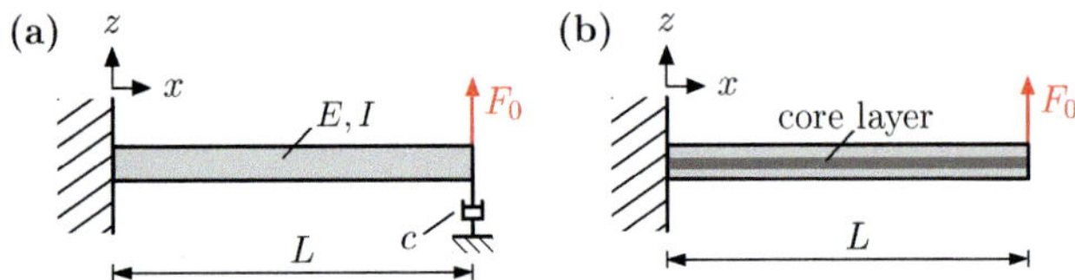

Fig. 8.4 Conceptual lightweight design: Different damping concepts **(a)** differential construction; **(b)** integral construction

- *Lightweight design: concepts*: Application of differential and integral construction (functional diversification or integration of a certain number of functions).

Figure 8.4 shows two different approaches within the framework of conceptual lightweight design. Configuration 8.4a follows the differential design and achieves improved damping behavior through the use of an external damper. In configuration 8.4b, the damping behavior is improved within the framework of the integral design by using a cellular material as the core material (see [1,9]).

- *Lightweight design: manufacturing*: Consideration of the different manufacturing methods, production and assembly processes.

Here, the lightweight potential can be improved by using the latest processes, such as generative or additive manufacturing processes (3D printing, see [3]).

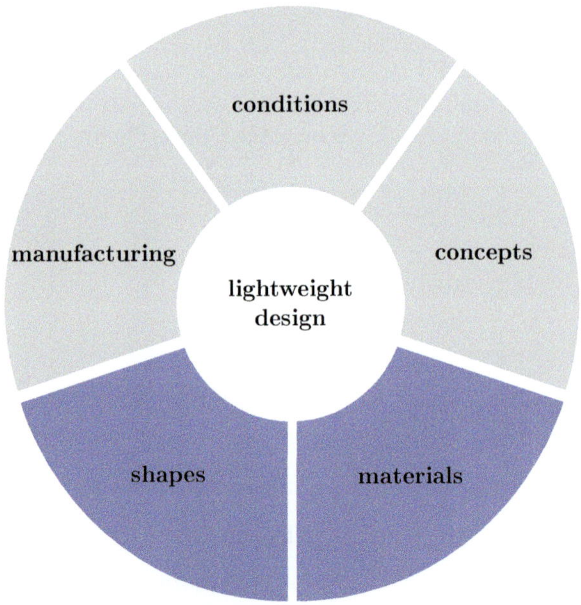

Fig. 8.5 Classic lightweight design concepts. Based on [4]

This book provides a simple and comprehensive introduction to the different lightweight design concepts. The focus was on the concepts that have a direct connection to applied mechanics and strength of materials. The examples of one-dimensional structural elements were intended to ensure that the theory is also easily accessible for students of a bachelor's degree in engineering. Figure 8.5 summarizes the different lightweight design concepts, with the concepts highlighted in blue being explained in more detail in this book.

8.2   Numerical Methods

There is no question that real engineering constructions often have to be approximated by more complex models. In this case, the definition of the lightweight index according to Eq. (3.1) can quickly reach certain limits. Load cases in which different loads act simultaneously (individual forces and moments, distributed loads) can, under certain circumstances, no longer be combined into a single external force (F_0). Therefore, the specific energy absorption was introduced as an alternative concept.

The analytical solutions for the different beam theories according to Eq. (2.15), (2.22)–(2.31) and (2.31)–(2.32) can only be applied to simple problems. For more complex structures and questions, numerical approximation methods have become established, with the finite element method being the standard tool in the field of structural mechanics. The basic idea of this approximation is that the basic equations of continuum mechanics are no longer fulfilled for every point of the continuum, but only averaged for a so-called finite element. Deformations are only calculated at a finite number of points, the so-called nodes, and interpolation is simply carried out between them. These nodes are placed at least at the ends/corners of the elements and allow individual elements to be connected to form a coherent structure. A structure is thus approximated by a mesh of connected elements and the system's degrees of freedom are reduced to the nodes. More information on the finite element method can be found in the relevant literature, e.g. [7, 8, 10–12]. Numerical optimization methods are often used in the same context.

For example, a three-dimensional beam can be approximated by a one-dimensional beam element as part of a finite element analysis, see Fig. 8.6.

The beam element itself is described by two nodes and a connecting line. Cross-sectional and material properties are only assigned as numerical input in a finite

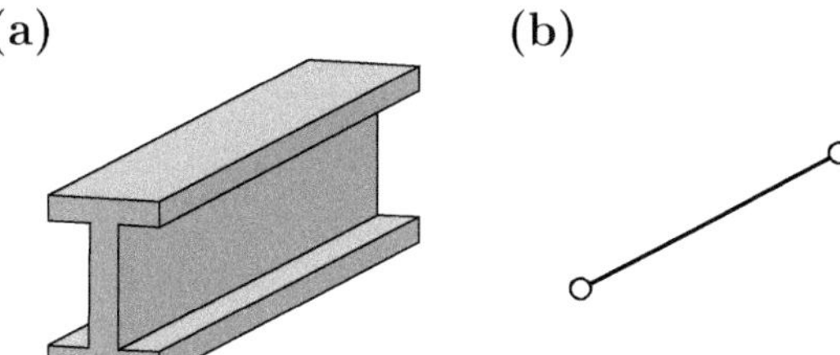

Fig. 8.6 (a) Three-dimensional beam structure with I-section; (b) Approximation using a single one-dimensional beam element. The element nodes are symbolized by circles (o)

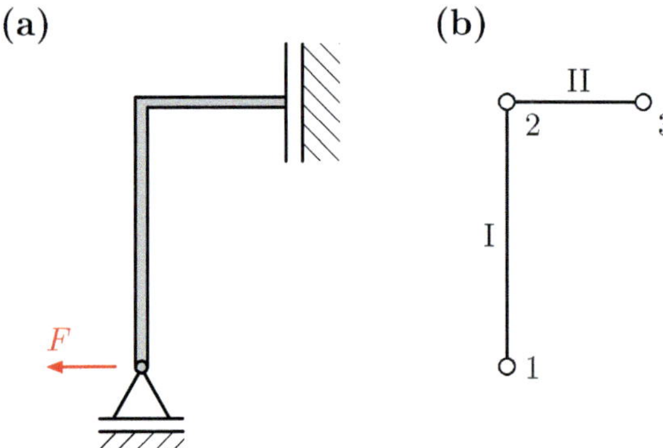

Fig. 8.7 (**a**) Mechanical model of a plane frame structure; (**b**) finite element mesh consisting of two one-dimensional elements I and II

element program and do not have to be taken into account in the mesh. This results in an extremely simplified generation of a beam structure in a finite element program.

The following briefly looks at the example of a plane frame structure. As part of the so-called modeling, a real structure must first be converted into a mechanical model by means of simplifications and assumptions, see Fig. 8.7a. The simplified structure is then converted into a finite element mesh of one-dimensional beam elements (so-called discretization). The beam elements in commercial programs are so-called generalized beams, which can also deform in the axial direction or can be twisted along the longitudinal axis.

For each of these beam elements, the so-called principal finite element equation at element level $K^e u^e = f^e$ can be specified[1]. The so-called stiffness matrix K^e contains the information about the material and the geometry of the element. The column matrix of unknowns u^e contains the deformations, i.e. the displacements and rotations at the nodes, and the column matrix of the loads f^e contains the external forces and moments acting on the element. In general, the accuracy of a finite element analysis increases with the number of elements. The principal equations for the individual elements can be combined into a global system of equations in the form:

$$Ku = f .\tag{8.8}$$

After taking the support conditions into account, the solution for a linear system of equations is:

$$u = K^{-1}f .\tag{8.9}$$

The finite element method also allows other questions to be answered based on the same mesh. Taking into account the structural masses, which are summarized in a mass matrix M, the principal equation is:

$$M\ddot{u}(t) + Ku(t) = f(t) .\tag{8.10}$$

---

[1] The index "e" stands for element level.

The solution of this equation in the time domain is often carried out using classical finite difference methods and provides the deformation $\boldsymbol{u}(t)$, the velocity $\dot{\boldsymbol{u}}(t)$ and the acceleration $\ddot{\boldsymbol{u}}(t)$ of the transient problem. Considering the same finite element model allows the determination of the eigenfrequencies and eigenmodes by solving an eigenvalue problem:

$$\det\left(\boldsymbol{K} - \omega_i \boldsymbol{M}\right) = 0\,, \tag{8.11}$$

where ω_i represent the natural frequencies of the system. The natural modes, i.e. the deformed shape of the structure at certain natural frequencies, are defined by the following relationship:

$$\left(\boldsymbol{K} - \omega_i^2 \boldsymbol{M}\right)\boldsymbol{\Phi} = \boldsymbol{0}\,, \tag{8.12}$$

where $\boldsymbol{\Phi}$ represent the eigenforms of the system. Furthermore, the solution of another eigenvalue problem (with the same mesh) allows the determination of the buckling load as part of a stability analysis:

$$\det\left(\boldsymbol{K}\right) = \det\left(\boldsymbol{K}^{\text{el}} + \lambda \boldsymbol{K}^{\text{geo}}\right) = 0\,, \tag{8.13}$$

where $\boldsymbol{K}^{\text{el}}$ represents the stiffness matrix according to Eq. (8.9) for a linear-elastic problem and $\boldsymbol{K}^{\text{geo}}$ represents the geometric stiffness matrix including the external load, see [10]. The buckling load ultimately results from the relationship λf. The solutions to various structural mechanics questions described above can be achieved in a commercial finite element program simply by switching to the different solution modes—with the same mesh.

Finally, it should be noted here that the use of new materials or material combinations (for example based on nanomaterials, see [13]) or manufacturing processes (for example generative or additive manufacturing technology, see [3,5,6]) allow new designs with extraordinary properties.

References

1. Altenbach, H., Öchsner, A. (eds.): Cellular and Porous Materials in Structures and Processes. Springer, Wien (2010)
2. Ashby, M.F., Jones, D.R.H.: Engineering Materials 1: An Introduction to Properties, Applications and Design. Elsevier, Amsterdam (2005)
3. Fastermann, P.: 3D-Drucken: Wie die generative Fertigungstechnik funktioniert. Springer Vieweg, Berlin (2014)
4. Henning, F., Moeller, E.: Handbuch Leichtbau: Methoden, Werkstoffe, Fertigung. Hanser Verlag, München (2011)
5. Hitzler, L., Janousch, C., Schanz, J., Merkel, M., Heine, B., Mack, F., Hall, W., Öchsner, A.: Direction and location dependency of selective laser melted AlSi10Mg specimens. J. Mater. Process. Tech. 243, 48–61 (2017). https://doi.org/10.1016/j.jmatprotec.2016.11.029

6. Hitzler, L., Merkel, M., Hall, W., Öchsner, A.: A review of metal fabricated with laser- and powder-bed based additive manufacturing techniques: Process, nomenclature, materials, achievable properties, and its utilization in the medical sector. Adv. Eng. Mater. 20(5), 1700658 (2018). https://doi.org/10.1002/adem.201700658
7. Javanbakht, Z., Öchsner, A.: Advanced Finite Element Simulation with MSC Marc: Application of User Subroutines. Springer, Cham (2017)
8. Merkel, M., Öchsner, A.: Eindimensionale Finite Elemente: Ein Einstieg in die Methode. Springer Vieweg, Berlin (2014)
9. Öchsner, A., Augustin, C. (eds.): Multifunctional Metallic Hollow Sphere Structures: Manufacturing, Properties and Application. Springer, Berlin (2009)
10. Öchsner, A., Merkel, M.: One-Dimensional Finite Elements: An Introduction to the FE Method. Springer, Berlin (2013)
11. Öchsner, A.: Computational Statics and Dynamics: An Introduction Based on the Finite Element Method. Springer, Singapore (2016)
12. Reddy, J.N.: An Introduction to the Finite Element Method. McGraw Hill, Singapore (2006)
13. Yengejeh, S.I., Kazemi, S.A., Öchsner, A.: Carbon nanotubes as reinforcement in composites: A review of the analytical, numerical and experimental approaches. Comp. Mater. Sci. 136, 85–101 (2017). https://doi.org/10.1016/j.commatsci.2017.04.023

Short Solutions to the Calculations Problems

9

Abstract

In this chapter short solutions to the calculation problems are compiled. Here, not only final solutions, but also important intermediate steps are given so that the solutions can be understood.

9.1 Chapter 2

2.4.1 Simplified model of a tower under dead weight

The internal normal force distribution follows from the vertical force equilibrium as

$$N_x(x) = -\varrho g A(L - x),\tag{9.1}$$

or as a distributed load:

$$p_x(x) = -\frac{\mathrm{d}N_x(x)}{\mathrm{d}x} = -\varrho g A = p_0.\tag{9.2}$$

Alternatively, Eq. (9.2) can also be derived by normalizing the total deadweight of the tower by its length:

$$p_x(x) = -\frac{F_G}{L} = -\frac{mg}{L} = -\varrho g A = p_0.\tag{9.3}$$

Further results:

$$\sigma_x(x) = -\varrho g(L - x),\tag{9.4}$$

$$u_x(x) = \frac{1}{EA}\left(+\tfrac{1}{2}\varrho A g x^2 - \varrho A g L x\right),\tag{9.5}$$

$$L' = L + u_x(L) = L\left(1 - \tfrac{\varrho g L}{2E}\right),\tag{9.6}$$

$$L_{\max} = \frac{\sigma_{\max}}{\varrho g}.\tag{9.7}$$

A. Öchsner, *Lightweight Design*,
https://doi.org/10.1007/978-3-658-48162-9_9

2.4.2 Course of the bending line and maximum stress for a cantilever beam with different loads

Load case (a): (see also Fig. 9.1)

$$u_z(x) = \frac{F_0 L^3}{EI} \left(\frac{1}{6} \left[\frac{x}{L} \right]^3 - \frac{1}{2} \left[\frac{x}{L} \right]^2 \right), \tag{9.8}$$

$$M_y(x) = F_0 L \left(- \left[\frac{x}{L} \right] + 1 \right), \tag{9.9}$$

$$\sigma_{x,\max} = \frac{6 F_0 L}{a^3} \left(- \left[\frac{x}{L} \right] + 1 \right). \tag{9.10}$$

Fig. 9.1 Course of the **a** bending line and **b** maximum stress for a cantilever beam with a point load

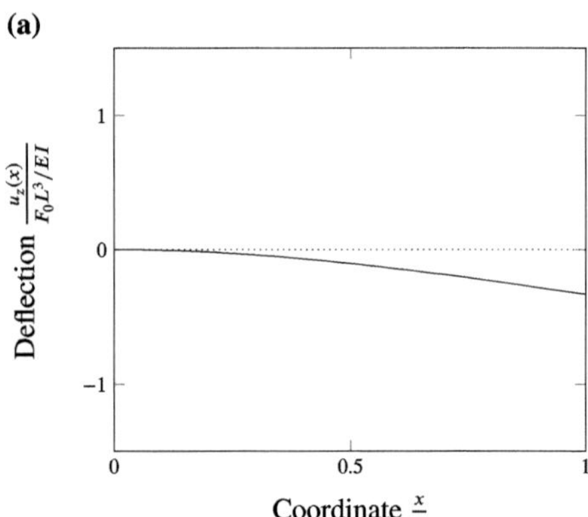

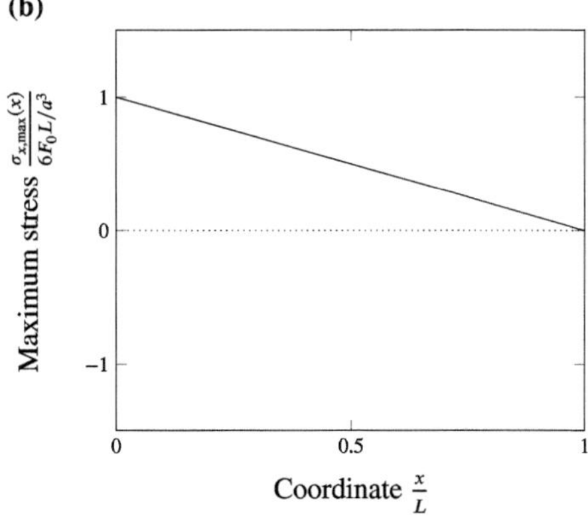

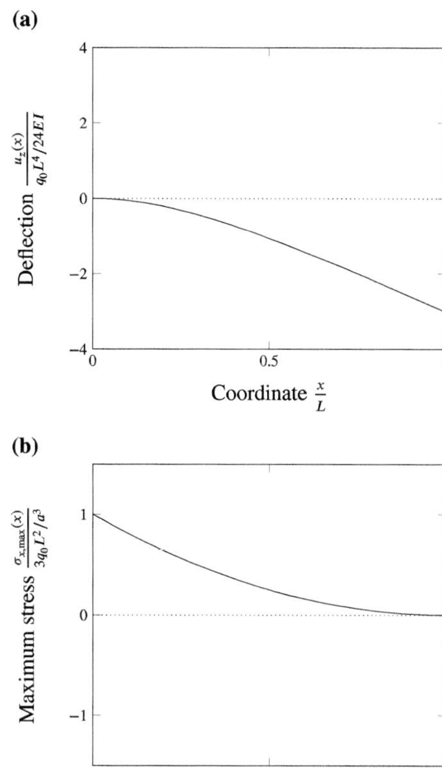

Fig. 9.2 Course of the **a** bending line and **b** maximum stress for a cantilever beam with a constant distributed load

Load case (b): (see also Fig. 9.2)

$$u_z(x) = -\frac{q_0 L^4}{24EI}\left(\left[\frac{x}{L}\right]^4 - 4\left[\frac{x}{L}\right]^3 + 6\left[\frac{x}{L}\right]^2\right),\qquad(9.11)$$

$$M_y(x) = \frac{q_0 L^2}{2}\left(\left[\frac{x}{L}\right]^2 - 2\left[\frac{x}{L}\right] + 1\right),\qquad(9.12)$$

$$\sigma_{x,\max} = \frac{3q_0 L^2}{a^3}\left(\left[\frac{x}{L}\right]^2 - 2\left[\frac{x}{L}\right] + 1\right).\qquad(9.13)$$

2.4.3 Course of bending line, stress and strain for a bi-material cantilever beam

- Boundary and transition conditions for determining the integration constants

$$u_{z_\mathrm{I}}(x_\mathrm{I} = 0) = 0\,, \quad M_{y_\mathrm{II}}(x_\mathrm{II} = \tfrac{L}{2}) = 0\,, \tag{9.14}$$

$$\varphi_{y_\mathrm{I}}(x_\mathrm{I} = 0) = 0\,, \quad Q_{z_\mathrm{II}}(x_\mathrm{II} = \tfrac{L}{2}) = -F_0\,, \tag{9.15}$$

$$u_{z_\mathrm{I}}(x_\mathrm{I} = \tfrac{L}{2}) = u_{z_\mathrm{II}}(x_\mathrm{II} = 0)\,, \tag{9.16}$$

$$\varphi_{y_\mathrm{I}}(x_\mathrm{I} = \tfrac{L}{2}) = \varphi_{y_\mathrm{II}}(x_\mathrm{II} = 0)\,, \tag{9.17}$$

$$Q_{z_\mathrm{I}}(x_\mathrm{I} = \tfrac{L}{2}) = Q_{z_\mathrm{II}}(x_\mathrm{II} = 0)\,, \tag{9.18}$$

$$M_{y_\mathrm{I}}(x_\mathrm{I} = \tfrac{L}{2}) = M_{y_\mathrm{II}}(x_\mathrm{II} = 0)\,. \tag{9.19}$$

- Integration constants

$$c_1 = F_0\,, c_2 = -F_0 L\,, c_3 = c_4 = 0\,, c_5 = F_0\,, c_6 = -\tfrac{F_0 L}{2}\,, \tag{9.20}$$

$$c_7 = -\frac{3L^2 E_\mathrm{II} F_0}{8 E_\mathrm{I}}\,, c_8 = -\frac{5L^3 E_\mathrm{II} F_0}{48 E_\mathrm{I}}\,. \tag{9.21}$$

- Bending line (see Fig. 9.3a)

$$u_{z_\mathrm{I}}(x_\mathrm{I}) = \frac{F_0 L^3}{E_\mathrm{I} I_y}\left(\frac{1}{6}\left[\frac{x_\mathrm{I}}{L}\right]^3 - \frac{1}{2}\left[\frac{x_\mathrm{I}}{L}\right]^2\right), \tag{9.22}$$

$$u_{z_\mathrm{II}}(x_\mathrm{II}) = \frac{F_0 L^3}{E_\mathrm{II} I_y}\left(\frac{1}{6}\left[\frac{x_\mathrm{II}}{L}\right]^3 - \frac{1}{4}\left[\frac{x_\mathrm{II}}{L}\right]^2 - \frac{3\,E_\mathrm{II}}{8\,E_\mathrm{I}}\left[\frac{x_\mathrm{II}}{L}\right]^1 - \frac{5\,E_\mathrm{II}}{48\,E_\mathrm{I}}\right). \tag{9.23}$$

- Stresses (see Fig. 9.3b)

$$\sigma_{z_\mathrm{I},\mathrm{max}}(x_\mathrm{I}) = \frac{F_0 L a}{2 I_y}\left(-\left[\frac{x_\mathrm{I}}{L}\right] + 1\right), \tag{9.24}$$

$$\sigma_{z_\mathrm{II},\mathrm{max}}(x_\mathrm{II}) = \frac{F_0 L a}{2 I_y}\left(-\left[\frac{x_\mathrm{II}}{L}\right] + \frac{1}{2}\right). \tag{9.25}$$

- Strains (see Fig. 9.3c)

$$\varepsilon_{z\mathrm{I},\max}(x_\mathrm{I}) = \frac{F_0 L a}{2 E_\mathrm{I} I_y}\left(-\left[\frac{x_\mathrm{I}}{L}\right] + 1\right), \tag{9.26}$$

$$\varepsilon_{z\mathrm{II},\max}(x_\mathrm{II}) = \frac{F_0 L a}{2 E_\mathrm{II} I_y}\left(-\left[\frac{x_\mathrm{II}}{L}\right] + \frac{1}{2}\right). \tag{9.27}$$

2.4.4 Comparison of the maximum deflections of a cantilever beam with constant line load according to different beam theories

See Table 9.1 for details on the integration constants.

Timoshenko:

$$\frac{u_{z,\max}}{-\dfrac{q_0 L^4}{E I_y}} = \frac{1}{8} + \frac{1+v}{10}\left(\frac{h}{L}\right)^2. \tag{9.28}$$

Levinson:

$$\frac{u_{z,\max}}{-\dfrac{q_0 L^4}{E I_y}} = \frac{1}{8} + \frac{3(1+v)}{20}\left(\frac{h}{L}\right)^2. \tag{9.29}$$

Table 9.1 Integration constants and normalized deflection according to the various beam theories for a cantilever with a constant line load (q_0 in negative z-direction)

Theory	c_1	c_2	c_3	c_4	$\dfrac{u_z(L)}{-\dfrac{q_0 L^4}{8 E I_y}}$
Euler-Bernoulli	$q_0 L$	$-\dfrac{q_0 L^2}{2}$	0	0	1
Timoshenko	$q_0 L$	$-\dfrac{q_0 L^2}{2} + \dfrac{q_0 E I_y}{k_s A G}$	$-\dfrac{q_0 L E I_y}{k_s A G}$	0	$1 + \dfrac{4(1+v)}{5}\left(\dfrac{h}{L}\right)^2$
Levinson	$q_0 L$	$-\dfrac{q_0 L^2}{2} + \dfrac{6 q_0 E I_y}{5 A G}$	$-\dfrac{3 q_0 L E I_y}{2 A G}$	0	$1 + \dfrac{6(1+v)}{5}\left(\dfrac{h}{L}\right)^2$

Fig. 9.3 Course of **a** bending line, **b** stress (no influence of E_i) and **c** strain for bimaterial cantilever beam

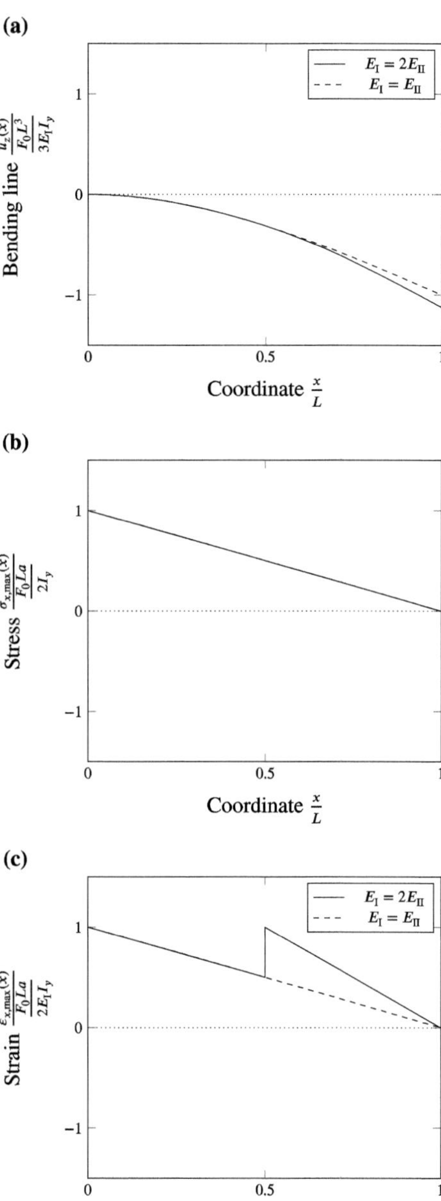

2.4.5 Second moment of area for a circular ring

- Thick wall ring I:

$$I = \frac{\pi}{4}\left(r_a^4 - r_i^4\right) = \frac{\pi}{4}\left(r_a^2 + r_i^2\right)\left(r_a^2 - r_i^2\right) \tag{9.30}$$

$$= \frac{\pi}{4}\left(r_a^2 + r_i^2\right)\left(r_a - r_i\right)\left(r_a + r_i\right) . \tag{9.31}$$

With $r_m = \frac{1}{2}\left(r_a + r_i\right), s = \left(r_a - r_i\right)$ and $r_a = r_m + \frac{s}{2}$ or $r_i = r_m - \frac{s}{2}$ respectively

$$\left(r_a^2 + r_i^2\right) = \left(r_m + \frac{s}{2}\right)^2 + \left(r_m - \frac{s}{2}\right)^2 \tag{9.32}$$

$$= r_m^2\left(1 + \frac{s}{2r_m}\right)^2 + r_m^2\left(1 - \frac{s}{2r_m}\right)^2 = 2r_m^2\left(1 + \frac{s^2}{4r_m^2}\right) , \tag{9.33}$$

finally follows for the second moment of area

$$I = \pi r_m^3 s\left(1 + \frac{s^2}{4r_m^2}\right) = \frac{\pi}{8} d_m^3 s\left(1 + \frac{s^2}{d_m^2}\right) , \tag{9.34}$$

where $d_m = 2r_m$ denotes the mean diameter.

- Thin wall ring I:

For $s \ll r_m$:

$$I \approx \pi r_m^3 s = \frac{\pi}{8} d_m^3 s . \tag{9.35}$$

Relative error of simplification less than 1%

$$\frac{\pi r_m^3 s \times \dfrac{s^2}{4r_m^2}}{\pi r_m^3 s\left(1 + \dfrac{s^2}{4r_m^2}\right)} \leq 0.01 \quad \Rightarrow \quad \frac{r_m^2}{s^2} \geq \frac{1 - 0.01}{4 \times 0.01} = 24.75 \tag{9.36}$$

$$\Rightarrow \quad \frac{r_m}{s} \geq 4.97 . \tag{9.37}$$

- Cross-sectional area A:

$$A = \pi \left(r_a^2 - r_i^2 \right) = \pi \left(r_a - r_i \right) \left(r_a + r_i \right) \tag{9.38}$$

$$= 2\pi r_m s = \pi d_m s \quad \text{(exact)}, \tag{9.39}$$

or

$$A = \pi \left(r_a + r_i \right) s , \tag{9.40}$$

or for $s \ll r_m$ with $r_a \approx r_i$:

$$A \approx 2\pi r_i s \approx 2\pi r_a s . \tag{9.41}$$

2.4.6 Second moment of area for a square box profile

- Thick profile I:

$$I = \frac{1}{12} \left(a_a^4 - a_i^4 \right) = \frac{1}{12} \left(a_a^2 + a_i^2 \right) \left(a_a^2 - a_i^2 \right) \tag{9.42}$$

$$= \frac{1}{12} \left(a_a^2 + a_i^2 \right) \left(a_a - a_i \right) \left(a_a + a_i \right) . \tag{9.43}$$

With $a_m = \frac{1}{2} \left(a_a + a_i \right)$, $s = \frac{1}{2} \left(a_a - a_i \right)$ and $a_a = a_m + s$ or $a_i = a_m - s$ respectively

$$\left(a_a^2 + a_i^2 \right) = \left(a_m + s \right)^2 + \left(a_m - s \right)^2 \tag{9.44}$$

$$= a_m^2 \left(1 + \frac{s}{a_m} \right)^2 + a_m^2 \left(1 - \frac{s}{a_m} \right)^2 = 2 a_m^2 \left(1 + \frac{s^2}{a_m^2} \right), \tag{9.45}$$

finally follows for the second moment of area

$$I = \frac{2}{3} a_m^3 s \left(1 + \frac{s^2}{a_m^2} \right) . \tag{9.46}$$

- Thin profile I:
  For $s \ll a_m$:

$$I \approx \frac{2}{3} a_m^3 s . \tag{9.47}$$

Relative error of simplification less than 1%

$$\frac{\frac{2}{3} a_{\mathrm{m}}^3 s \times \frac{s^2}{a_{\mathrm{m}}^2}}{\frac{2}{3} a_{\mathrm{m}}^3 s \left(1 + \frac{s^2}{a_{\mathrm{m}}^2}\right)} \le 0.01 \quad \Rightarrow \quad \frac{a_{\mathrm{m}}^2}{s^2} \ge \frac{1 - 0.01}{0.01} = 99 \tag{9.48}$$

$$\Rightarrow \quad \frac{a_{\mathrm{m}}}{s} \ge 9.95 \, . \tag{9.49}$$

- Cross-sectional area A:

$$A = a_{\mathrm{a}}^2 - a_{\mathrm{i}}^2 = (a_{\mathrm{a}} - a_{\mathrm{i}}) \, (a_{\mathrm{a}} + a_{\mathrm{i}}) \tag{9.50}$$
$$= 4 a_{\mathrm{m}} s \ \text{(exact)} \, . \tag{9.51}$$

2.4.7 General solution of the differential equations of the Timoshenko beamusing a computer algebra systems

Coefficient comparison with respect to the constant factors of $u(x)$ and $\phi(x)$ results in:

$$C1 = -\frac{kAG \left.\frac{\mathrm{d}u}{\mathrm{d}x}\right|_{x=0}}{EI} - \frac{kAG\phi(0)}{EI}, \tag{9.52}$$

$$C2 = -\left.\frac{\mathrm{d}\phi}{\mathrm{d}x}\right|_{x=0} - \frac{q}{kAG}, \tag{9.53}$$

$$C3 = \left.\frac{\mathrm{d}u(x)}{\mathrm{d}x}\right|_{x=0}, \tag{9.54}$$

$$C4 = u(0) \, . \tag{9.55}$$

2.4.8 Maximum difference between von Mises and Tresca

Maximum absolute error:

$$\left| \frac{1}{\sqrt{3}} k_{\mathrm{t}} - \frac{1}{2} k_{\mathrm{t}} \right| = \frac{2 - \sqrt{3}}{2\sqrt{3}} k_{\mathrm{t}} = 0.07735 k_{\mathrm{t}} \approx 7.7\% k_{\mathrm{t}} \, . \tag{9.56}$$

Maximum relative error (option 1):

$$\left| \frac{\frac{1}{\sqrt{3}} k_{\mathrm{t}} - \frac{1}{2} k_{\mathrm{t}}}{\frac{1}{\sqrt{3}} k_{\mathrm{t}}} \right| = \frac{2 - \sqrt{3}}{2} = 0.13397 \approx 13.4\% \, . \tag{9.57}$$

Maximum relative error (option 2):

$$\left| \frac{\frac{1}{2} k_t - \frac{1}{\sqrt{3}} k_t}{\frac{1}{2} k_t} \right| = \left| \frac{\sqrt{3} - 2}{\sqrt{3}} \right| = 0.15470 \approx 15.5\% \,. \tag{9.58}$$

2.4.9 Bar under tensile and torsional loading

$$\sigma_{\text{eff}} = \sqrt{\sigma^2 + 3\tau^2} = \frac{4}{\pi} \times \sqrt{\left(\frac{F_0}{h^2} \right)^2 + 3 \left(\frac{4M_t}{h^3} \right)^2} \,. \tag{9.59}$$

9.2 Chapter 3

3.4.1 Finite element calculation of a stepped tensile bar

Information on the individual steps of a finite element analysis can be found in [1,3,5].

Reduced system of equations:

$$\frac{EA}{\frac{1}{3} L} \begin{bmatrix} 5 & -2 & 0 \\ -2 & 3 & -1 \\ 0 & -1 & 1 \end{bmatrix} \begin{bmatrix} u_{2X} \\ u_{3X} \\ u_{4X} \end{bmatrix} = \begin{bmatrix} 0 \\ 0 \\ F_0 \end{bmatrix} \,. \tag{9.60}$$

Solution:

$$\begin{bmatrix} u_{2X} \\ u_{3X} \\ u_{4X} \end{bmatrix} = \frac{\frac{1}{3} L F_0}{EA} \begin{bmatrix} \frac{1}{3} \\ \frac{5}{6} \\ \frac{11}{6} \end{bmatrix} \,. \tag{9.61}$$

Lightweight index:

$$M = \frac{R_{p0.2}}{2 \varrho g L} \,. \tag{9.62}$$

3.4.2 Calculation of the lightweight index for a cantilever beam with constant distributed load

- Internal reactions:
  The internal reactions for the problem in Fig. 3.34 are shown in Fig. 9.4.
- Maximum bending moment:

$$M_{y,\max} = M_y(0) = \frac{q_0 L^2}{2}.$$ (9.63)

- Lightweight index:
  Since there is no single force in this case, the definition equation of the lightweight index is evaluated using the resultant of the distributed load:

$$M = \frac{F_0}{F_G} = \frac{q_0 L}{F_G} = \frac{1}{3} \times \frac{R_{p0.2}}{\varrho g \frac{L^2}{h}}.$$ (9.64)

Numerical values:

$$M_{St} = 1712.015 \; ; \; M_{Al} = 4497.575 \; ; \; M_{Ti} = 5422.170.$$ (9.65)

3.4.3 Calculation of the lightweight index for a simply supported beam with constant distributed load

- Maximum bending moment (see Fig. 9.38):

$$M_{y,\max} = M_y(\tfrac{L}{2}) = -\frac{q_0 L^2}{8}.$$ (9.66)

- Lightweight index:
  Since there is no single force in this case, the definition equation of the lightweight index is evaluated using the resultant of the distributed load:

$$M = \frac{F_0}{F_G} = \frac{q_0 L}{F_G} = \frac{4}{3} \times \frac{R_{p0.2}}{\varrho g \frac{L^2}{h}}.$$ (9.67)

Numerical values:

$$M_{St} = 6848.062 \; ; \; M_{Al} = 17990.3001 \; ; \; M_{Ti} = 21688.681.$$ (9.68)

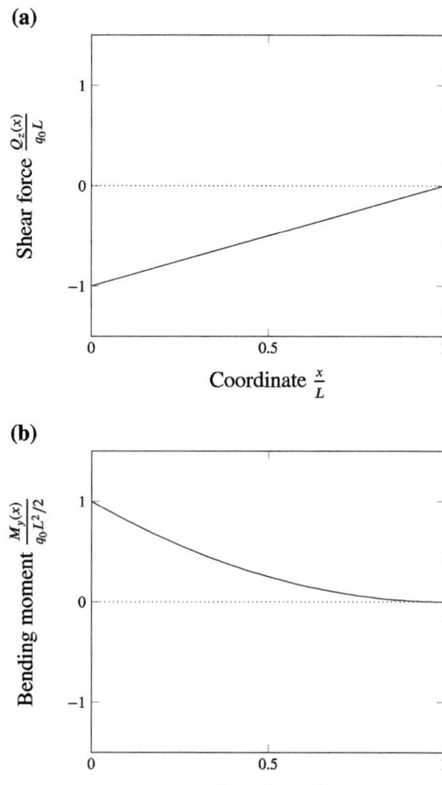

Fig. 9.4 Internal reactions for a cantilever beam with constant distributed load according to Fig. 3.34: **a** shear force and **b** bending moment

3.4.4 Calculation of the lightweight index for a simply supported beam with point load

- Maximum bending moment (see Fig. 9.41):

$$M_{y,\mathrm{max}} = M_y\left(\tfrac{L}{2}\right) = -\frac{F_0 L}{4}. \tag{9.69}$$

- Lightweight index:

$$M = \frac{F_0}{F_\mathrm{G}} = \frac{2}{3} \times \frac{R_{\mathrm{p}0.2}}{\varrho g \frac{L^2}{h}}. \tag{9.70}$$

Numerical values:

$$M_\mathrm{St} = 3424.031 \; ; \;\; M_\mathrm{Al} = 8995.150 \; ; \;\; M_\mathrm{Ti} = 10844.340 . \tag{9.71}$$

3.4.5 Calculation of the specific energy absorption for a cantilever beam with constant distributed load

- Bending moment distribution:

$$M_y(x) = \frac{q_0 L^2}{2}\left(\frac{1}{L^2}x^2 - \frac{2}{L}x + 1\right).$$
(9.72)

- Total strain energy:

$$\Pi = \frac{q_0^2 L^5}{40 EI}.$$
(9.73)

- Specific energy absorption:

$$SEA = \frac{3q_0^2 L^4}{10 b^2 h^4 E\varrho} = \frac{R_{\text{p0.2}}^2}{30 E\varrho}.$$
(9.74)

3.4.6 Calculation of the specific energy absorption for a simply supported beamwith constant distributed load

- Bending moment distribution:

$$M_y(x) = \frac{q_0 L^2}{2}\left(\frac{1}{L^2}x^2 - \frac{1}{L}x\right).$$
(9.75)

- Total strain energy:

$$\Pi = \frac{q_0^2 L^5}{240 EI}.$$
(9.76)

- Specific energy absorption:

$$SEA = \frac{q_0^2 L^4}{20 b^2 h^4 E\varrho} = \frac{4 R_{\text{p0.2}}^2}{45 E\varrho}.$$
(9.77)

3.4.7 Calculation of the specific energy absorption for a simply supported beam with point load

- Bending moment distribution:

$$M_y(x) = -\frac{F_0 x}{2} \quad \text{for} \quad 0 \leq x \leq \frac{L}{2}.$$
(9.78)

- Total strain energy:

$$\Pi = \frac{F_0^2 L^3}{8Ebh^3} = \frac{F_0^2 L^3}{96EI}. \tag{9.79}$$

- Specific energy absorption:

$$SEA = \frac{F_0^2 L^2}{8b^2 h^4 E\varrho} = \frac{R_{p0.2}^2}{18E\varrho}. \tag{9.80}$$

3.4.8 Calculation of specific energy absorption for a bimaterial cantilever beam

- Total strain energy:

$$\Pi = \frac{F_0^2 L^3}{48 I_y} \times \left(\frac{7}{E_{\mathrm{I}}} + \frac{1}{E_{\mathrm{II}}}\right). \tag{9.81}$$

- Mass:

$$m = \frac{a^2 L}{2} \times (\varrho_{\mathrm{I}} + \varrho_{\mathrm{II}}). \tag{9.82}$$

- Specific energy absorption:

$$SEA = \frac{\Pi}{m} = \frac{F_0^2 L^2}{2a^6} \times \frac{\frac{7}{E_{\mathrm{I}}} + \frac{1}{E_{\mathrm{II}}}}{\varrho_{\mathrm{I}} + \varrho_{\mathrm{II}}}. \tag{9.83}$$

9.3   Chapter 4

4.4.1 Analysis of the lightweight index for a cantilever beam with end force and tubular cross section

$$M(d, s) = \frac{F_0(d, s)}{F_G(d, s)} = \frac{R_{p0.2}}{4\varrho g L^2} \times \frac{\frac{1}{d}\left(-2s^4 + d^3 s + 4ds^3 - 3d^2 s^2\right)}{ds - s^2} \tag{9.84}$$

$$= \frac{R_{p0.2}}{4\varrho g L^2} \times \frac{F_0^*(d, s)}{F_G^*(d, s)} \tag{9.85}$$

$$= \frac{R_{p0.2}}{4\varrho g L^2} \times \frac{-2s^3 + d^3 + 4ds^2 - 3d^2 s}{d(d - s)}. \tag{9.86}$$

The functions $F_0(d, s)$, $F_G(d, s)$ and $M(d, s) = F_0(d, s)/F_G(d, s)$ are shown in Fig. 9.5.

4.4.2 Implementation of the concept of lightweight design shapes: Optimization of a box profile

- Second moment of area of the box profile:

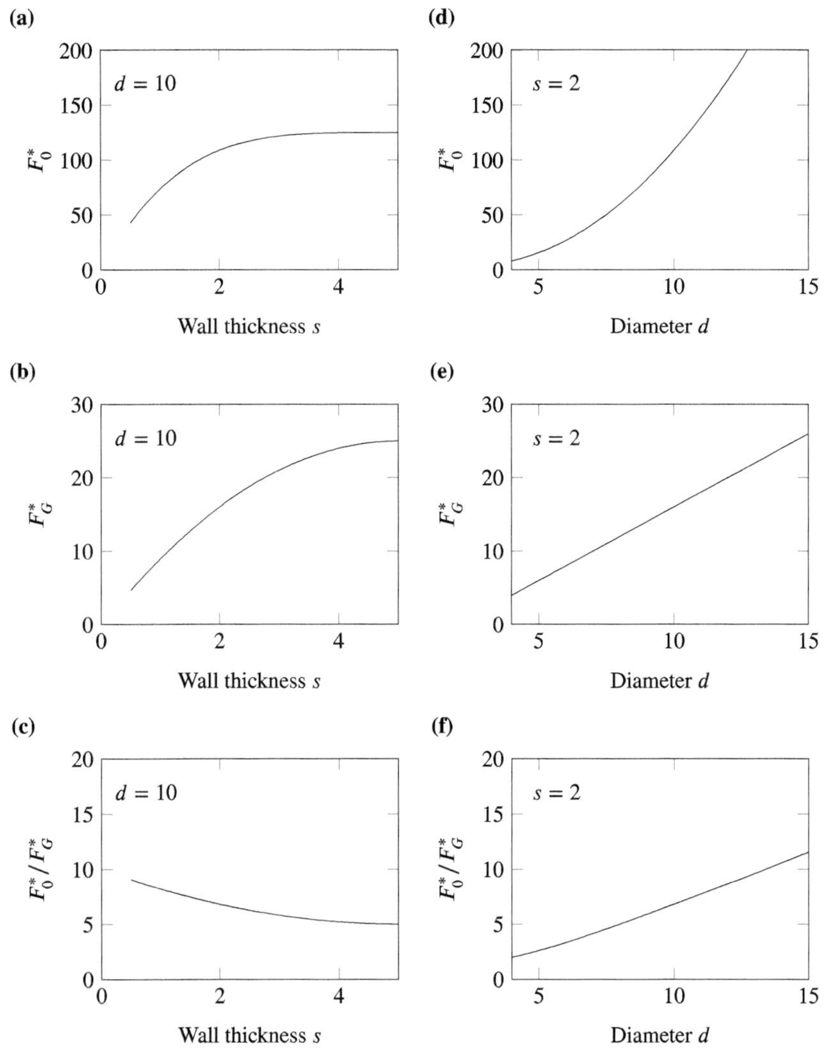

Fig. 9.5 Functions $F_0(d, s)$, $F_G(d, s)$ and $M(d, s) = F_0(d, s)/F_G(d, s)$ depending on s at $d = 10\,\text{mm} = \text{const.}$ (**a–c**) and on d at $s = 2\,\text{mm} = \text{const.}$ (**d–f**)

$$I_y^{\square} = \frac{b\,a^3}{12} - \frac{1}{12}\,(b - 2w)\,(a - 2w)^3\,, \tag{9.87}$$

$$= -\frac{4w^4}{3} + \frac{2b\,w^3}{3} + 2a\,w^3 - ab\,w^2 - a^2\,w^2 + \frac{a^2 bw}{2} + \frac{a^3 w}{6}\,, \tag{9.88}$$

or considering $w = \frac{h}{5}$

$$I_y^{\square} = -\frac{4h^4}{1875} + \frac{2b\,h^3}{375} + \frac{2a\,h^3}{125} - \frac{ab\,h^2}{25} - \frac{a^2\,h^2}{25} + \frac{a^2 bh}{10} + \frac{a^3 h}{30}\,, \tag{9.89}$$

or using $b = \frac{29h}{10} - a$

$$I_y^{\square} = \frac{h^4}{75} - \frac{79a\,h^3}{750} + \frac{29a^2\,h^2}{100} - \frac{a^3 h}{15}\,, \tag{9.90}$$

or in normalized representation:

$$\frac{I_y^{\square}}{h^4} = \frac{1}{75} - \frac{79}{750}\left(\frac{a}{h}\right) + \frac{29}{100}\left(\frac{a}{h}\right)^2 - \frac{1}{15}\left(\frac{a}{h}\right)^3. \tag{9.91}$$

The relationship between the relations $\frac{a}{h}$ and $\frac{a}{b}$ is obtained from the equivalence of the cross-sectional areas, i.e. $b = \frac{29h}{10} - a$, as:

$$\frac{a}{h} = \frac{29}{10}\left(\frac{b}{a} + 1\right)^{-1}. \tag{9.92}$$

The course of the second moment of area is shown in Fig. 9.6.
- Lightweight index:

$$M = \frac{F_0}{F_G} = \frac{R_{St}}{\varrho_{St} g L^2 h^2} \times \frac{I_y^{\square}}{\frac{a}{2}} \tag{9.93}$$

$$= \frac{R_{St}}{\varrho_{St} g L^2 h^2} \times \frac{\frac{h^4}{75} - \frac{79ah^3}{750} + \frac{29a^2 h^2}{100} - \frac{a^3 h}{15}}{\frac{a}{2}} \tag{9.94}$$

$$= \frac{R_{St}}{\varrho_{St} g L^2 h^2} \times \frac{\left(\frac{I_y^{\square}}{h^4}\right) \times h^4}{\frac{1}{2}\left(\frac{a}{h}\right) \times h}. \tag{9.95}$$

The course of the lightweight index is shown in Fig. 9.7.

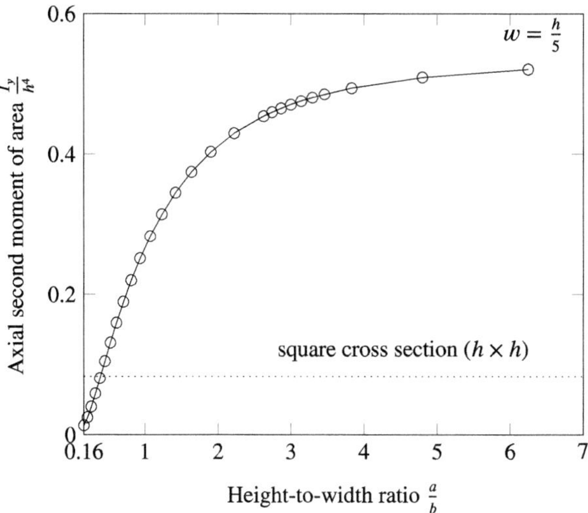

Fig. 9.6 Lightweight index for different box profiles and stress criterion ($\sigma_{\max} = R_{\mathrm{p0.2}}$). Same cross-sectional area ($A^{\boxtimes} = A^{\square}$) and thus same mass ($m^{\boxtimes} = m^{\square}$) assumed

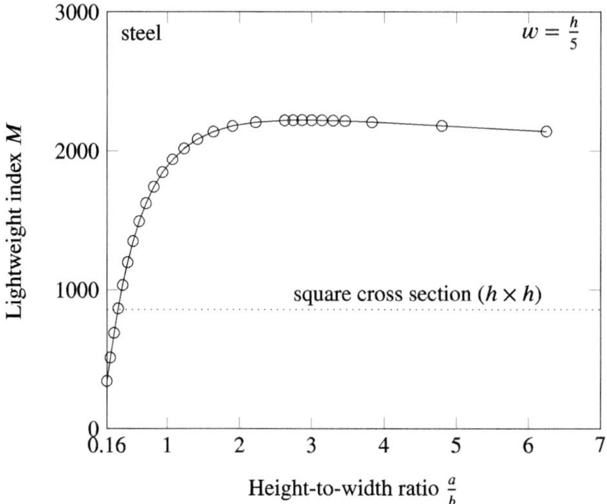

Fig. 9.7 Axial second moment of area for different box profiles. Same cross-sectional area ($A^{\boxtimes} = A^{\square}$) and thus same mass ($m^{\boxtimes} = m^{\square}$) assumed

4.4.3 Optimization of a beam in bending along its axis

- Reference configuration:

$$A_0^{\circledcirc} = \frac{\pi}{4}\left(d_0^2 - d_i^2\right) = \frac{\pi}{4}\left(d_0^2 - (d_0 - 2s_0)^2\right) = \frac{\pi}{4}\left(d_0^2 - (d_0 - \tfrac{2d_0}{5})^2\right)$$
$$= \frac{4\pi}{25}d_0^2 \,. \tag{9.96}$$

$$I_{y,0}^{\circledcirc} = \frac{\pi}{64}\left(d_0^4 - d_i^4\right) = \frac{\pi}{64}\left(d_0^4 - (d_0 - 2s_0)^4\right) = \frac{\pi}{64}\left(d_0^4 - (d_0 - \tfrac{2d_0}{5})^4\right)$$
$$= \frac{17}{1250}\pi d_0^4 \,. \tag{9.97}$$

$$m_0^{\circledcirc} = \varrho V_0^{\circledcirc} = \varrho A_0^{\circledcirc} L = \varrho \frac{4\pi}{25}d_0^2 L \,. \tag{9.98}$$

Using the stress criterion, i.e.,

$$\sigma_{x,\max} = \frac{M_{y,\max}}{I_{y,0}^{\circledcirc}} \times \frac{d_0}{2} \overset{!}{=} R_{p0.2} \,, \tag{9.99}$$

one obtains the maximum force as:

$$F_0 = \frac{R_{p0.2} I_{y,0}^{\circledcirc}}{\frac{1}{2}d_0 L} \,. \tag{9.100}$$

Thus, the lightweight index of the reference configuration results as:

$$M_0 = \frac{F_0}{F_G} = \frac{17}{100} \times \frac{R_{p0.2}}{\varrho g \frac{L^2}{d_0}} \,. \tag{9.101}$$

The total strain energy of the reference configuration is based on the following integration:

$$\Pi_0 = \int_0^L \frac{M_y(x)^2}{2EI_{y,0}^{\circledcirc}}\,\mathrm{d}x = \frac{F_0^2}{2EI_{y,0}^{\circledcirc}}\int_0^L \left(L^2 - 2Lx + x^2\right)\mathrm{d}x = \frac{F_0^2 L^3}{6EI_{y,0}^{\circledcirc}} \,. \tag{9.102}$$

Using Eq. (9.100) this gives:

$$\Pi_0 = \frac{L^3}{6EI_{y,0}^{\circledcirc}} \times \frac{R_{p0.2}^2 I_{y,0}^{\circledcirc,2}}{\frac{1}{4}d_0^2 L^2} = \frac{L R_{p0.2}^2 I_{y,0}^{\circledcirc}}{\frac{3}{2}E d_0^2} \,, \tag{9.103}$$

$$SEA_0 = \frac{\Pi_0}{m_0} = \frac{L\,R_{\text{p0.2}}^2\,I_{y,0}^{\circledcirc}}{\frac{3}{2}E d_0^2} \times \frac{1}{\varrho\,\frac{4\pi}{25}\,d_0^2\,L} = \frac{17}{300} \times \frac{R_{\text{p0.2}}^2}{E\varrho}. \tag{9.104}$$

- Optimized configuration:
  Using the stress criterion, i.e.,

$$\sigma_{x,\max} = \frac{M_y(x)}{I_y^{\circledcirc}(x)} \times \frac{d(x)}{2} = \frac{F_0 L\left(1 - \frac{x}{L}\right)}{\frac{17}{1250}\pi d^4(x)} \times \frac{d(x)}{2} \overset{!}{=} R_{\text{p0.2}}, \tag{9.105}$$

one obtains

$$d^3(x) = \frac{F_0 L}{\frac{17}{1250}\pi\,2R_{\text{p0.2}}} \times \left(1 - \frac{x}{L}\right) = d_0^3\left(1 - \frac{x}{L}\right). \tag{9.106}$$

Thus, the outer circle diameter of the beam cross-section is (see Fig. 9.8):

$$d(x) = d_0 \times \sqrt[3]{1 - \frac{x}{L}}. \tag{9.107}$$

The volume of the optimized configuration is obtained based on the following integration:

$$V = \int_0^L A(x)\mathrm{d}x = \int_0^L \frac{4\pi}{25}d^2(x)\mathrm{d}x = \frac{4\pi d_0^2}{25}\int_0^L \left(1 - \frac{x}{L}\right)^{\frac{2}{3}}\mathrm{d}x = \frac{12\pi}{125}d_0^2 L, \tag{9.108}$$

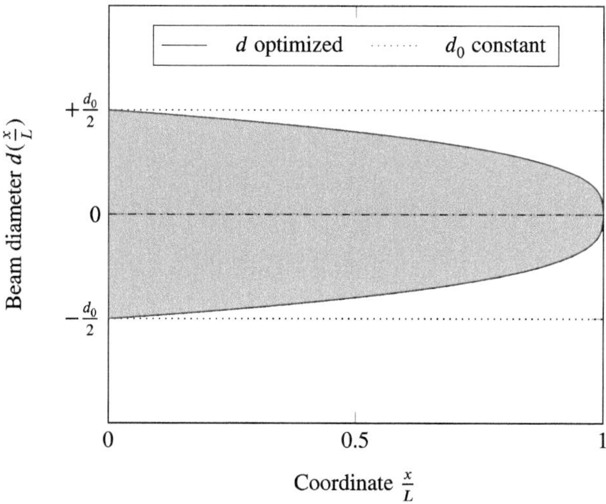

Fig. 9.8 Beam diameter (outer) along the beam axis

or converted to the mass of the optimized configuration:

$$m = \varrho V = \frac{12\pi}{125} d_0^2 L \varrho \,. \qquad (9.109)$$

Thus, the lightweight index of the optimized configuration results as:

$$M = \frac{F_0}{F_G} = \frac{R_{p0.2} I_{y,0}^{\circledcirc}}{\frac{1}{2} d_0 L} \times \frac{1}{\varrho g \frac{12\pi}{125} d_0^2 L} = \frac{17}{60} \times \frac{R_{p0.2}}{\varrho g \frac{L^2}{d_0}} \,. \qquad (9.110)$$

The total strain energy of the optimized configuration is based on the following integration:

$$\Pi = \int_0^L \frac{M_y(x)^2}{2EI_y^{\circledcirc}(x)} \mathrm{d}x = \frac{F_0^2 L^2}{2EI_{y,0}^{\circledcirc}} \int_0^L \frac{(1 - \frac{x}{L})^2}{(1 - \frac{x}{L})^{\frac{4}{3}}} \mathrm{d}x = \frac{F_0^2 L^2}{2EI_{y,0}^{\circledcirc}} \int_0^L (1 - \frac{x}{L})^{\frac{2}{3}} \mathrm{d}x$$

$$= \frac{3}{10} \times \frac{F_0^2 L^3}{EI_{y,0}^{\circledcirc}} \,. \qquad (9.111)$$

Using Eq. (9.100) this gives:

$$\Pi = \frac{3}{10} \frac{L^3}{EI_{y,0}^{\circledcirc}} \times \frac{R_{p0.2}^2 I_{y,0}^{\circledcirc,2}}{\frac{1}{4} d_0^2 L^2} = \frac{6}{5} \times \frac{L R_{p0.2}^2 I_{y,0}^{\circledcirc}}{E d_0^2} \,, \qquad (9.112)$$

$$SEA = \frac{\Pi}{m} = \frac{6}{5} \times \frac{L R_{p0.2}^2 I_{y,0}^{\circledcirc}}{E d_0^2} \times \frac{1}{\frac{12\pi}{125} d_0^2 L \varrho} = \frac{17}{100} \times \frac{R_{p0.2}^2}{E \varrho} \,. \qquad (9.113)$$

- Ratios:

$$\frac{m}{m_0} = \frac{3}{5} \approx 0.6 \,, \qquad (9.114)$$

$$\frac{M}{M_0} = \frac{5}{3} \approx 1.667 \,, \qquad (9.115)$$

$$\frac{SEA}{SEA_0} = 3.0 \,. \qquad (9.116)$$

$$SEA_0 = \frac{\Pi_0}{m_0} = \frac{L\,R_{\mathrm{p0.2}}^2 I_{y,0}^{\odot}}{\frac{3}{2}Ed_0^2} \times \frac{1}{\varrho\,\frac{4\pi}{25}d_0^2 L} = \frac{17}{300} \times \frac{R_{\mathrm{p0.2}}^2}{E\varrho}. \tag{9.104}$$

- Optimized configuration:
  Using the stress criterion, i.e.,

$$\sigma_{x,\max} = \frac{M_y(x)}{I_y^{\odot}(x)} \times \frac{d(x)}{2} = \frac{F_0 L(1 - \frac{x}{L})}{\frac{17}{1250}\pi d^4(x)} \times \frac{d(x)}{2} \stackrel{!}{=} R_{\mathrm{p0.2}}, \tag{9.105}$$

one obtains

$$d^3(x) = \frac{F_0 L}{\frac{17}{1250}\pi\,2R_{\mathrm{p0.2}}} \times \left(1 - \frac{x}{L}\right) = d_0^3\left(1 - \frac{x}{L}\right). \tag{9.106}$$

Thus, the outer circle diameter of the beam cross-section is (see Fig. 9.8):

$$d(x) = d_0 \times \sqrt[3]{1 - \frac{x}{L}}. \tag{9.107}$$

The volume of the optimized configuration is obtained based on the following integration:

$$V = \int_0^L A(x)\mathrm{d}x = \int_0^L \frac{4\pi}{25}d^2(x)\mathrm{d}x = \frac{4\pi d_0^2}{25}\int_0^L\left(1 - \frac{x}{L}\right)^{\frac{2}{3}}\mathrm{d}x = \frac{12\pi}{125}d_0^2 L, \tag{9.108}$$

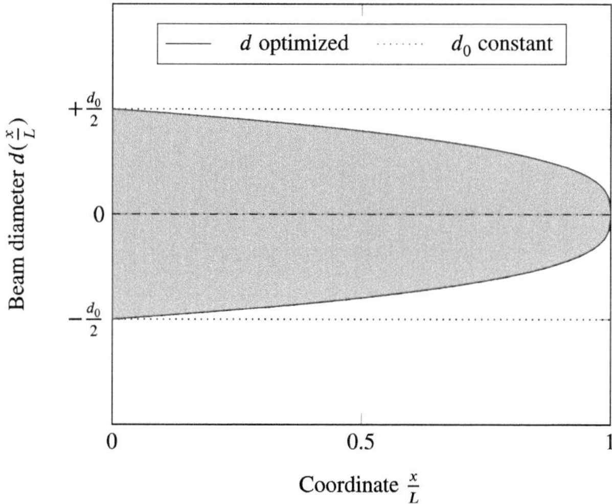

Fig. 9.8 Beam diameter (outer) along the beam axis

or converted to the mass of the optimized configuration:

$$m = \varrho V = \frac{12\pi}{125} d_0^2 L \varrho \,. \tag{9.109}$$

Thus, the lightweight index of the optimized configuration results as:

$$M = \frac{F_0}{F_G} = \frac{R_{p0.2} I_{y,0}^{\odot}}{\frac{1}{2} d_0 L} \times \frac{1}{\varrho g \frac{12\pi}{125} d_0^2 L} = \frac{17}{60} \times \frac{R_{p0.2}}{\varrho g \frac{L^2}{d_0}} \,. \tag{9.110}$$

The total strain energy of the optimized configuration is based on the following integration:

$$\Pi = \int_0^L \frac{M_y(x)^2}{2E I_y^{\odot}(x)} \mathrm{d}x = \frac{F_0^2 L^2}{2E I_{y,0}^{\odot}} \int_0^L \frac{(1 - \frac{x}{L})^2}{(1 - \frac{x}{L})^{\frac{4}{3}}} \mathrm{d}x = \frac{F_0^2 L^2}{2E I_{y,0}^{\odot}} \int_0^L (1 - \frac{x}{L})^{\frac{2}{3}} \mathrm{d}x$$

$$= \frac{3}{10} \times \frac{F_0^2 L^3}{E I_{y,0}^{\odot}} \,. \tag{9.111}$$

Using Eq. (9.100) this gives:

$$\Pi = \frac{3}{10} \frac{L^3}{E I_{y,0}^{\odot}} \times \frac{R_{p0.2}^2 I_{y,0}^{\odot,2}}{\frac{1}{4} d_0^2 L^2} = \frac{6}{5} \times \frac{L R_{p0.2}^2 I_{y,0}^{\odot}}{E d_0^2} \,, \tag{9.112}$$

$$SEA = \frac{\Pi}{m} = \frac{6}{5} \times \frac{L R_{p0.2}^2 I_{y,0}^{\odot}}{E d_0^2} \times \frac{1}{\frac{12\pi}{125} d_0^2 L \varrho} = \frac{17}{100} \times \frac{R_{p0.2}^2}{E \varrho} \,. \tag{9.113}$$

- Ratios:

$$\frac{m}{m_0} = \frac{3}{5} \approx 0.6 \,, \tag{9.114}$$

$$\frac{M}{M_0} = \frac{5}{3} \approx 1.667 \,, \tag{9.115}$$

$$\frac{SEA}{SEA_0} = 3.0 \,. \tag{9.116}$$

4.4.4 Optimization of a cantilevered beam loaded with two single forces of the same magnitude

- Optimized diameter:

$$d(x) = d_0 \quad \text{for} \quad 0 \le x \le \frac{L}{2}, \tag{9.117}$$

$$d(x) = d_0 \left(2 - \frac{2x}{L} \right)^{\frac{1}{3}} \quad \text{for} \quad \frac{L}{2} \le x \le L, \tag{9.118}$$

or as a *single* equation using the Föppl-symbol[1] (see Fig. 9.9):

$$d(x) = d_0 \left(1 + \left(1 - \frac{2x}{L} \right) \left\langle x - \frac{L}{2} \right\rangle^0 \right)^{\frac{1}{3}}. \tag{9.119}$$

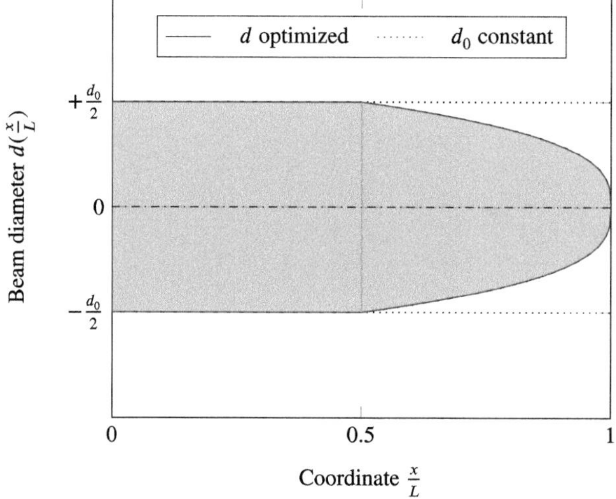

Fig. 9.9 Beam diameter (outer) along the beam axis

---

[1] In the English literature, this is also related to the British mathematician William Herrick Macaulay.

- Axial second moment of area:

$$I_y^{\circledcirc} = \frac{17}{1250} \times \pi d_0^4 \quad \text{for} \quad 0 \le x \le \frac{L}{2} \,, \tag{9.120}$$

$$I_y^{\circledcirc} = \frac{17}{1250} \times \pi d_0^4 (16)^{\frac{1}{3}} \left(1 - \frac{x}{L}\right)^{\frac{4}{3}} \quad \text{for} \quad \frac{L}{2} \le x \le L \,, \tag{9.121}$$

or as a *single* equation using the Föppl-symbol:

$$I_y^{\circledcirc} = \frac{17}{1250} \times \pi d_0^4 \left(1 + \left(1 - \frac{2x}{L}\right)\left\langle x - \frac{L}{2}\right\rangle^0 \right)^{\frac{4}{3}} . \tag{9.122}$$

- Total strain energy:

$$\Pi = \int_0^{\frac{L}{2}} \frac{M_y^2}{2EI_y^{\circledcirc}}\,\mathrm{d}x + \int_{\frac{L}{2}}^{L} \frac{M_y(x)^2}{2EI_y^{\circledcirc}(x)}\,\mathrm{d}x \tag{9.123}$$

$$= \frac{125}{17} \times \frac{F_0^2 L^3}{\pi d_0^4 E} . \tag{9.124}$$

- Volume:

$$V = \int_0^{\frac{L}{2}} A_0\,\mathrm{d}x + \int_{\frac{L}{2}}^{L} A(x)\,\mathrm{d}x \tag{9.125}$$

$$= \frac{16}{125} \times \pi d_0^2 L . \tag{9.126}$$

- Specific energy absorption:

$$SEA = \frac{\Pi}{m} = \frac{125^2}{16 \times 17} \times \frac{F_0^2 L^2}{\pi^2 d_0^6 \varrho E} = \frac{17}{100} \times \frac{R_{\text{p0.2}}^2}{\varrho E} . \tag{9.127}$$

4.4.5 Optimization of a beam under 3-point bending: rectangular cross section

- Optimized diameter (see also Fig. 9.10):

$$h(x) = h_0 \left(\frac{2x}{L} \right)^{\frac{1}{2}} \quad \text{for} \quad 0 \le x \le \frac{L}{2}, \tag{9.128}$$

$$h(x) = h_0 \left(2 \left(1 - \frac{x}{L} \right) \right)^{\frac{1}{2}} \quad \text{for} \quad \frac{L}{2} \le x \le L, \tag{9.129}$$

where the reference diameter in the middle of the beam is as follows:

$$d_0 = \left(\frac{3 F_0 L}{2b R_{\mathrm{p0.2}}} \right)^{\frac{1}{2}}. \tag{9.130}$$

- Total strain energy:

$$\Pi = 2 \times \int_0^{\frac{L}{2}} \frac{M_y(x)^2}{2 E I_y} \, \mathrm{d}x = \frac{2}{2^3} \times \frac{F_0^2 L^3}{b h_0^3 E}. \tag{9.131}$$

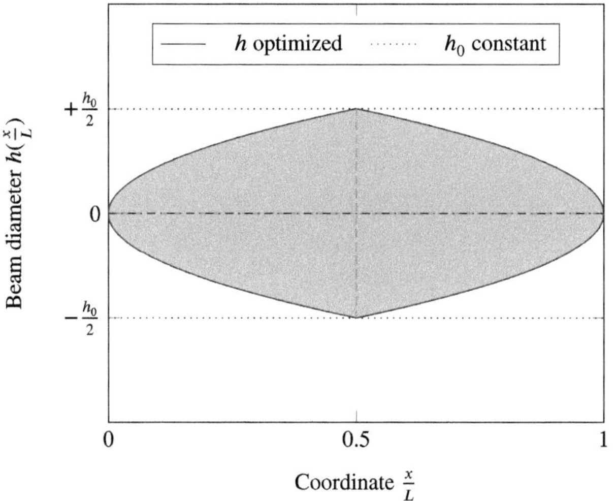

Fig. 9.10 Beam diameter (outer) along the beam axis

- Volume:

$$V = 2 \times \int_0^{\frac{L}{2}} A(x)\, \mathrm{d}x = \frac{2}{3} \times bh_0 L\,. \tag{9.132}$$

- Specific energy absorption:

$$SEA = \frac{\Pi}{m} = \frac{R_{\mathrm{p0.2}}^2}{6\varrho E}\,. \tag{9.133}$$

- Lightweight index:

$$M = \frac{F_0}{F_\mathrm{G}} = \frac{F_0}{V\varrho g} = \frac{R_{\mathrm{p0.2}} h_0}{L^2 \varrho g}\,. \tag{9.134}$$

- Values for the beam with constant diameter::

$$\Pi_0 = 2 \times \int_0^{\frac{L}{2}} \frac{M_y(x)^2}{2EI_{y,0}}\, \mathrm{d}x = \frac{1}{8} \times \frac{F_0^2 L^3}{Ebh_0^3}\,, \tag{9.135}$$

$$V_0 = bh_0 L\,, \tag{9.136}$$

$$SEA_0 = \frac{R_{\mathrm{p0.2}}^2}{18\varrho E}\,. \tag{9.137}$$

- Comparison:

$$\frac{SEA}{SEA_0} = 3.0\,, \tag{9.138}$$

$$\frac{M}{M_0} = 1.5\,. \tag{9.139}$$

4.4.6 Optimization of a beam under 3-point bending: tubular cross section

- Optimized diameter (see also Fig. 9.11):

$$d(x) = d_0 \left(\frac{2x}{L} \right)^{\frac{1}{3}} \quad \text{for} \quad 0 \le x \le \frac{L}{2}, \tag{9.140}$$

$$d(x) = d_0 \left(2 \left(1 - \frac{x}{L} \right) \right)^{\frac{1}{3}} \quad \text{for} \quad \frac{L}{2} \le x \le L, \tag{9.141}$$

where the reference diameter in the middle of the beam is as follows:

$$d_0 = \left(\frac{F_0 L}{2 \times \frac{68}{1250} \pi R_{\mathrm{p}0.2}} \right)^{\frac{1}{3}} . \tag{9.142}$$

- Total strain energy:

$$\Pi = 2 \times \int_0^{\frac{L}{2}} \frac{M_y(x)^2}{2 E I_y^{\odot}} \, \mathrm{d}x = \frac{375}{272} \times \frac{F_0^2 L^3}{\pi d_0^4 E} \approx 1379 \times \frac{F_0^2 L^3}{\pi d_0^4 E}. \tag{9.143}$$

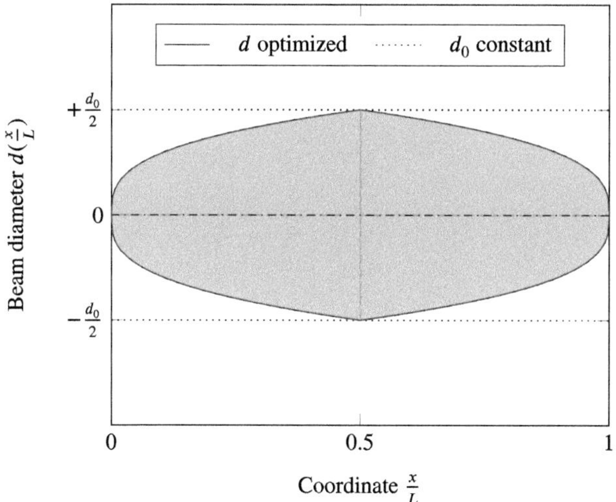

Fig. 9.11 Beam diameter (outer) along the beam axis

- Volume:

$$V = 2 \times \int_0^{\frac{L}{2}} A(x)\, \mathrm{d}x = \frac{12}{125} \times \pi d_0^2 L \,. \tag{9.144}$$

- Specific energy absorption:

$$SEA = \frac{\Pi}{m} = \frac{125 \times 375}{12 \times 272} \times \frac{F_0^2 L^2}{\pi^2 d_0^6 \varrho E} \approx 14.361 \frac{F_0^2 L^2}{\pi^2 d_0^6 \varrho E} \,. \tag{9.145}$$

- Lightweight index:

$$M = \frac{F_0}{F_G} = \frac{F_0}{V \varrho g} = \frac{125}{12} \times \frac{F_0}{\pi d_0^2 L \varrho g} \approx 10.417 \times \frac{F_0}{\pi d_0^2 L \varrho g} \,. \tag{9.146}$$

- Values for beam with constant diameter:

$$\Pi_0 = 2 \times \int_0^{\frac{L}{2}} \frac{M_y(x)^2}{2E I_{y,0}^{\odot}}\, \mathrm{d}x = \frac{1250}{1632} \times \frac{F_0^2 L^3}{\pi d_0^4 E} \,, \tag{9.147}$$

$$V_0 = \frac{4}{25} \times \pi d_0^2 L \,, \tag{9.148}$$

$$SEA_0 = \frac{31{,}250}{6528} \times \frac{F_0^2 L^2}{\pi^2 d_0^6 \varrho E} \approx 4.787 \times \frac{F_0^2 L^2}{\pi^2 d_0^6 \varrho E} \,, \tag{9.149}$$

$$M_0 = \frac{25 F_0}{4\pi d_0^2 L \varrho g} \,. \tag{9.150}$$

- Comparison:

$$\frac{SEA}{SEA_0} = 3.00 \,, \tag{9.151}$$

$$\frac{M}{M_0} = 1.67 \,. \tag{9.152}$$

4.4.7 Optimization of a beam under asymmetric 3-point bending

- Bending moment distribution:

$$M_y(x) = -F_0 \left(1 - \frac{b}{L} \right) x \ \text{ for } x \le b \,, \tag{9.153}$$

$$M_y(x) = -F_0 \left(1 - \frac{x}{L} \right) b \ \text{ for } x \ge b \,. \tag{9.154}$$

- Optimized diameter (see also Fig. 9.12):

$$d(x) = d_0 \left(\frac{x}{b} \right)^{\frac{1}{3}} \ \text{ for } x \le b \,, \tag{9.155}$$

$$d(x) = d_0 \left(\frac{1 - \frac{x}{L}}{1 - \frac{b}{L}} \right)^{\frac{1}{3}} \ \text{ for } x \ge b \,, \tag{9.156}$$

with $d_0 = b^{1/3} \left(\frac{486 F_0 (1 - b/L)}{65 R_{p0.2}} \right)^{1/3}$.

- Reference cross section:

$$A_0 = \frac{5}{9} d_0^2 \,, \tag{9.157}$$

$$I_0 = \frac{65}{972} d_0^4 \,, \tag{9.158}$$

$$F_0 = \frac{65 d_0^3 R_{p0.2}}{486 b (1 - b/L)} \,, \tag{9.159}$$

$$M_0 = \frac{13 d_0 R_{p0.2}}{54 \varrho g L b (1 - b/L)} \,, \tag{9.160}$$

$$\Pi_0 = \frac{65 d_0^2 L R_{p0.2}^2}{1458 E} \,, \tag{9.161}$$

$$SEA_0 = \frac{13 R_{p0.2}^2}{162 \varrho E} \,. \tag{9.162}$$

- Optimized section:

$$A(x) = \frac{5}{9}d(x)^2 \,, \tag{9.163}$$

$$I(x) = \frac{65}{972}d(x)^4 \,, \tag{9.164}$$

$$V = \frac{1}{3}d_0^2 L \,, \tag{9.165}$$

$$F_0 = \frac{65d_0^3 R_{\mathrm{p0.2}}}{486b(1 - b/L)} \,, \tag{9.166}$$

$$M = \frac{65d_0 R_{\mathrm{p0.2}}}{162\varrho g L b(1 - b/L)} \,, \tag{9.167}$$

$$\Pi = \frac{13d_0^2 L R_{\mathrm{p0.2}}^2}{162E} \,, \tag{9.168}$$

$$SEA = \frac{13 R_{\mathrm{p0.2}}^2}{54\varrho E} \,. \tag{9.169}$$

- Comparison:

$$\frac{m}{m_0} = \frac{3}{5} \,, \tag{9.170}$$

$$\frac{M}{M_0} = \frac{5}{3} \,, \tag{9.171}$$

$$\frac{SEA}{SEA_0} = 3 \,. \tag{9.172}$$

Numerical values for particular cases of b are shown in Table 9.2.

Table 9.2 Comparison of the lightweight index and the specific energy absorption for different values of b

$\dfrac{b}{L}$	$\dfrac{M_0}{\dfrac{d_0 R_{\mathrm{p0.2}}}{\varrho g L^2}}$	$\dfrac{M}{\dfrac{d_0 R_{\mathrm{p0.2}}}{\varrho g L^2}}$	$\dfrac{SEA_0}{\dfrac{R_{\mathrm{p0.2}}^2}{E\varrho}}$	$\dfrac{SEA}{\dfrac{R_{\mathrm{p0.2}}^2}{E\varrho}}$	$\dfrac{M_0}{M}$	$\dfrac{SEA_0}{SEA}$
$\dfrac{1}{4}$	$\dfrac{104}{81}$	$\dfrac{520}{243}$	$\dfrac{13}{162}$	$\dfrac{13}{54}$	$\dfrac{5}{3}$	3
$\dfrac{1}{3}$	$\dfrac{13}{12}$	$\dfrac{65}{36}$	$\dfrac{13}{162}$	$\dfrac{13}{54}$	$\dfrac{5}{3}$	3
$\dfrac{1}{2}$	$\dfrac{26}{27}$	$\dfrac{130}{81}$	$\dfrac{13}{162}$	$\dfrac{13}{54}$	$\dfrac{5}{3}$	3

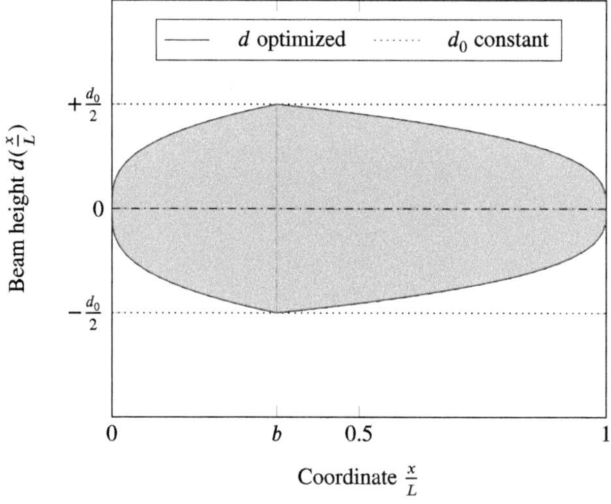

Fig. 9.12 Beam height along the beam axis

4.4.8 Optimization of a simply supported beam under constant distributed load: rectangular cross section

- Moment distribution:

$$M_y(x) = \frac{q_0 L^2}{2}\left(\left[\frac{x}{L}\right]^2 - \left[\frac{x}{L}\right]^1\right) < 0.\tag{9.173}$$

- Optimized diameter (see also Fig. 9.13):

$$h(x) = 2h_0\left(\left[\frac{x}{L}\right] - \left[\frac{x}{L}\right]^2\right)^{\frac{1}{2}}.\tag{9.174}$$

- Optimized diameter:

$$V = \frac{b\pi L h_0}{4},\tag{9.175}$$

$$\Pi = \frac{3\pi L^5 q_0^2}{128 E b h_0^3},\tag{9.176}$$

$$SEA = \frac{3}{32} \times \frac{L^4 q_0^2}{E b^2 h_0^4 \varrho} = \frac{1}{6} \times \frac{R_{\mathrm{p}0.2}^2}{E\varrho}.\tag{9.177}$$

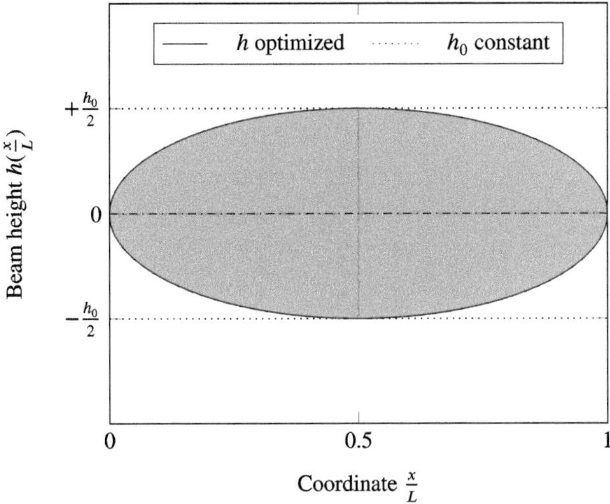

Fig. 9.13 Beam height along the beam axis

- Reference cross section:

$$V_0 = b h_0 L \,, \tag{9.178}$$

$$\Pi_0 = \frac{L^5 q_0^2}{20 E b h_0^3} \,, \tag{9.179}$$

$$SEA_0 = \frac{1}{20} \times \frac{L^4 q_0^2}{E b^2 h_0^4 \varrho} = \frac{4}{45} \times \frac{R_{\mathrm{p}0.2}^2}{E \varrho} \,. \tag{9.180}$$

- Comparison:

$$\frac{SEA}{SEA_0} = \frac{15}{8} \approx 1.875 \,. \tag{9.181}$$

4.4.9 Optimization of a simply supported beam under constant distributed load: tubular cross section (numerical integration)

- Moment distribution:

$$M_y(x) = \frac{q_0 L^2}{2} \left(\left[\frac{x}{L} \right]^2 - \left[\frac{x}{L} \right]^1 \right) < 0 \,. \tag{9.182}$$

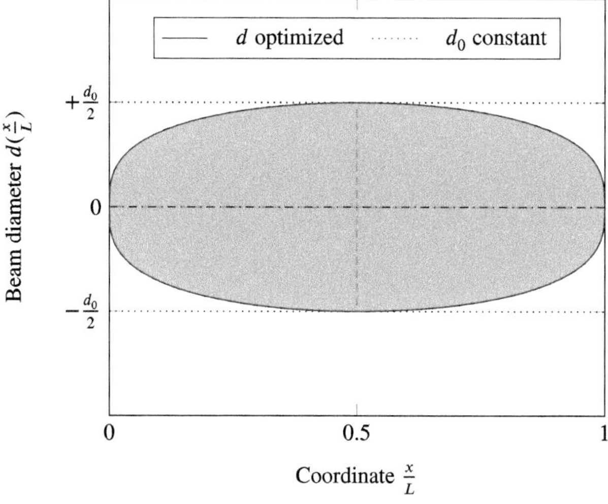

Fig. 9.14 Beam diameter along the beam axis

- Optimized diameter (see also Fig. 9.14):

$$d(x) = 4^{\frac{1}{3}} d_0 \left(-\left[\frac{x}{L}\right]^2 + \left[\frac{x}{L}\right]^1 \right)^{\frac{1}{3}}, \tag{9.183}$$

with

$$d_0 = \left(\frac{324 q_0 L^2}{65\pi R_{\mathrm{p}0.2}} \right)^{\frac{1}{3}} = 35.17\,\mathrm{mm}. \tag{9.184}$$

$$\Pi = 161921.87\,\mathrm{Nmm}, \tag{9.185}$$

$$V = 1013496.90\,\mathrm{mm}^3. \tag{9.186}$$

- Constant diameter:

$$\Pi_0 = 116831.68\,\mathrm{Nmm}, \tag{9.187}$$

$$V_0 = 1371130.53\,\mathrm{mm}^3. \tag{9.188}$$

- Ratio:

$$\frac{SEA}{SEA_0} = 1.875 \,. \tag{9.189}$$

4.4.10 Optimization of a beam with a square box profile along the longitudinal axis under the influence of a constant distributed load

- Moment distribution:

$$M_y(x) = \frac{q_0 L^2}{2} \left(\left[\frac{x}{L}\right]^2 - \left[\frac{x}{L}\right]^1 \right) \; < 0 \,. \tag{9.190}$$

- Reference beam with constant cross section:

$$A_0 = \frac{5 d_0^2}{9} \,, \tag{9.191}$$

$$I_0 = \frac{65 d_0^4}{972} \,, \tag{9.192}$$

$$q_0 = \frac{260 d_0^3 R_{\text{p0.2}}}{243 L^2} \,, \tag{9.193}$$

$$M_0 \approx \frac{52}{27} \times \frac{d_0 R_{\text{p0.2}}}{\varrho g L^2} \,, \tag{9.194}$$

$$\approx 1.92593 \times \frac{d_0 R_{\text{p0.2}}}{\varrho g L^2} \,, \tag{9.195}$$

$$\Pi_0 = \frac{52}{729} \times \frac{L d_0^2 R_{\text{p0.2}}^2}{E} \,, \tag{9.196}$$

$$SEA_0 = \frac{52}{405} \times \frac{R_{\text{p0.2}}^2}{\varrho E} \,, \tag{9.197}$$

$$\approx 0.12840 \times \frac{R_{\text{p0.2}}^2}{\varrho E} \,. \tag{9.198}$$

- Beams with optimized cross section:

$$d(x) = 4^{\frac{1}{3}} d_0 \left(-\left[\frac{x}{L}\right]^2 + \left[\frac{x}{L}\right]^1 \right)^{\frac{1}{3}}, \tag{9.199}$$

$$d_0 = \frac{1}{4^{\frac{1}{3}}} \left(\frac{243}{65} \times \frac{q_0 L^2}{R_{\text{p0.2}}} \right)^{\frac{1}{3}}, \tag{9.200}$$

$$V = \frac{5}{9} \times 4^{\frac{2}{3}} \times 0.293342 d_0^2 L, \tag{9.201}$$

$$\approx 0.41 \times d_0^2 L, \tag{9.202}$$

$$M \approx \frac{52}{27} \times \frac{1}{4^{\frac{2}{3}} \times 0.293342} \times \frac{d_0 R_{\text{p0.2}}}{\varrho g L^2}, \tag{9.203}$$

$$\approx 2.60551 \times \frac{d_0 R_{\text{p0.2}}}{\varrho g L^2}, \tag{9.204}$$

$$\Pi = \frac{2 \times 52 \times 260 \times 0.293342}{4^{\frac{7}{3}} \times 13 \times 243} \times \frac{L d_0^2 R_{\text{p0.2}}^2}{E}, \tag{9.205}$$

$$SEA = \frac{13}{54} \times \frac{R_{\text{p0.2}}^2}{\varrho E} \approx 0.24074 \times \frac{R_{\text{p0.2}}^2}{\varrho E}. \tag{9.206}$$

The optimized beam is shown in Fig. 9.15.

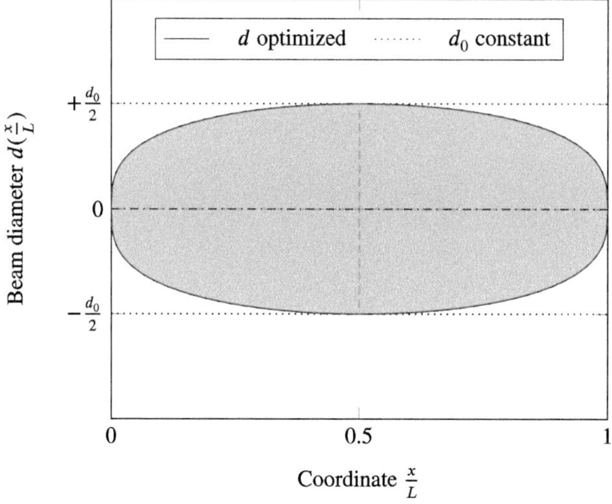

Fig. 9.15 Beam diameter along the beam axis

4.4.11 Optimization of a beam with a rectangular box profile along the longitudinal axis under the influence of a constant distributed load

- Moment distribution:

$$M_y(x) = \frac{q_0 L^2}{2}\left(\left[\frac{x}{L}\right]^2 - \left[\frac{x}{L}\right]^1\right) < 0. \tag{9.207}$$

- Reference beam with constant cross section:

$$A_0 = \frac{8a_0^2}{9}, \tag{9.208}$$

$$I_0 = \frac{199a_0^4}{486}, \tag{9.209}$$

$$q_0 = \frac{796a_0^3 R_{p0.2}}{243L^2}, \tag{9.210}$$

$$M_0 \approx \frac{199}{54} \times \frac{a_0 R_{p0.2}}{\varrho g L^2}, \tag{9.211}$$

$$\approx 3.685185 \times \frac{a_0 R_{p0.2}}{\varrho g L^2}, \tag{9.212}$$

$$\Pi_0 = \frac{398}{3645} \times \frac{La_0^2 R_{p0.2}^2}{E}, \tag{9.213}$$

$$\approx 0.109191 \times \frac{La_0^2 R_{p0.2}^2}{E}, \tag{9.214}$$

$$SEA_0 = \frac{199}{1620} \times \frac{R_{p0.2}^2}{\varrho E}, \tag{9.215}$$

$$\approx 0.122840 \times \frac{R_{p0.2}^2}{\varrho E}. \tag{9.216}$$

- Beams with optimized cross section:

$$a(x) = 4^{\frac{1}{3}} a_0 \left(-\left[\frac{x}{L}\right]^2 + \left[\frac{x}{L}\right]^1 \right)^{\frac{1}{3}}, \tag{9.217}$$

$$a_0 = \left(\frac{243}{796} \times \frac{q_0 L^2}{R_{\mathrm{p}0.2}} \right)^{\frac{1}{3}}, \tag{9.218}$$

$$V = \frac{8}{9} \times 4^{\frac{2}{3}} \times 0.293342 a_0^2 L, \tag{9.219}$$

$$\approx 0.657 \times a_0^2 L, \tag{9.220}$$

$$M \approx \frac{199}{54} \times \frac{1}{4^{\frac{2}{3}} \times 0.293342} \times \frac{a_0 R_{\mathrm{p}0.2}}{\varrho g L^2}, \tag{9.221}$$

$$\approx 4.985431 \times \frac{a_0 R_{\mathrm{p}0.2}}{\varrho g L^2}, \tag{9.222}$$

$$\Pi = \frac{199 \times 0.293342}{4^{\frac{1}{3}} \times 243} \times \frac{L a_0^2 R_{\mathrm{p}0.2}^2}{E}, \tag{9.223}$$

$$\approx 0.151333 \times \frac{L a_0^2 R_{\mathrm{p}0.2}^2}{E}, \tag{9.224}$$

$$SEA = \frac{199}{864} \times \frac{R_{\mathrm{p}0.2}^2}{\varrho E} \approx 0.230324 \times \frac{R_{\mathrm{p}0.2}^2}{\varrho E}. \tag{9.225}$$

The optimized beam is shown in Fig. 9.16.

4.4.12 Optimization of a beam with I profile along the longitudinal axis under the influence of a constant distributed load

- Moment distribution:

$$M_y(x) = \frac{q_0 L^2}{2} \left(\left[\frac{x}{L}\right]^2 - \left[\frac{x}{L}\right]^1 \right) < 0. \tag{9.226}$$

▶ Case $\alpha = 2$:

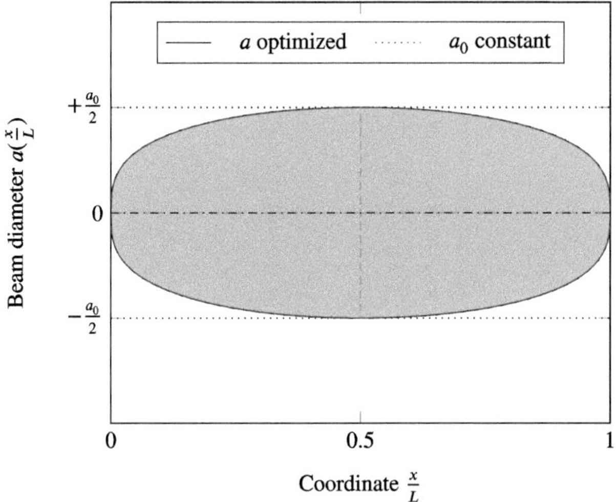

Fig. 9.16 Beam diameter along the beam axis

- Reference beam with constant cross section:

$$A_0 = \frac{11a_0^2}{18}, \tag{9.227}$$

$$I_0 = \frac{671a_0^4}{1944}, \tag{9.228}$$

$$q_0 = \frac{671a_0^3 R_{p0.2}}{243L^2}, \tag{9.229}$$

$$M_0 = \frac{122}{27} \times \frac{a_0 R_{p0.2}}{\varrho g L^2}, \tag{9.230}$$

$$\approx 4.518519 \times \frac{a_0 R_{p0.2}}{\varrho g L^2}, \tag{9.231}$$

$$\Pi_0 = \frac{671}{7290} \times \frac{L a_0^2 R_{p0.2}^2}{E}, \tag{9.232}$$

$$\approx 0.0920439 \times \frac{L a_0^2 R_{p0.2}^2}{E}, \tag{9.233}$$

$$SEA_0 = \frac{61}{405} \times \frac{R_{p0.2}^2}{\varrho E}, \tag{9.234}$$

$$\approx 0.150617 \times \frac{R_{p0.2}^2}{\varrho E}. \tag{9.235}$$

- Beam with optimized cross section:

$$a(x) = 4^{\frac{1}{3}} a_0 \left(-\left[\frac{x}{L}\right]^2 + \left[\frac{x}{L}\right]^1 \right)^{\frac{1}{3}}, \tag{9.236}$$

$$a_0 = \left(\frac{1}{4} \times \frac{972}{671} \times \frac{q_0 L^2}{R_{p0.2}} \right)^{\frac{1}{3}}, \tag{9.237}$$

$$sV \approx \frac{11}{18} \times 4^{\frac{2}{3}} \times 0.293342 a_0^2 L, \tag{9.238}$$

$$\approx 0.452 \times a_0^2 L, \tag{9.239}$$

$$M \approx \frac{122}{27} \times \frac{1}{4^{\frac{2}{3}} \times 0.293342} \times \frac{a_0 R_{p0.2}}{\varrho g L^2}, \tag{9.240}$$

$$\approx 6.112917 \times \frac{a_0 R_{p0.2}}{\varrho g L^2}, \tag{9.241}$$

$$\Pi \approx \frac{671 \times 0.293342}{243 \times 4^{\frac{4}{3}}} \times \frac{L a_0^2 R_{p0.2}^2}{E}, \tag{9.242}$$

$$\approx 0.127569 \times \frac{L a_0^2 R_{p0.2}^2}{E}, \tag{9.243}$$

$$SEA \approx 0.282407 \times \frac{R_{p0.2}^2}{\varrho E}. \tag{9.244}$$

The optimized beam is shown in Fig. 9.17.

▶ Case $\alpha = 3$:

- Reference beam with constant cross section:

$$A_0 = \frac{7 a_0^2}{9}, \tag{9.245}$$

$$I_0 = \frac{907 a_0^4}{972}, \tag{9.246}$$

$$q_0 = \frac{3628 a_0^3 R_{p0.2}}{729 L^2}, \tag{9.247}$$

$$M_0 = \frac{3628}{567} \times \frac{a_0 R_{p0.2}}{\varrho g L^2}, \tag{9.248}$$

$$\approx 6.398589 \times \frac{a_0 R_{p0.2}}{\varrho g L^2}, \tag{9.249}$$

$$\Pi_0 = \frac{3628}{32,805} \times \frac{L a_0^2 R_{p0.2}^2}{E}, \tag{9.250}$$

$$\approx 0.110593 \times \frac{L a_0^2 R_{p0.2}^2}{E}, \tag{9.251}$$

$$SEA_0 = \frac{3628}{25,515} \times \frac{R_{p0.2}^2}{\varrho E}, \tag{9.252}$$

$$\approx 0.142191 \times \frac{R_{p0.2}^2}{\varrho E}. \tag{9.253}$$

- Beam with optimized cross section:

$$a(x) = 4^{\frac{1}{3}} a_0 \left(-\left[\frac{x}{L}\right]^2 + \left[\frac{x}{L}\right]^1 \right)^{\frac{1}{3}}, \tag{9.254}$$

$$a_0 = \left(\frac{1}{4} \times \frac{729}{907} \times \frac{q_0 L^2}{R_{p0.2}} \right)^{\frac{1}{3}}, \tag{9.255}$$

$$V \approx \frac{7}{9} \times 4^{\frac{2}{3}} \times 0.293342 a_0^2 L, \tag{9.256}$$

$$\approx 0.575 \times a_0^2 L, \tag{9.257}$$

$$M \approx \frac{3628}{567} \times \frac{1}{4^{\frac{2}{3}} \times 0.293342} \times \frac{a_0 R_{p0.2}}{\varrho g L^2}, \tag{9.258}$$

$$\approx 8.656387 \times \frac{a_0 R_{p0.2}}{\varrho g L^2}, \tag{9.259}$$

$$\Pi \approx \frac{7256 \times 0.293342}{2187 \times 4^{\frac{4}{3}}} \times \frac{L a_0^2 R_{p0.2}^2}{E}, \tag{9.260}$$

$$\approx 0.153277 \times \frac{L a_0^2 R_{p0.2}^2}{E}, \tag{9.261}$$

$$SEA \approx 0.266608 \times \frac{R_{p0.2}^2}{\varrho E}. \tag{9.262}$$

The optimized beam is shown in Fig. 9.17.

4.4.13 Optimization of a beam with I-profile along the longitudinal axis and the height-to-width ratio under the influence of a constant distributed load

The evaluation of the specific energy absorption SEA of the optimized cross-section for different values of α is shown in Fig. 9.18.

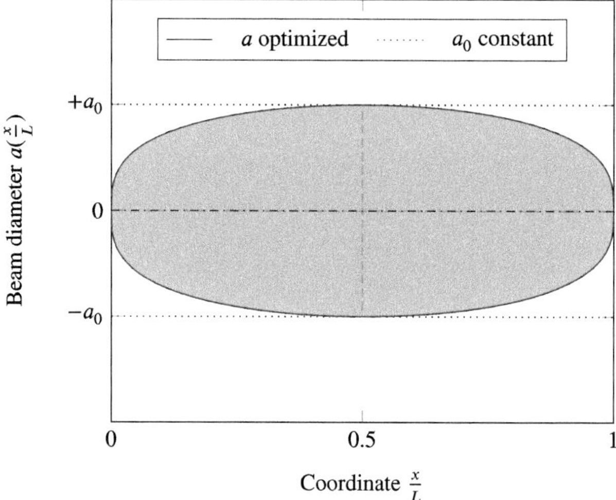

Fig. 9.17 Beam diameter along the beam axis

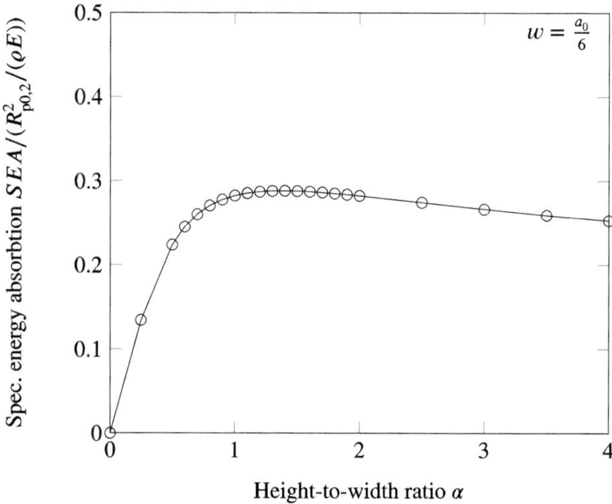

Fig. 9.18 Specific energy absorbtion for different values of α

4.4.14 Optimization of a beam along the longitudinal axis under the influence of two different single forces

- Moment distribution:

$$M_{y,\mathrm{I}}(x) = \frac{3F_0 L}{4}\left(1 - \frac{2x}{3L}\right) \quad \text{for} \quad 0 \le x \le \frac{L}{2}, \tag{9.263}$$

$$M_{y,\mathrm{II}}(x) = F_0 L\left(1 - \frac{x}{L}\right) \quad \text{for} \quad \frac{L}{2} \le x \le L. \tag{9.264}$$

- Optimized diameter (see also Fig. 9.19):

$$d_{\mathrm{I}}(x) = d_0 \times \sqrt[3]{1 - \frac{2x}{3L}} \quad \text{for} \quad 0 \le x \le \frac{L}{2}, \tag{9.265}$$

$$d_{\mathrm{II}}(x) = \sqrt[3]{\frac{4}{3}} \times d_0 \times \sqrt[3]{1 - \frac{x}{L}} \quad \text{for} \quad \frac{L}{2} \le x \le L, \tag{9.266}$$

where the reference diameter at the fixed support is as follows:

$$d_0 = \sqrt[3]{\frac{\frac{3F_0 L}{4}}{\frac{272}{1875}\,R_{\mathrm{p0.2}}}}. \tag{9.267}$$

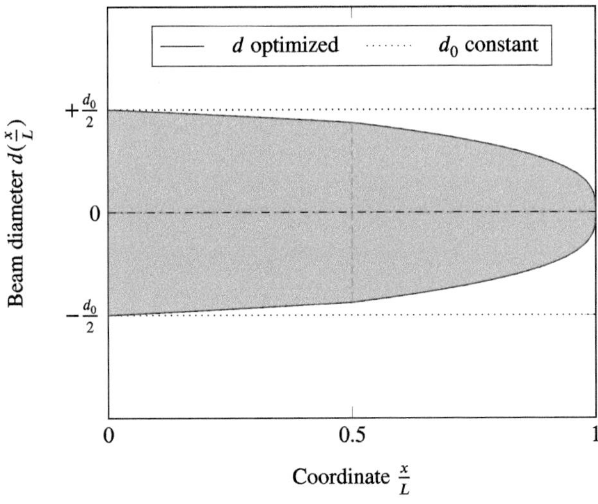

Fig. 9.19 Beam diameter (outer) along the beam axis

- Total strain energy:

$$\Pi = \int\limits_0^L \frac{(M_y(x))^2}{2EI}\,\mathrm{d}x = \int\limits_0^{\frac{L}{2}} \frac{M_{y,\mathrm{I}}(x)^2}{2EI_{\mathrm{I}}}\,\mathrm{d}x + \int\limits_{\frac{L}{2}}^{L} \frac{(M_{y,\mathrm{II}}(x))^2}{2EI_{\mathrm{II}}}\,\mathrm{d}x \tag{9.268}$$

$$= \frac{16{,}875 \times F_0^2 L^2 \left(\dfrac{9L}{10} - \dfrac{2^{\frac{2}{3}} 3^{\frac{1}{3}} L}{5}\right)}{4352 \times E d_0^4} + \frac{125 \times 3^{\frac{10}{3}} F_0^2 L^3}{17 \times 2^{\frac{5}{3}} 4^{\frac{10}{3}} E d_0^4} \tag{9.269}$$

$$\approx 2.602043 \times \frac{F_0^2 L^3}{E d_0^4}. \tag{9.270}$$

- Total mass:

$$m = \int\limits_0^L \varrho A(x)\mathrm{d}x = \int\limits_0^{\frac{L}{2}} \varrho A_{\mathrm{I}}(x)\mathrm{d}x + \int\limits_{\frac{L}{2}}^{L} \varrho A_{\mathrm{II}}(x)\mathrm{d}x \tag{9.271}$$

$$= \frac{16}{25} \times \left(\frac{9L}{10} - \frac{2^{\frac{2}{3}} 3^{\frac{1}{3}} L}{5}\right) d_0^2 \varrho + \frac{3^{\frac{1}{3}} 4^{\frac{8}{3}} L d_0^2 \varrho}{125 \times 2^{\frac{5}{3}}} \tag{9.272}$$

$$\approx 0.429477 \times L\, d_0^{\,2} \varrho. \tag{9.273}$$

- Specific energy absorption:

$$SEA = \frac{\Pi}{m} = 6.058637 \times \frac{F_0^2 L^2}{E d_0^6 \varrho}. \tag{9.274}$$

4.4.15 Optimization of a beam along the longitudinal axis under the influence of a single force and a single moment

- Geometric dimensions of the cross section:

$$I(x) = \frac{61a(x)^4}{972}, \tag{9.275}$$

$$A(x) = \frac{4a(x)^2}{9}. \tag{9.276}$$

- Beam contour $a = a(x)$ along the beam axis (see Fig. 9.20):

$$a(x) = \left(\frac{6}{7}\right)^{\frac{1}{3}} a_0 \left(\frac{7}{6} - \frac{x}{L}\right)^{\frac{1}{3}} \approx 0.9499 \times a_0 \left(\frac{7}{6} - \frac{x}{L}\right)^{\frac{1}{3}} , \qquad (9.277)$$

where the following dimension results for the reference cross-section:

$$a_0 = \left(\frac{567 F_0 L}{61 R_{\mathrm{p}0.2}}\right)^{\frac{1}{3}} . \qquad (9.278)$$

- Constant cross section:

$$\Pi_0 = \frac{513}{122} \times \frac{F_0^2 L^3}{E a_0^4} \approx 4.2049 \times \frac{F_0^2 L^3}{E a_0^4} \approx 0.0487 \times \frac{R_{\mathrm{p}0.2}^2 L a_0^2}{E} , \qquad (9.279)$$

$$SEA_0 \approx 0.1095 \times \frac{R_{\mathrm{p}0.2}^2}{E \varrho} . \qquad (9.280)$$

- Optimized cross section:

$$\Pi \approx 7.2946 \times \frac{F_0^2 L^3}{E a_0^4} \approx 0.0844 \times \frac{R_{\mathrm{p}0.2}^2 L a_0^2}{E} , \qquad (9.281)$$

$$SEA \approx 0.2824 \times \frac{R_{\mathrm{p}0.2}^2}{E \varrho} . \qquad (9.282)$$

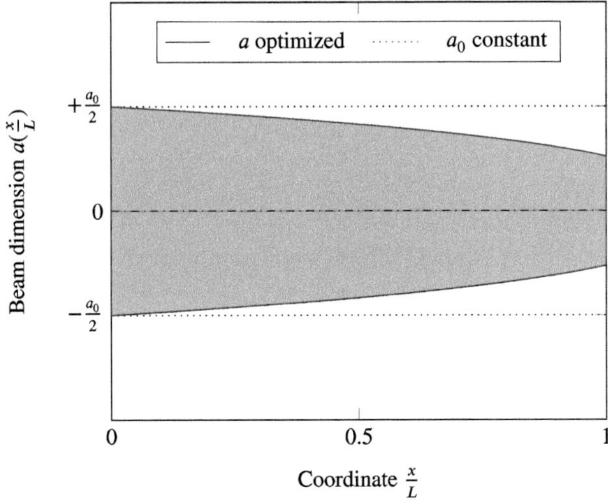

Fig. 9.20 Beam dimension along the beam axis

- Ratio:

$$\frac{SEA}{SEA_0} = \frac{0.282407}{0.109505} \approx 2.5789 \,. \tag{9.283}$$

4.4.16 Optimization of a cantilever beam along the longitudinal axis under the influence of a constant distributed load

- Moment distribution:

$$M_y(x) = \frac{q_0 L^2}{2}\left(1 - \frac{x}{L}\right)^2 \,. \tag{9.284}$$

- Constant cross section:

$$A_0 = \frac{8a_0{}^2}{9}\,, \tag{9.285}$$

$$I_0 = \frac{199a_0{}^4}{486}\,, \tag{9.286}$$

$$m_0 = \frac{8\,L\,a_0{}^2\varrho}{9}\,, \tag{9.287}$$

$$\Pi_0 = \frac{243L^5\,q_0{}^2}{3980E\,a_0{}^4}\,, \tag{9.288}$$

$$SEA_0 = \frac{2187L^4\,q_0{}^2}{31{,}840E\,a_0{}^6\varrho}\,. \tag{9.289}$$

- Optimized cross section: (see Fig. 9.21):

$$a(x) = a_0\left(1 - \frac{x}{L}\right)^{\frac{2}{3}}\,, \tag{9.290}$$

$$A(x) = \frac{8}{9}\mathrm{a}(x)^2\,, \tag{9.291}$$

$$I(x) = \frac{199}{486}\mathrm{a}(x)^4\,, \tag{9.292}$$

$$m = \frac{8\,L\,a_0{}^2\varrho}{21}\,, \tag{9.293}$$

$$\Pi = \frac{729L^5\,q_0{}^2}{5572E\,a_0{}^4}\,, \tag{9.294}$$

$$SEA = \frac{2187L^4\,q_0{}^2}{6368E\,a_0{}^6\varrho}\,. \tag{9.295}$$

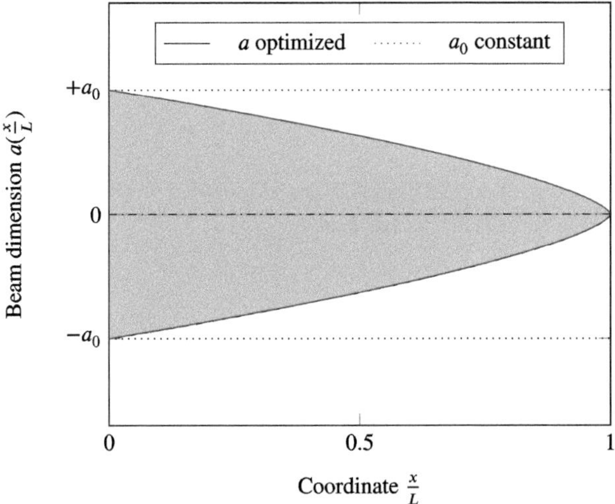

Fig. 9.21 Beam dimension along the beam axis

- Ratios:

$$\frac{m}{m_0} = \frac{3}{7}, \tag{9.296}$$

$$\frac{SEA}{SEA_0} = 5. \tag{9.297}$$

In order for the 'optimized' beam to have the same cross-section at every point x, there should be a constant bending moment distribution. This could be achieved by a single moment at the free end.

4.4.17 Optimization of a simply supported beam under constant distributed load: box cross section

- Moment distribution:

$$M_y(x) = \frac{q_0 L^2}{2} \left(\left[\frac{x}{L}\right]^2 - \left[\frac{x}{L}\right]^1 \right) < 0. \tag{9.298}$$

- Beam contour $a = a(x)$ along the beam axis (see Fig. 9.22):

$$a(x) = (4)^{\frac{1}{3}} a_0 \left(-\left(\frac{x}{L}\right)^2 + \frac{x}{L} \right)^{\frac{1}{3}} \approx 1.587 \times a_0 \left(-\left(\frac{x}{L}\right)^2 + \frac{x}{L} \right)^{\frac{1}{3}} > 0, \tag{9.299}$$

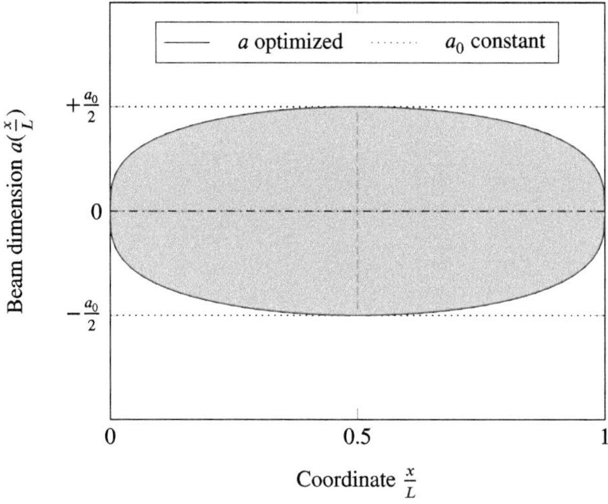

Fig. 9.22 Beam dimension a along the beam axis

where the reference cross section has the following dimensions:

$$a_0 = \left(\frac{486 q_0 L^2}{4 \times 73 R_{\mathrm{p0.2}}} \right)^{\frac{1}{3}}.$$ (9.300)

4.4.18 Optimization of a simply supported bending under constant distributed load: box cross section. Numerical calculation of the specific energy absorption

- Constant cross section:

$$A_0 = \frac{7 a_0{}^2}{18},$$ (9.301)

$$I_0 = \frac{73 a_0{}^4}{1944},$$ (9.302)

$$m_0 = \frac{8 L a_0{}^2 \varrho}{9},$$ (9.303)

$$\Pi_0 = \frac{81 L^5 q_0{}^2}{730 E a_0{}^4},$$ (9.304)

$$SEA_0 = \frac{729 L^4 q_0{}^2}{2555 E a_0{}^6 \varrho}.$$ (9.305)

- Optimized cross section (see Fig. 9.22)
  Numerical integration of the integral yields ($n = 4112$):

$$\int_0^L \left(-\left(\frac{x}{L}\right)^2 + \frac{x}{L} \right)^{\frac{2}{3}} dx \approx 0.2933415685302212 \times L. \tag{9.306}$$

$$a(x) = (4)^{\frac{1}{3}}\, a_0 \left(-\left(\frac{x}{L}\right)^2 + \frac{x}{L} \right)^{\frac{1}{3}}, \tag{9.307}$$

$$A(x) = \frac{7}{18} a(x)^2, \tag{9.308}$$

$$I(x) = \frac{73}{1944} a(x)^4, \tag{9.309}$$

$$V = \frac{7}{18} \times 4^{\frac{2}{3}} \times a_0^2 \times 0.2933415685302212\, L, \tag{9.310}$$

$$m = \frac{7}{18} \times 4^{\frac{2}{3}} \times a_0^2 \times \varrho \times 0.2933415685302212\, L, \tag{9.311}$$

$$\Pi = \frac{0.1537837221422918 L^5 q_0^2}{E\, a_0^4}, \tag{9.312}$$

$$SEA = \frac{0.5349804305283756 L^4 q_0^2}{E\, a_0^6 \varrho}. \tag{9.313}$$

- Ratios:

$$\frac{m}{m_0} = 0.739, \tag{9.314}$$

$$\frac{SEA}{SEA_0} = 1.875. \tag{9.315}$$

4.4.19 Optimization of a simply supported beam under the influence of a single moment and a single force

- Moment distribution:

$$M_y(x) = -\frac{F_0 L}{2}\left(1 + \frac{x}{L} \right) \quad \text{for} \ \ 0 \leq x \leq \frac{L}{2}. \tag{9.316}$$

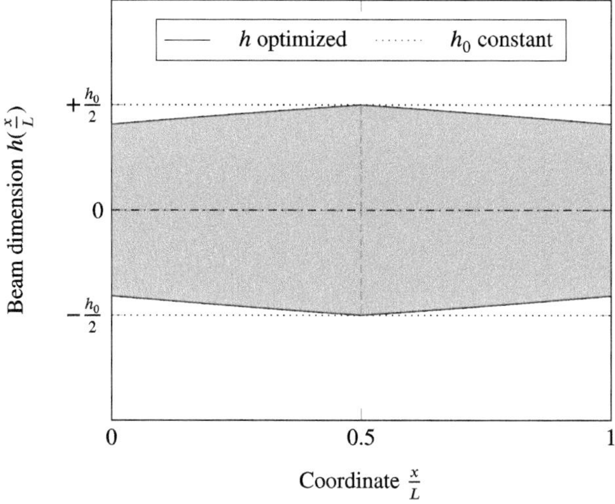

Fig. 9.23 Beam dimension h along the beam axis

- Beam height $h = h(x)$ along the beam axis (see Fig. 9.23):

$$h(x) = \left(\frac{2}{3}\right)^{\frac{1}{2}} h_0 \left(1 + \frac{x}{L}\right)^{\frac{1}{2}}, \tag{9.317}$$

where the following dimension results for the reference cross-section:

$$h_0^2 = \frac{9 F_0 L}{2b \times R_{\text{p0.2}}}. \tag{9.318}$$

- Constant cross section:

$$\Pi_0 = \frac{27}{8} \frac{F_0^2 L^3}{Ebh_0^3} = \frac{19}{162} \frac{Lh_0 b R_{\text{p0.2}}^2}{E}, \tag{9.319}$$

$$SEA_0 = \frac{19}{162} \frac{R_{\text{p0.2}}^2}{\varrho E}. \tag{9.320}$$

- Optimized cross section:

$$V = 2 \left(\frac{2}{3}\right)^{(1/2)} \left(\frac{3^{(1/2)}}{2^{(1/2)}} - \frac{2}{3}\right) Lbh_0, \tag{9.321}$$

$$\Pi = \frac{3}{\left(\frac{2}{3}\right)^{(3/2)}} \left(\frac{3^{(1/2)}}{2^{(1/2)}} - \frac{2}{3}\right) \frac{4}{81} \frac{Lh_0 b R_{\text{p0.2}}^2}{E}, \tag{9.322}$$

$$SEA = \frac{1}{6} \frac{R_{\text{p0.2}}^2}{\varrho E}. \tag{9.323}$$

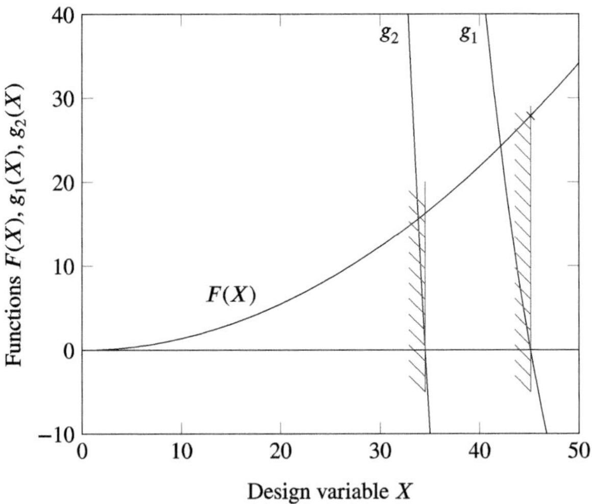

Fig. 9.24 Optimal design of a cantilever beam with: $L = 2540\,\mathrm{mm}$, $F_0 = 2667\,\mathrm{N}$, $E = 68{,}948\,\mathrm{MPa}$, $R_{\mathrm{p}0.2} = 247\,\mathrm{MPa}$, $\varrho = 2.691 \times 10^{-6}\,\mathrm{kg/mm^3}$, $r_1 = 0.03$ (original set of Eqs. (9.324)–(9.326)). Exact solution for the minimum: $X_{\mathrm{extr}} = 45.160\,\mathrm{mm}$

4.4.20 Optimal design of a cantilever beam (classical optimization problem))

The objective function of the optimization problem, i.e. the mass of the beam, can be represented as a function of the design variable $X = a$ as follows:

$$F(X) = m(X) = 2\varrho L X^2 , \tag{9.324}$$

where the minimization is to be carried out taking into account the following inequality constraints (see Fig. 9.24), see [4]:

$$g_1(X) = \frac{F_0 L^3}{2EX^4} - r_1 L \leq 0 \quad \text{(displacement)} , \tag{9.325}$$

$$g_2(X) = \frac{3F_0 L}{2X^3} - R_{\mathrm{p}0,2} \leq 0 \quad \text{(stress)} . \tag{9.326}$$

Analytical solution based on the graphical representation in Fig. 9.24:

$$X_{\mathrm{extr}} = \left(\frac{F_0 L^2}{2E r_1} \right)^{\frac{1}{4}} = 45.16\,\mathrm{mm} . \tag{9.327}$$

As an alternative to Eqs. (9.324)–(9.326), one can also specify a normalized representation, i.e. without a specific unit. By introducing the normalized geometric dimension $a^n = a/L = X^n$, one arrives at the corresponding equations of the optimization problem:

$$F^n(X^n) = F^n(a^n) = \frac{m}{\varrho \times L^3} = 2 \times \left(a^n\right)^2 = 2 \times \left(X^n\right)^2 , \tag{9.328}$$

$$g_1^n(X^n) = g_1^n(a^n) = \frac{F_0}{2EL^2 r_1 (a^n)^4} - 1 = \frac{F_0}{2EL^2 r_1 (X^n)^4} - 1 \le 0, \tag{9.329}$$

$$g_2^n(X^n) = g_2^n(a^n) = \frac{3F_0}{2L^2 R_{p0.2}(a^n)^3} - 1 = \frac{3F_0}{2L^2 R_{p0.2}(X^n)^3} - 1 \le 0. \tag{9.330}$$

The graphical representation of these normalized equations is shown in Fig. 9.25a. It can be seen that both inequality constraints again have a single root in the considered region of the abscissa. However, both slopes are extremely steep. Thus, a scaling factor β can be introduced to modify the inequality restrictions as follows:

$$g_1^n(X^n)/\beta = \frac{1}{\beta}\left(\frac{F_0}{2EL^2 r_1(X^n)^4} - 1 \le 0\right) , \tag{9.331}$$

$$g_2^n(X^n)/\beta = \frac{1}{\beta}\left(\frac{3F_0}{2L^2 R_{p0.2}(X^n)^3} - 1 \le 0\right) . \tag{9.332}$$

It is obvious that the scaling does not change the position of the roots and a similar course of all curves is maintained, see Fig. 9.25b.

4.4.21 Optimal design of a compression strut (classical optimization problem)
The objective function of the optimization problem, i.e. the mass of the compression strut, can be represented as a function of the design variable $X = a$ as follows:

$$F(X) = m(X) = 2\varrho L X^2 , \tag{9.333}$$

where the minimization is to be carried out taking into account the following inequality constraints (see Fig. 9.26), see [4]:

$$g_1(X) = \frac{F_0}{2X^2} - R_{p0.2} \le 0 \quad \text{(stress)}, \tag{9.334}$$

$$g_2(X) = F_0 - \frac{\pi^2 E X^4}{24L^2} \le 0 \quad \text{(buckling)}. \tag{9.335}$$

4.4.22 Optimal design of a short cantilever beam (classical optimization problem)
The objective function of the optimization problem, i.e. the mass of the beam, can be represented as a function of the design variable $X = a$ as follows:

$$F(X) = m(X) = 2\varrho L X^2 , \tag{9.336}$$

(a)

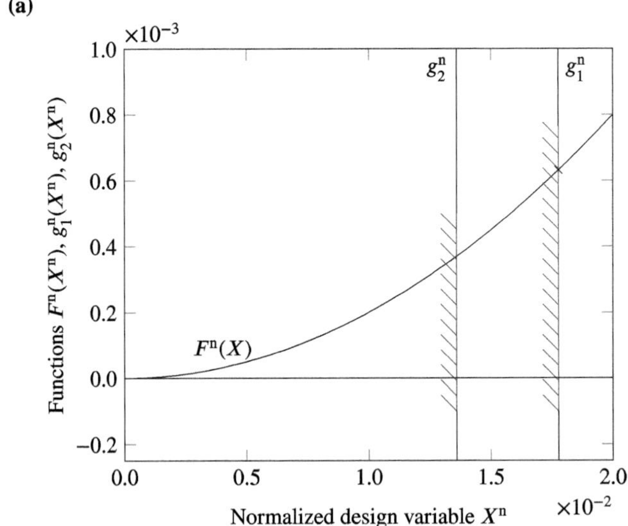

(b)

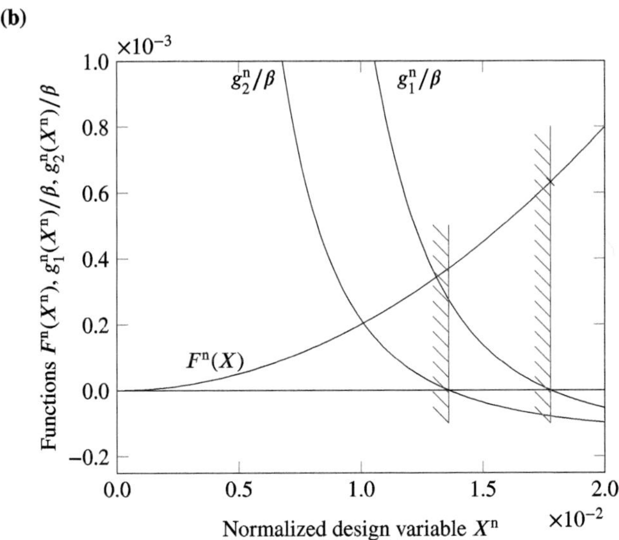

Fig. 9.25 Optimal design of a cantilever beam: **a** normalized equations and **b** normalized equations with scaled inequality constraints ($\beta = 7000$). Exact solution for the minimum: $X_{\text{extr}}^n = 0.01778$

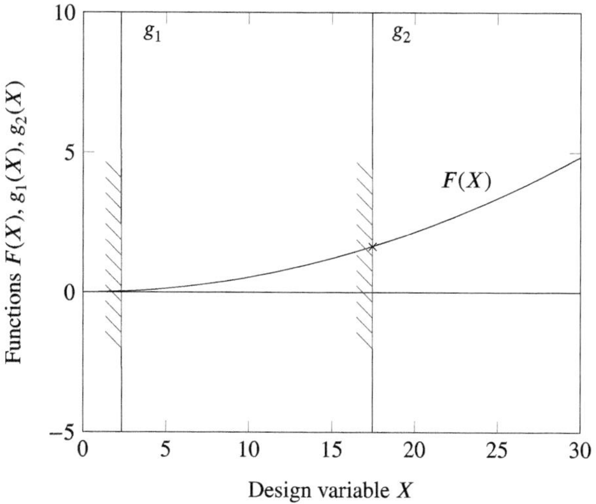

Fig. 9.26 Optimal design of a compression strut: $L = 1000\,\text{mm}$, $F_0 = 2667\,\text{N}$, $E = 70{,}000\,\text{MPa}$, $R_{\text{p0.2}} = 247\,\text{MPa}$, $\varrho = 2.691 \times 10^{-6}\,\text{kg/mm}^3$ (original set of Eqs. (9.333)–(9.335)). Exact solution for the minimum: $X_{\text{extr}} = 17.447\,\text{mm}$

where the minimization is to be carried out taking into account the following inequality constraints (see Fig. 9.27), see [4]:

$$g_1(X) = \frac{3F_0 L}{2X^3} - R_{\text{p0.2}} \le 0 \quad \text{(normal stress)}, \tag{9.337}$$

$$g_2(X) = \frac{3F_0}{4X^2} - \frac{R_{\text{p0.2}}}{2} \le 0 \quad \text{(shear stress)}. \tag{9.338}$$

4.4.23 Optimal design of a cantilever beam: constant rectangular cross section (classical optimization problem)

The objective function of the optimization problem, i.e. the mass of the beam, can be represented as a function of the two design variables $X_1 = a$ aund $X_2 = b$ as follows:

$$F(X_1, X_2) = m(X_1, X_2) = \varrho V = \varrho L X_1 X_2, \tag{9.339}$$

where the minimization is to be carried out taking into account the following three inequality constraints (see Fig. 9.28), see [4]:

$$g_1(X_1, X_2) = \frac{4F_0 L^3}{E X_1 X_2^3} - r_1 L \le 0 \quad \text{(displacement)}, \tag{9.340}$$

$$g_2(X_1, X_2) = \frac{6F_0 L}{X_1 X_2^2} - R_{\text{p0.2}} \le 0 \quad \text{(normal stress)}, \tag{9.341}$$

$$g_3(X_1, X_2) = X_2 - 20X_1 \le 0 \quad \text{(height-to-width ratio)}. \tag{9.342}$$

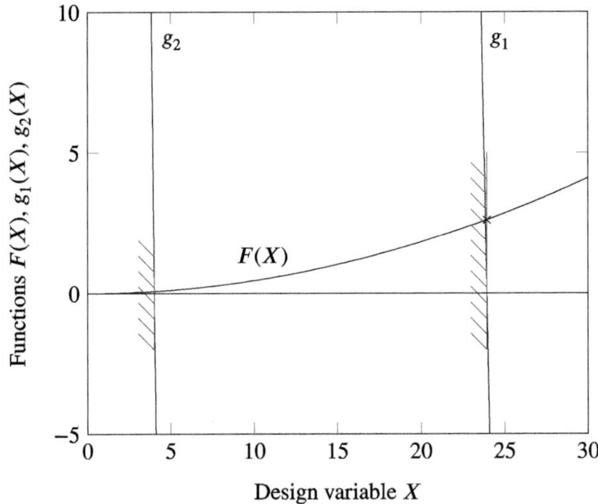

Fig. 9.27 Optimal design of a short cantilever beam: $L = 846.67\,\text{mm}$, $F_0 = 2667\,\text{N}$, $E = 68{,}948\,\text{MPa}$, $R_{\text{p0.2}} = 247\,\text{MPa}$, $\varrho = 2.691 \times 10^{-6}\,\text{kg/mm}^3$, (original set of Eqs. (9.336)–(9.338)). Exact solution for the minimum: $X_{\text{extr}} = 23.936\,\text{mm}$

From Fig. 9.28 one can see that the minimum—taking into account the given inequality constraints—is the intersection point of the curves g_1 and g_3. A simple calculation gives the analytical solution to:

$$X_{1,\text{extr}} = (20)^{-\frac{3}{4}} \times \left(\frac{4F_0L^2}{Er_1}\right)^{\frac{1}{4}} = 8.031\,\text{mm}\,, \tag{9.343}$$

$$X_{2,\text{extr}} = 20 \times X_{1,\text{extr}} = 160.614\,\text{mm}\,. \tag{9.344}$$

The result of problem 4.4.20, i.e. $a = 45.160$ and $2a = 90.3$, is marked by the $\star$ marker in Fig. 9.28. For the given geometric dimensions this results in a total mass of $F(a, 2a) = m(a, 2a) = 27.9$ kg, while the optimization for a variable width and height results in $F(X_{1,\text{extr}}, X_{2,\text{extr}}) = m(X_{1,\text{extr}}, X_{2,\text{extr}}) = 8.8$ kg.

4.4.24 Optimal design of a simply supported beam: constant rectangular cross section (classical optimization problem)

The objective function of the optimization problem, i.e. the mass of the beam, can be represented as a function of the two design variables $X_1 = a$ and $X_2 = b$ as follows:

$$F(X_1, X_2) = m(X_1, X_2) = \varrho V = \varrho L X_1 X_2\,, \tag{9.345}$$

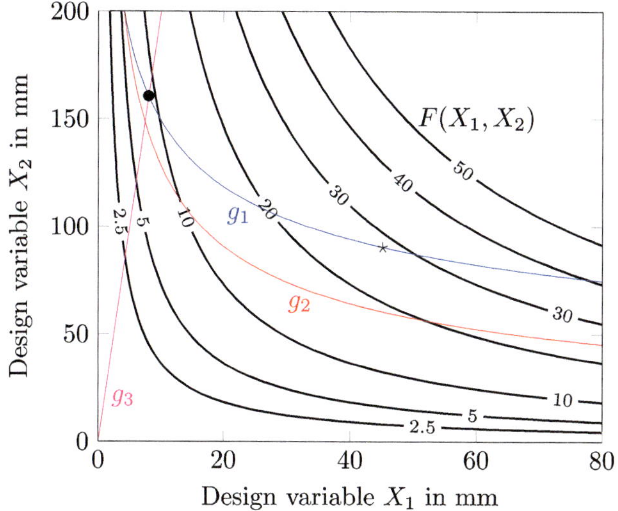

Fig. 9.28 Optimal design of a cantilever beam with: $L = 2540\,\text{mm}$, $F_0 = 2667\,\text{N}$, $E = 68{,}948\,\text{MPa}$, $R_{\text{p0.2}} = 247\,\text{MPa}$, $\varrho = 2.691 \times 10^{-6}\,\text{kg/mm}^3$, $r_1 = 0.03$ (original set of Eqs. (9.339)–(9.342)). Exact solution for the minimum: $X_{1,\text{extr}} = 8.031\,\text{mm}$, $X_{2,\text{extr}} = 160.614\,\text{mm}$ (indicated by the ● marker)

where the minimization is to be carried out taking into account the following four inequality constraints (see Fig. 9.29), see [4]:

$$g_1(X_1, X_2) = \frac{3F_0 L}{2X_1 X_2^2} - R_{\text{p0.2}} \leq 0 \quad \text{(normal stress)}, \tag{9.346}$$

$$g_2(X_1, X_2) = \frac{3F_0}{4X_1 X_2} - \frac{R_{\text{p0.2}}}{2} \leq 0 \quad \text{(shear stress)}, \tag{9.347}$$

$$g_3(X_1, X_2) = \frac{F_0 L^3}{4E X_1 X_2^3} - r_1 L \leq 0 \quad \text{(displacement)}, \tag{9.348}$$

$$g_4(X_1, X_2) = X_2 - 20X_1 \leq 0 \quad \text{(height-to-width ratio)}. \tag{9.349}$$

From Fig. 9.29 one can see that the minimum—taking into account the given inequality constraints—is the intersection of the curves g_1 and g_4 for the case $r_1 = 0.03$ and the intersection of the curves g_3 and g_4 for the case $r_1 = 0.003$. A simple calculation gives the analytical solution to:

$$X_{1,\text{extr}}\big|_{r_1=0.03} = (20)^{-\frac{2}{3}} \times \left(\frac{3F_0 L}{2R_{\text{p0.2}}}\right)^{\frac{1}{3}} = 4.685\,\text{mm}, \tag{9.350}$$

$$X_{2,\text{extr}}\big|_{r_1=0.03} = 20 \times X_{1,\text{extr}} = 93.704\,\text{mm}, \tag{9.351}$$

(a)

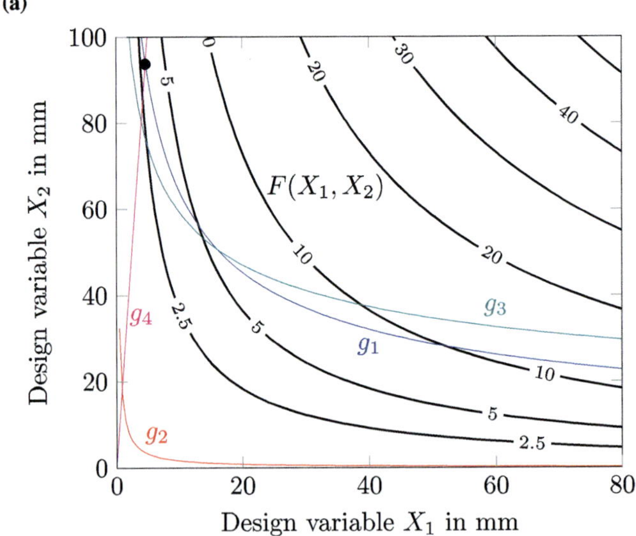

(b)

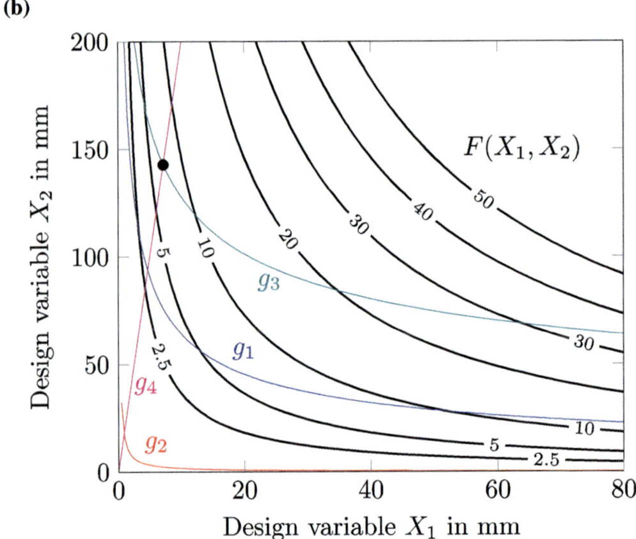

Fig. 9.29 Optimal design of a simply supported beam with $L = 2540\,\text{mm}$, $F_0 = 2667\,\text{N}$, $E = 68{,}948\,\text{MPa}$, $R_{p0.2} = 247\,\text{MPa}$, $\varrho = 2.691 \times 10^{-6}\,\text{kg/mm}^3$ (original set of Eqs. (9.345)–(9.349)): **a** $r_1 = 0.03$, exact solution for the minimum: $X_{1,\text{extr}} = 4.685\,\text{mm}$, $X_{2,\text{extr}} = 93.704\,\text{mm}$ (indicated by the ● marker), **b** $r_1 = 0.003$, exact solution for the minimum: $X_{1,\text{extr}} = 7.140\,\text{mm}$, $X_{2,\text{extr}} = 142.809\,\text{mm}$ (indicated by the ● marker)

or

$$X_{1,\text{extr}}\big|_{r_1=0.003} = (20)^{-\frac{3}{4}} \times \left(\frac{F_0 L^2}{4 E r_1}\right)^{\frac{1}{4}} = 7.140\,\text{mm}\,, \tag{9.352}$$

$$X_{2,\text{extr}}\big|_{r_1=0.003} = 20 \times X_{1,\text{extr}} = 142.809\,\text{mm}\,. \tag{9.353}$$

4.4.25 Optimal design of a compression strut: constant rectangular cross section (classical optimization problem)

The objective function of the optimization problem, i.e. the mass of the compression strut, can be represented as a function of the two design variables $X_1 = a$ and $X_2 = b$ as follows:

$$F(X_1, X_2) = m(X_1, X_2) = \varrho V = \varrho L X_1 X_2\,, \tag{9.354}$$

where the minimization is to be carried out taking into account the following three inequality constraints (see Fig. 9.30), see [4]:

$$g_1(X_1, X_2) = \frac{F_0}{X_1 X_2} - R_{\text{p}0.2} \leq 0 \quad \text{(normal stress)}\,, \tag{9.355}$$

$$g_2(X_1, X_2) = F_0 - \frac{\pi^2 E X_1 X_2^3}{48 L^2} \leq 0 \quad \text{(buckling around the y-axis)}\,, \tag{9.356}$$

$$g_3(X_1, X_2) = F_0 - \frac{\pi^2 E X_2 X_1^3}{48 L^2} \leq 0 \quad \text{(buckling around the z-axis)}\,. \tag{9.357}$$

From Fig. 9.30 one can see that the minimum—taking into account the given inequality constraints—is the intersection point of the curves g_2 and g_3. A simple calculation gives the analytical solution:

$$X_{1,\text{extr}} = \left(\frac{48 F_0 L^2}{\pi^2 E}\right)^{\frac{1}{4}} = 20.748\,\text{mm}\,, \tag{9.358}$$

$$X_{2,\text{extr}} = \frac{48 F_0 L^2}{\pi^2 E X_{1,\text{extr}}^3} = 20.748\,\text{mm}\,. \tag{9.359}$$

The result of problem 4.4.21, i.e. $a = 17.447$ and $2a = 34.894$, is marked by the $\star$ marker in Fig. 9.30. For the given geometric dimensions this results in a total mass of $F(a, 2a) = m(a, 2a) = 1.638$ kg, while the optimization for a variable width and height results in $F(X_{1,\text{extr}}, X_{2,\text{extr}}) = m(X_{1,\text{extr}}, X_{2,\text{extr}}) = 1.158$ kg.

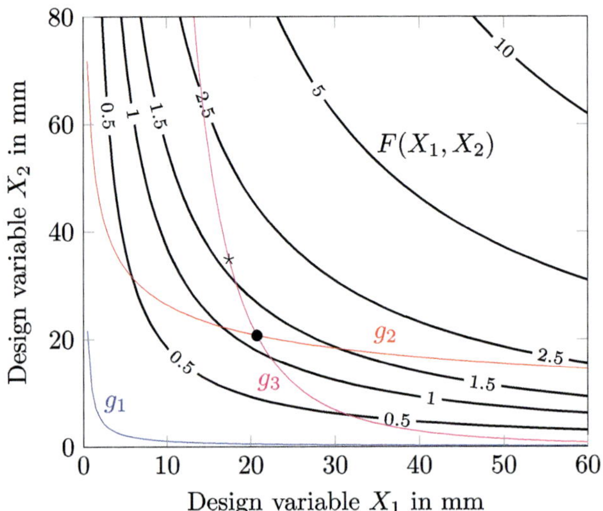

Fig. 9.30 Optimal design of a compression strut: $L = 1000\,\text{mm}$, $F_0 = 2667\,\text{N}$, $E = 70{,}000\,\text{MPa}$, $R_{\text{p0.2}} = 247\,\text{MPa}$, $\varrho = 2.691 \times 10^{-6}\,\text{kg/mm}^3$, (original set of Eqs. (9.354)–(9.357)). Exact solution for the minimum: $X_{1,\text{extr}} = 20.748\,\text{mm}$, $X_{2,\text{extr}} = 20.748\,\text{mm}$ (indicated by the • marker)

4.4.26 Optimal design of a short cantilever beam: constant rectangular cross section (classical optimization problem)

The objective function of the optimization problem, i.e. the mass of the beam, can be represented as a function of the two design variables $X_1 = a$ and $X_2 = b$ as follows:

$$F(X_1, X_2) = m(X_1, X_2) = \varrho V = \varrho L X_1 X_2 \,, \tag{9.360}$$

where the minimization is to be carried out taking into account the following three inequality constraints (see Fig. 9.31), see [4]:

$$g_1(X_1, X_2) = \frac{6 F_0 L}{X_1 X_2^2} - R_{\text{p0,2}} \leq 0 \quad \text{(normal stress)} \,, \tag{9.361}$$

$$g_2(X_1, X_2) = \frac{3 F_0}{2 X_1 X_2} - \frac{R_{\text{p0,2}}}{2} \leq 0 \quad \text{(shear stress)} \,, \tag{9.362}$$

$$g_3(X_1, X_2) = X_2 - 20 X_1 \leq 0 \quad \text{(height-to-width ratio)} \,. \tag{9.363}$$

From Fig. 9.31 one can see that the minimum—taking into account the given inequality constraints—is the intersection point of the curves g_1 and g_3. A simple calculation gives the analytical solution:

$$X_{1,\text{extr}} = 20^{-\frac{2}{3}} \times \left(\frac{6 F_0 L}{R_{\text{p0.2}}} \right)^{\frac{1}{3}} = 5.157\,\text{mm} \,, \tag{9.364}$$

$$X_{2,\text{extr}} = 20 \times X_{1,\text{extr}} = 103.135\,\text{mm} \,. \tag{9.365}$$

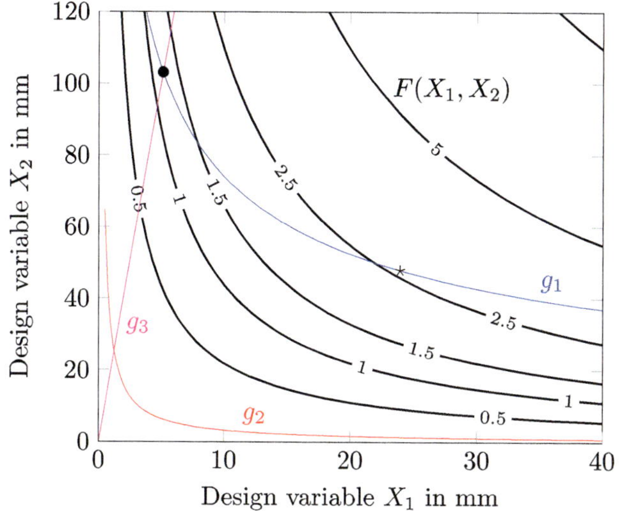

Fig. 9.31 Optimal design of a short cantilever beam: $L = 846.67\,\text{mm}$, $F_0 = 2667\,\text{N}$, $E = 68{,}948\,\text{MPa}$, $R_{\text{p0.2}} = 247\,\text{MPa}$, $\varrho = 2.691 \times 10^{-6}\,\text{kg/mm}^3$, (original set of Eqs. (9.360)–(9.363)). Exact solution for the minimum: $X_{1,\text{extr}} = 5.157\,\text{mm}$, $X_{2,\text{extr}} = 103.135\,\text{mm}$ (indicated by the ● marker)

The result of problem 4.4.22, i.e. $a = 23.936$ and $2a = 47.872$, is marked by the ★ marker in Fig. 9.31. For the given geometric dimensions this results in a total mass of $F(a, 2a) = m(a, 2a) = 2.611$ kg, while the optimization for a variable width and height results in $F(X_{1,\text{extr}}, X_{2,\text{extr}}) = m(X_{1,\text{extr}}, X_{2,\text{extr}}) = 1.212$ kg.

4.4.27 Optimal design of a stepped cantilever beam with two sections (classical optimization problem)

The problem requires the calculation of both the deformations and the stress distributions. This task can be approached by a 'hand calculation' based on the finite element method. A single stiffness matrix of an Euler-Bernoulli beam element can be given as follows (see [5,6]):

$$
K_i^{\text{e}} = \frac{E_i I_i}{L_i^3}
\begin{bmatrix}
12 & -6L_i & -12 & -6L_i \\
-6L_i & 4L_i^2 & 6L_i & 2L_i^2 \\
-12 & 6L_i & 12 & 6L_i \\
-6L_i & 2L_i^2 & 6L_i & 4L_i^2
\end{bmatrix} .
\tag{9.366}
$$

The assembly of the two element stiffness matrices K_i^{e}, taking into account $E_{\text{I}} = E_{\text{II}} = E$, $L_{\text{I}} = L_{\text{II}} = \frac{L}{2}$, $I_{\text{I}} = \frac{1}{12}ab_{\text{I}}^3$ and $I_{\text{II}} = \frac{1}{12}ab_{\text{II}}^3$, results in the following global (reduced) system of equations:

$$
E \begin{bmatrix}
\left(\dfrac{8a\,b_{\mathrm{II}}^3}{L^3}+\dfrac{8a\,b_{\mathrm{I}}^3}{L^3}\right) & \left(\dfrac{2a\,b_{\mathrm{I}}^3}{L^2}-\dfrac{2a\,b_{\mathrm{II}}^3}{L^2}\right) & -\dfrac{8a\,b_{\mathrm{II}}^3}{L^3} & -\dfrac{2a\,b_{\mathrm{II}}^3}{L^2} \\[2ex]
\left(\dfrac{2a\,b_{\mathrm{I}}^3}{L^2}-\dfrac{2a\,b_{\mathrm{II}}^3}{L^2}\right) & \left(\dfrac{2a\,b_{\mathrm{II}}^3}{3L}+\dfrac{2a\,b_{\mathrm{I}}^3}{3L}\right) & \dfrac{2a\,b_{\mathrm{II}}^3}{L^2} & \dfrac{a\,b_{\mathrm{II}}^3}{3L} \\[2ex]
-\dfrac{8a\,b_{\mathrm{II}}^3}{L^3} & \dfrac{2a\,b_{\mathrm{II}}^3}{L^2} & \dfrac{8a\,b_{\mathrm{II}}^3}{L^3} & \dfrac{2a\,b_{\mathrm{II}}^3}{L^2} \\[2ex]
-\dfrac{2a\,b_{\mathrm{II}}^3}{L^2} & \dfrac{a\,b_{\mathrm{II}}^3}{3L} & \dfrac{2a\,b_{\mathrm{II}}^3}{L^2} & \dfrac{2a\,b_{\mathrm{II}}^3}{3L}
\end{bmatrix}
\begin{bmatrix} u_{2z} \\ \varphi_{2y} \\ u_{3z} \\ \varphi_{3y} \end{bmatrix}
=
\begin{bmatrix} 0 \\ 0 \\ -F_0 \\ 0 \end{bmatrix}.
$$

$$(9.367)$$

Equation (9.367) already takes into account that the two degrees of freedom (i.e. u_{1z} and φ_{1y}) at node 1 are zero. The solution of the system of equations can be obtained, for example, by inverting the global stiffness matrix and then multiplying with the right-hand side, i.e. $u = K^{-1}f$. This gives the column matrix of the unknown node values as:

$$
\begin{bmatrix} u_{2z} \\ \varphi_{2y} \\ u_{3z} \\ \varphi_{3y} \end{bmatrix}
=
\begin{bmatrix}
-\dfrac{5F_0 L^3}{4Ea\,b_{\mathrm{I}}^3} \\[2ex]
\dfrac{9F_0 L^2}{2Ea\,b_{\mathrm{I}}^3} \\[2ex]
-\dfrac{F_0 L^3\left(7b_{\mathrm{II}}^3+b_{\mathrm{I}}^3\right)}{2Ea\,b_{\mathrm{I}}^3 b_{\mathrm{II}}^3} \\[2ex]
\dfrac{3F_0 L^2\left(3b_{\mathrm{II}}^3+b_{\mathrm{I}}^3\right)}{2Ea\,b_{\mathrm{I}}^3 b_{\mathrm{II}}^3}
\end{bmatrix}.
$$

$$(9.368)$$

The bending moment and shear force distributions within a single element of length L_i can generally be expressed using (the index '1' refers to the start node while the index '2' refers to the end node of an element) the following relations:

$$
M_y^{\mathrm{e}}(x) = EI_y \left(\left[+\frac{6}{L^2}-\frac{12x}{L^3}\right] u_{1z} + \left[-\frac{4}{L}+\frac{6x}{L^2}\right] \varphi_{1y} \right.
$$
$$
\left. + \left[-\frac{6}{L^2}+\frac{12x}{L^3}\right] u_{2z} + \left[-\frac{2}{L}+\frac{6x}{L^2}\right] \varphi_{2y} \right),
$$

$$(9.369)$$

$$
Q_z^{\mathrm{e}}(x) = EI_y \left(\left[-\frac{12}{L^3}\right] u_{1z} + \left[+\frac{6}{L^2}\right] \varphi_{1y} + \left[+\frac{12}{L^3}\right] u_{2z} + \left[+\frac{6}{L^2}\right] \varphi_{2y} \right).
$$

$$(9.370)$$

Thus, the normal and shear stress distributions can generally be calculated as follows:

$$
\sigma_x^{\mathrm{e}}(x,z) = \frac{M_y^{\mathrm{e}}(x)}{I_y} \times z,
$$

$$(9.371)$$

$$
\tau_{xz}^{\mathrm{e}}(x,z) = \frac{Q_z^{\mathrm{e}}(x)}{2I_y}\left[\left(\frac{b}{2}\right)^2 - z^2\right].
$$

$$(9.372)$$

The nodal values of the internal reactions for calculating the normal and shear stresses are summarized in Table 9.3. It is noted here that in this case the nodal values based on the finite element calculation are identical to the analytical solution.

From Table 9.3 it can be concluded that the maximum normal stress in each element is reached at the left node and that the shear stress is constant in each element. Thus, the critical stresses in each element can be calculated as follows:

$$\sigma_{x,\mathrm{I}} = \frac{6F_0L}{a\,b_{\mathrm{I}}^2}, \tag{9.373}$$

$$\sigma_{x,\mathrm{II}} = \frac{3F_0L}{a\,b_{\mathrm{II}}^2}, \tag{9.374}$$

or for the shear stresses:

$$\tau_{xz,\mathrm{I}} = -\frac{3F_0}{2a\,b_{\mathrm{I}}}, \tag{9.375}$$

$$\tau_{xz,\mathrm{II}} = -\frac{3F_0}{2a\,b_{\mathrm{II}}}. \tag{9.376}$$

The objective function of the optimization problem, i.e. the mass of the stepped beam, can be represented as a function of the two design variables $X_1 = b_{\mathrm{I}}$ and $X_2 = b_{\mathrm{II}}$ as follows:

$$F\,(X_1,\,X_2) = \varrho L_{\mathrm{I}}aX_1 + \varrho L_{\mathrm{II}}aX_2 = \frac{\varrho La}{2}\,(X_1 + X_2)\,, \tag{9.377}$$

where the minimization is to be carried out under consideration of the following seven inequality constraints:

$$g_1(X_1,\,X_2) = +\frac{F_0L^3(X_1^3 + 7X_2^3)}{2EaX_1^3X_2^3 r_1 L} - 1 \le 0 \quad (\text{max. deflection})\,, \tag{9.378}$$

$$g_2(X_1,\,X_2) = \frac{6F_0L}{aX_1^2} - R_{\mathrm{p0.2}} \le 0 \quad (\text{normal stress in I})\,, \tag{9.379}$$

$$g_3(X_1,\,X_2) = \frac{3F_0L}{aX_2^2} - R_{\mathrm{p0.2}} \le 0 \quad (\text{normal stress in II})\,, \tag{9.380}$$

Table 9.3 Nodal values of the internal reactions bending moment (M_y) and shear force (Q_z) at each node

	Element I		Element II	
	Left node	Right node	Left node	Right node
M_y	F_0L	$\frac{F_0L}{2}$	$\frac{F_0L}{2}$	0
Q_z	$-F_0$	$-F_0$	$-F_0$	$-F_0$

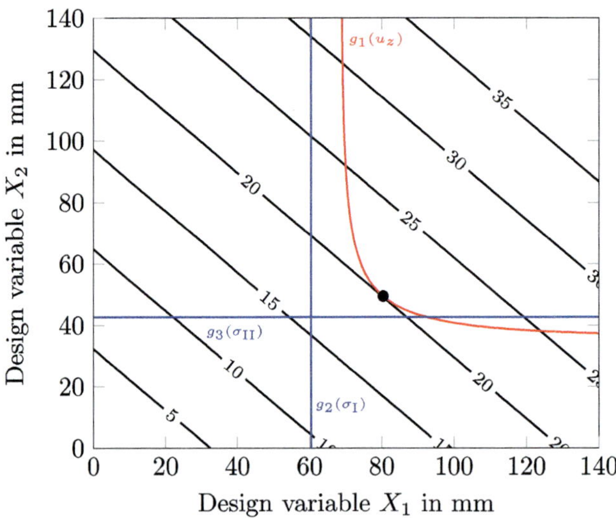

Fig. 9.32 Optimal design of a stepped cantilever beam with two sections: $L = 2540\,\text{mm}$, $a = 45.16\,\text{mm}$, $r_1 = 0.06$, $F_0 = 2667\,\text{N}$, $E = 68{,}948\,\text{MPa}$, $R_{\text{p0.2}} = 247\,\text{MPa}$, $\varrho = 2.691 \times 10^{-6}\,\text{kg/mm}^3$, (original set of Eqs. (9.377)–(9.384)). Exact solution for the minimum: $X_{1,\text{extr}} = 80.442\,\text{mm}$, $X_{2,\text{extr}} = 49.455\,\text{mm}$ (indicated by the $\bullet$ marker)

$$g_4(X_1, X_2) = \frac{3F_0}{2aX_1} - \frac{R_{\text{p0.2}}}{2} \leq 0 \quad \text{(shear stress in I)}, \tag{9.381}$$

$$g_5(X_1, X_2) = \frac{3F_0}{2aX_2} - \frac{R_{\text{p0.2}}}{2} \leq 0 \quad \text{(shear stress in II)}, \tag{9.382}$$

$$g_6(X_1, X_2) = X_1 - 20a \leq 0 \quad \text{(height-to-width ratio in I)}, \tag{9.383}$$

$$g_7(X_1, X_2) = X_2 - 20a \leq 0 \quad \text{(height-to-width ratio in II)}. \tag{9.384}$$

The graphical representation of the objective function and the inequality constraints in the X_1-X_2 design space is shown in Figs. 9.32–9.33.

From Figs. 9.32–9.33 it can be concluded that considering g_1, g_2, and g_3 can be sufficient if a reasonable range of the design variables is chosen. Furthermore, from Fig. 9.32 it can be concluded that the minimum is achieved if g_1 is tangent to the objective function F. In order to derive an analytical solution for the minimum, one must take into account that the objective function F, in the representation $X_2(X_1) = \frac{2c}{\varrho La} - X_1$, represents straight lines with the slope of -1. Thus, the condition for the minimum is that the curve g_1 assumes a slope of -1. Based on Eq. (9.379), the expression for g_1 can be rewritten:

$$X_2 = \left(-\frac{F_0 L^3 X_1^3}{7F_0 L^3 - 2Ear_1 LX_1^3} \right)^{\frac{1}{3}}, \tag{9.385}$$

(a)

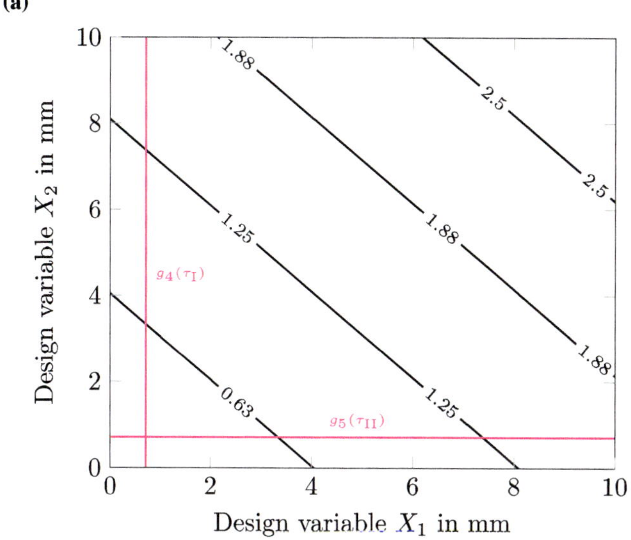

(b)

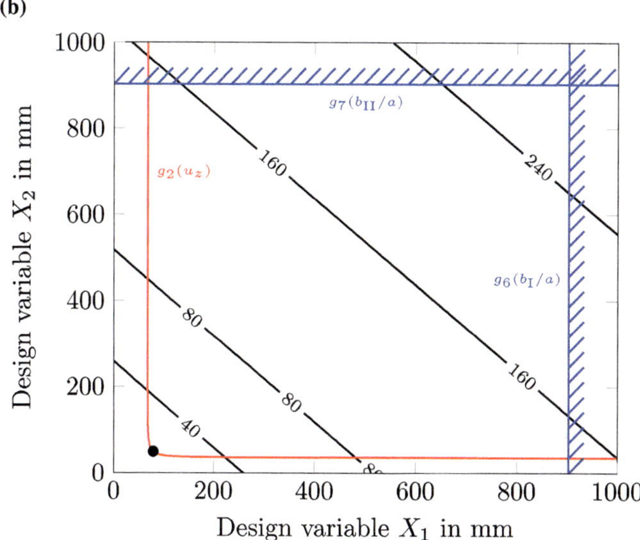

Fig. 9.33 Optimal design of a stepped cantilever beam with two sections: **a** magnification for smaller values of the design variables and **b** global view

which, after a short calculation, gives the expression of the first-order derivative as follows:

$$\left.\frac{\mathrm{d}X_2}{\mathrm{d}X_1}\right|_{g_1} = \frac{-7F_0^{\frac{4}{3}}L^{\frac{8}{3}}}{\left(7F_0L^2 - 2Ear_1X_1^3\right)^{\frac{4}{3}}} \overset{!}{=} -1. \tag{9.386}$$

The last equation can be rearranged so that one side has a root, and Newton's method can be used to determine the root. It is important that the root is sought in a reasonable interval, e.g. $70 \leq X_{1,\,\mathrm{extr}} \leq 85$.

4.4.28 Optimal design of a simply supported beam with three sections (classical optimization problem)

The problem requires the calculation of both the deformations and the stress distributions. This task can be approached by a 'hand calculation' based on the finite element method. A single stiffness matrix of an Euler-Bernoulli beam element can be given as follows (see [5,6]):

$$\boldsymbol{K}_i^{\mathrm{e}} = \frac{E_i I_i}{L_i^3} \begin{bmatrix} 12 & -6L_i & -12 & -6L_i \\ -6L_i & 4L_i^2 & 6L_i & 2L_i^2 \\ -12 & 6L_i & 12 & 6L_i \\ -6L_i & 2L_i^2 & 6L_i & 4L_i^2 \end{bmatrix}. \tag{9.387}$$

This special problem allows the consideration of symmetry in terms of geometry and loading. Thus, the reduced mechanical system according to Fig. 9.34 allows a faster calculation than when considering the entire structure.

The assembly of the two element stiffness matrices $\boldsymbol{K}_i^{\mathrm{e}}$, taking into account $E_{\mathrm{I}} = E_{\mathrm{II}} = E$, $L_{\mathrm{I}} = L_{\mathrm{II}} = \frac{L}{3}$, $I_{\mathrm{I}} = \frac{1}{12}ab_{\mathrm{I}}^3$, and $I_{\mathrm{II}} = \frac{1}{12}ab_{\mathrm{II}}^3$, results in the following global (reduced) system of equations:

$$E \begin{bmatrix} \frac{a b_{\mathrm{I}}^3}{L} & \frac{9a b_{\mathrm{I}}^3}{2L^2} & \frac{a b_{\mathrm{I}}^3}{2L} & 0 \\[6pt] \frac{9a b_{\mathrm{I}}^3}{2L^2} & \left(\frac{216a b_{\mathrm{II}}^3}{L^3} + \frac{27a b_{\mathrm{I}}^3}{L^3}\right) & \left(\frac{9a b_{\mathrm{I}}^3}{2L^2} - \frac{18a b_{\mathrm{II}}^3}{L^2}\right) & -\frac{216a b_{\mathrm{II}}^3}{L^3} \\[6pt] \frac{a b_{\mathrm{I}}^3}{2L} & \left(\frac{9a b_{\mathrm{I}}^3}{2L^2} - \frac{18a b_{\mathrm{II}}^3}{L^2}\right) & \left(\frac{2a b_{\mathrm{II}}^3}{L} + \frac{a b_{\mathrm{I}}^3}{L}\right) & \frac{18a b_{\mathrm{II}}^3}{L^2} \\[6pt] 0 & -\frac{216a b_{\mathrm{II}}^3}{L^3} & \frac{18a b_{\mathrm{II}}^3}{L^2} & \frac{216a b_{\mathrm{II}}^3}{L^3} \end{bmatrix} \begin{bmatrix} \varphi_{1y} \\ u_{2z} \\ \varphi_{2y} \\ u_{3z} \end{bmatrix} = \begin{bmatrix} 0 \\ 0 \\ 0 \\ -\frac{F_0}{2} \end{bmatrix}. \tag{9.388}$$

The last equation already takes into account that the translational degree of freedom at node 1 and the rotational degree of freedom at node 3 are zero. The solution of the system of equations can be obtained, for example, by inverting the global stiffness matrix and then multiplying with the right-hand side, i.e. $\boldsymbol{u} = \boldsymbol{K}^{-1}\boldsymbol{f}$. This gives the

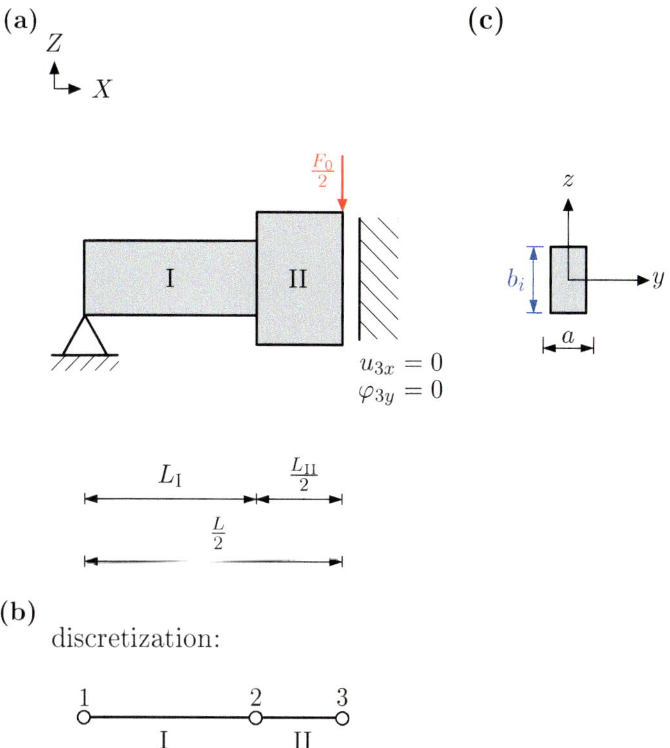

Fig. 9.34 Simply supported beam with three sections under consideration of symmetry: **a** general configuration, **b** discretization, and **c** cross-section of element i

column matrix of the unknown node values as:

$$
\begin{bmatrix} \varphi_{1y} \\ u_{2z} \\ \varphi_{2y} \\ u_{3z} \end{bmatrix} =
\begin{bmatrix}
\dfrac{F_0 L^2 \left(4 b_{\mathrm{II}}^{3} + 5 b_{\mathrm{I}}^{3}\right)}{12 E a\, b_{\mathrm{I}}^{3} b_{\mathrm{II}}^{3}} \\[2mm]
-\dfrac{F_0 L^3 \left(8 b_{\mathrm{II}}^{3} + 15 b_{\mathrm{I}}^{3}\right)}{108 E a\, b_{\mathrm{I}}^{3} b_{\mathrm{II}}^{3}} \\[2mm]
\dfrac{5 F_0 L^2}{12 E a\, b_{\mathrm{II}}^{3}} \\[2mm]
-\dfrac{F_0 L^3 \left(8 b_{\mathrm{II}}^{3} + 19 b_{\mathrm{I}}^{3}\right)}{108 E a\, b_{\mathrm{I}}^{3} b_{\mathrm{II}}^{3}}
\end{bmatrix} .
\tag{9.389}
$$

The bending moment and shear force distributions within a single element of length L_i can be generally expressed using (the index '1' refers to the start node, while the index '2' refers to the end node of an element) the following relations:

$$
M_y^{\mathrm{e}}(x) = E I_y \left(\left[+\frac{6}{L^2} - \frac{12x}{L^3} \right] u_{1z} + \left[-\frac{4}{L} + \frac{6x}{L^2} \right] \varphi_{1y} \right.
$$
$$
\left. + \left[-\frac{6}{L^2} + \frac{12x}{L^3} \right] u_{2z} + \left[-\frac{2}{L} + \frac{6x}{L^2} \right] \varphi_{2y} \right) ,
\tag{9.390}
$$

$$Q_z^e(x) = E I_y \left(\left[-\frac{12}{L^3} \right] u_{1z} + \left[+\frac{6}{L^2} \right] \varphi_{1y} + \left[+\frac{12}{L^3} \right] u_{2z} + \left[+\frac{6}{L^2} \right] \varphi_{2y} \right).$$
(9.391)

Thus, the normal and shear stress distributions can generally be calculated as follows:

$$\sigma_x^e(x, z) = \frac{M_y^e(x)}{I_y} \times z,$$
(9.392)

$$\tau_{xz}^e(x, z) = \frac{Q_z^e(x)}{2 I_y} \left[\left(\frac{b}{2} \right)^2 - z^2 \right].$$
(9.393)

The nodal values of the internal reactions for calculating the normal and shear stresses are summarized in Table 9.4. It is noted here that in this case the nodal values based on the finite element calculation are identical to the analytical solution.

From Table 9.4 it can be concluded that the maximum normal stress in each element is reached at the right node and that the shear stress is constant in each element. Thus, the critical stresses in each element can be calculated as follows:

$$\sigma_{x,\mathrm{I}} = -\frac{F_0 L}{a\, b_\mathrm{I}^2},$$
(9.394)

$$\sigma_{x,\mathrm{II}} = -\frac{3 F_0 L}{2a\, b_\mathrm{II}^2},$$
(9.395)

or for the shear stresses:

$$\tau_{xz,\mathrm{I}} = -\frac{3 F_0}{4a\, b_\mathrm{I}},$$
(9.396)

$$\tau_{xz,\mathrm{II}} = -\frac{3 F_0}{4a\, b_\mathrm{II}}.$$
(9.397)

The objective function of the optimization problem, i.e. the mass of the stepped beam, can be represented as a function of the two design variables $X_1 = b_\mathrm{I}$ and $X_2 = b_\mathrm{II}$

Table 9.4 Nodal values of the internal reactions bending moment (M_y) and shear force (Q_z) at each node

	Element I		Element II	
	Left node	Right node	Left node	Right node
M_y	0	$-\frac{F_0 L}{6}$	$-\frac{F_0 L}{6}$	$-\frac{F_0 L}{4}$
Q_z	$-\frac{F_0}{2}$	$-\frac{F_0}{2}$	$-\frac{F_0}{2}$	$-\frac{F_0}{2}$

as follows:

$$F(X_1, X_2) = \varrho L_{\mathrm{I}} a X_1 + \varrho L_{\mathrm{II}} a X_2 = \varrho L a \left(\frac{X_1}{3} + \frac{X_2}{6} \right), \qquad (9.398)$$

where the minimization is to be carried out under consideration of the following seven inequality constraints:

$$g_1(X_1, X_2) = +\frac{F_0 L^3 (19 X_1^3 + 8 X_2^3)}{108 E a X_1^3 X_2^3 r_1 L} - 1 \leq 0 \quad (\text{max. Durchbg.}), \qquad (9.399)$$

$$g_2(X_1, X_2) = \frac{F_0 L}{a X_1^2} - R_{\mathrm{p0.2}} \leq 0 \quad (\text{normal stress in I}), \qquad (9.400)$$

$$g_3(X_1, X_2) = \frac{3 F_0 L}{2 a X_2^2} - R_{\mathrm{p0.2}} \leq 0 \quad (\text{normal stress in II}), \qquad (9.401)$$

$$g_4(X_1, X_2) = \frac{3 F_0}{4 a X_1} - \frac{R_{\mathrm{p0.2}}}{2} \leq 0 \quad (\text{shear stress in I}), \qquad (9.402)$$

$$g_5(X_1, X_2) = \frac{3 F_0}{4 a X_2} - \frac{R_{\mathrm{p0.2}}}{2} \leq 0 \quad (\text{shear stress in II}), \qquad (9.403)$$

$$g_6(X_1, X_2) = X_1 - 20a \leq 0 \quad (\text{height-to-width ratio in I}), \qquad (9.404)$$

$$g_7(X_1, X_2) = X_2 - 20a \leq 0 \quad (\text{height-to-width ratio in II}). \qquad (9.405)$$

The graphical representation of the objective function and the inequality constraints (g_1, g_2, g_3) in the X_1–X_2 design space is shown in Fig. 9.35.

From Fig. 9.35 it can be concluded that considering g_1, g_2, and g_3 can be sufficient if a reasonable range of design variables is chosen. Furthermore, from Fig. 9.35 it can be concluded that the minimum is achieved for the case that g_1 is tangent to the objective function F. In order to derive an analytical solution for the minimum, one must take into account that the objective function F, in the representation $X_2(X_1) = \frac{6c}{\varrho L a} - 2 X_1$, represents straight lines with a slope of -2. Thus, the condition for the minimum is that the curve g_1 assumes a slope of -2. Based on Eq. (9.399), the expression for g_1 can be rewritten:

$$X_2 = \left(-\frac{19 F_0 L^3 X_1^3}{8 F_0 L^3 - 108 E a r_1 L X_1^3} \right)^{\frac{1}{3}}, \qquad (9.406)$$

which, after calculation, yields the expression of the first-order derivative as follows:

$$\left. \frac{\mathrm{d} X_2}{\mathrm{d} X_1} \right|_{g_1} = \frac{219^{\frac{1}{3}} F^{\frac{4}{3}} L^{\frac{8}{3}}}{\left(8 F L^2 - 108 E a r_1 X_1^3 \right)^{\frac{1}{3}} \left(27 E a r_1 X_1^3 - 2 F L^2 \right)} \overset{!}{=} -2 . \qquad (9.407)$$

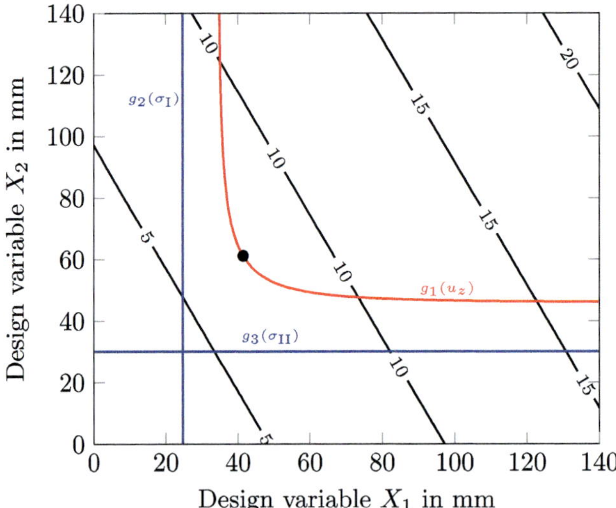

Fig. 9.35 Optimal design of a simply supported beam with three sections (consideration of symmetry): $L = 2540\,\text{mm}$, $a = 45.16\,\text{mm}$, $r_1 = 0.01$, $F_0 = 2667\,\text{N}$, $E = 68{,}948\,\text{MPa}$, $R_{\text{p0.2}} = 247\,\text{MPa}$, $\varrho = 2.691 \times 10^{-6}\,\text{kg/mm}^3$, (original set of Eqs. (9.398)–(9.405)). Exact solution for the minimum: $X_{1,\text{extr}} = 41.437\,\text{mm}$, $X_{2,\text{extr}} = 61.173\,\text{mm}$ (indicated by the • marker)

The last equation can be rearranged so that one side has a root, and Newton's method can be used to determine the root. It is important that the root is sought in a reasonable interval, e.g. $40 \le X_{1,\text{extr}} \le 60$.

9.4   Chapter 5

5.6.1 Average Bending Stiffness of a Sandwich
The generalization of Eq. (5.1) for different layer widths b^k results in:

$$\overline{EI}_y = \sum_{k=1}^{3} E^k \left[\tfrac{1}{12} b^k \left(\Delta h^k \right)^3 + b^k \Delta h^k \left(z_c^k \right)^2 \right]. \tag{9.408}$$

Application to our three-layer problem gives:

$$\overline{EI}_y = \frac{E^\text{F} b^\text{F} (\Delta h^\text{F})^3}{6} + \frac{E^\text{F} b^\text{F} \Delta h^\text{F} (\Delta h^\text{C} + \Delta h^\text{F})^2}{2} + \frac{E^\text{C} b^\text{C} (\Delta h^\text{C})^3}{12}. \tag{9.409}$$

For the special case $b^\text{C} = \frac{b^\text{F}}{3}$ and $\Delta h^\text{F} = \frac{\Delta h^\text{C}}{4}$ the simplified formula results in:

$$\overline{EI}_y = \frac{b^\text{F} (\Delta h^\text{F})^3}{9} \times \left[114 E^\text{F} + 16 E^\text{C} \right]. \tag{9.410}$$

Fig. 9.36 Circular
cross-section for deriving the
shear stress distribution τ_{xz}

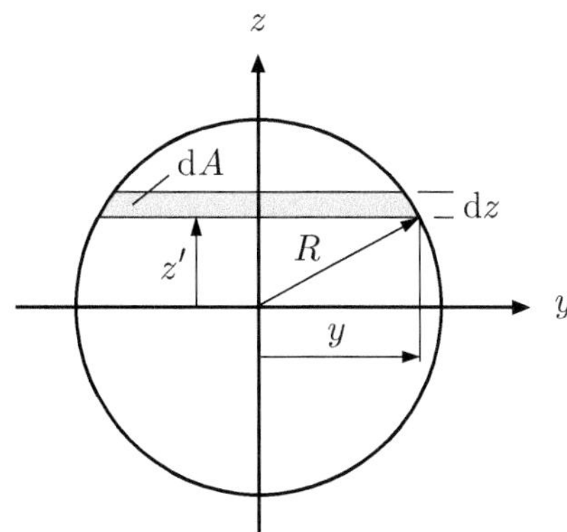

5.6.2 Simplification of the average bending stiffness for a homogeneous sandwich

Equation (5.1) simplifies for the homogeneous sandwich, i.e. $E^1 = E^2 = E^3 = E$ and $\Delta h^1 = \Delta h^2 = \Delta h^3 = \frac{1}{3}h$ to:

$$\overline{EI}_y = 2E\left(\frac{1}{12}b\frac{1}{27}h^3 + b\frac{1}{3}h\frac{1}{9}h^2\right) + E\left(\frac{1}{12}b\frac{1}{27}h^3\right) \tag{9.411}$$

$$= 2E\left(\frac{1}{12\times 27}bh^3 + \frac{12}{12\times 27}bh^3\right) + E\,\frac{1}{12\times 27}bh^3 \tag{9.412}$$

$$= \frac{1}{12}bh^3 = EI_y\,. \tag{9.413}$$

5.6.3 Calculation of the Shear Stress Distribution for a Circular Cross-Section

The starting point is again an infinitesimal beam element as in Fig. A.2. However, now the cross-section is as in Fig. 9.36.

The forces equilibrium in the x direction results in:

$$\int \sigma_x(x)\,\mathrm{d}A - \int\left(\sigma_x(x) + \frac{\mathrm{d}\sigma_x(x)}{\mathrm{d}x}\,\mathrm{d}x\right)\mathrm{d}A + \tau_{xz}2y\,\mathrm{d}x = 0\,. \tag{9.414}$$

Let us consider now Eq. (A.7) from the Appendix A.1.2:

$$\frac{\mathrm{d}\sigma_x(x)}{\mathrm{d}x} = +\frac{z}{I_y}\frac{\mathrm{d}M_y(x)}{\mathrm{d}x} = \frac{Q_z(x)\times z}{I_y}\,. \tag{9.415}$$

Consequently:

$$\tau_{xz} = \frac{Q_z(x)}{2yI_y}\int z\,\mathrm{d}A\,. \tag{9.416}$$

Taking into account $dA = 2y\,dz$ and $y = \sqrt{R^2 - z^2}$, the shear stress distribution results in:

$$\tau_{xz} = \frac{Q_z(x)}{3I_y}\left(R^2 - (z')^2\right).$$
(9.417)

The maximum shear stress results for $z' = 0$:

$$\tau_{xz,\max} = \frac{Q_z(x)R^2}{3I_y} = \frac{4Q_z(x)}{3\pi R^2} = \frac{4Q_z(x)}{3A}.$$
(9.418)

5.6.4 Limit Case of the Normalized Stiffness of a Sandwich

$$\frac{\overline{EI}_y}{EI_{y,\text{ref}}} = \left[\frac{1}{4} + \frac{3}{4}4 + \frac{1}{8}\right] = \frac{27}{8}.$$
(9.419)

Sandwich of height '1': $\overline{EI}_y = \frac{1}{12}b1^3 = \frac{b}{12}$; reference configuration of height '$\frac{2}{3}$': $\overline{EI}_{y,\text{ref}} = \frac{1}{12}b\left(\frac{2}{3}\right)^3 = \frac{2b}{81}$. Ratio: $\frac{\overline{EI}_y}{EI_{y,\text{ref}}} = \frac{81}{2\times12} = \frac{27}{8}$.

5.6.5 Deflection of a Sandwich Beam with Distributed Load Using the Partial Deflection Method

- General solution:

$$u_{z,b}(x) = \frac{1}{\overline{EI}_y}\left(-\frac{q_0x^4}{24} + \frac{c_1x^3}{6} + \frac{c_2x^2}{2} + c_3x + c_4\right),$$
(9.420)

$$u_{z,s}(x) = \frac{1}{AG^C}\left(\frac{q_0x^2}{2} - \frac{q_0Lx}{2}\right) + c_5.$$
(9.421)

- Particular solution:
  Using the boundary conditions $u_{z,b}(0) = u_{z,s}(0) = u_{z,b}(L) = 0$ and $M_y(0) = M_y(L) = 0$ the constants of integration result in $c_2 = c_4 = c_5 = 0$, $c_1 = \frac{q_0L}{2}$ and $c_3 = -\frac{q_0L^3}{24}$. This gives the particular solution:

$$u_z(x) = -\frac{q_0L^4}{24\overline{EI}_y}\left(\left(\frac{x}{L}\right)^4 - 2\left(\frac{x}{L}\right)^3 + \left(\frac{x}{L}\right)\right) - \frac{q_0L^2}{2AG^C}\left(-\left(\frac{x}{L}\right)^2 + \left(\frac{x}{L}\right)\right).$$
(9.422)

- Maximum value of deflection:

$$u_z\left(\frac{L}{2}\right) = -\frac{5q_0L^4}{384\overline{EI}_y} - \frac{q_0L^2}{8AG^C}.$$
(9.423)

- Ratio of the partial deflections:

$$\frac{u_{z,\text{b}}(\frac{L}{2})}{u_{z,\text{s}}(\frac{L}{2})} = \frac{5}{24} \times \frac{L^2 G^{\text{C}}}{E^{\text{F}} \Delta h^{\text{C}} \Delta h^{\text{F}}} = 0.165 = 16.5\% . \tag{9.424}$$

5.6.6 Deflection of a Sandwich Beam (Cantilever) with Distributed Load Usingthe Partial Deflection Method

- General solution:

$$u_{z,\text{b}}(x) = \frac{1}{\overline{EI_y}} \left(-\frac{q_0 x^4}{24} + \frac{c_1 x^3}{6} + \frac{c_2 x^2}{2} + c_3 x + c_4 \right) , \tag{9.425}$$

$$u_{z,\text{s}}(x) = \frac{1}{AG^{\text{C}}} \left(\frac{q_0 x^2}{2} - q_0 L x \right) + c_5 . \tag{9.426}$$

- Particular solution:
  Using the boundary conditions $u_{z,\text{b}}(0) = u_{z,\text{s}}(0) = 0$, $\frac{\mathrm{d} u_{z,\text{b}}(0)}{\mathrm{d}x} = 0$, $M_y(L) = 0$ and $Q_z(L) = 0$ the constants of integration result in $c_3 = c_4 = c_5 = 0, c_1 = q_0 L$ and $c_2 = -\frac{1}{2} q_0 L^2$. This gives the particular solution:

$$u_z(x) = -\frac{q_0 L^4}{2\overline{EI_y}} \left(\frac{1}{12} \left(\frac{x}{L} \right)^4 - \frac{1}{3} \left(\frac{x}{L} \right)^3 + \frac{1}{2} \left(\frac{x}{L} \right)^2 \right) - \frac{q_0 L^2}{AG^{\text{C}}} \left(-\frac{1}{2} \left(\frac{x}{L} \right)^2 + \left(\frac{x}{L} \right) \right) . \tag{9.427}$$

- Maximum value of deflection:

$$u_z(L) = -\frac{q_0 L^4}{8\overline{EI_y}} - \frac{q_0 L^2}{2AG^{\text{C}}} . \tag{9.428}$$

- Ratio of the partial deflections:

$$\frac{u_{z,\text{b}}(L)}{u_{z,\text{s}}(L)} = \frac{1}{4} \times \frac{L^2 G A^{\text{C}}}{\overline{EI_y}} = \frac{1}{2} \times \frac{L^2 G^{\text{C}}}{E^{\text{D}} \Delta h^{\text{C}} \Delta h^{\text{F}}} = 0.397 = 39.7\% . \tag{9.429}$$

5.6.7 Deflection of a Sandwich Beam (Cantilever) with Shear Force Using the Partial Deflection Method

- Internal reactions:
  The internal reactions for the problem according to Fig. 5.25 are shown in Fig. 9.37.

- General solution:

$$u_{z,\mathrm{b}}(x) = \frac{1}{EI_y}\left(-F_0\left(\frac{Lx^2}{2}-\frac{x^3}{6}\right)+c_1 x + c_2\right), \tag{9.430}$$

$$u_{z,\mathrm{c}}(x) = -\frac{F_0 x}{AG^{\mathrm{C}}} + c_3. \tag{9.431}$$

- Particular solution:
  Using the boundary conditions $u_{z,\mathrm{b}}(0) = u_{z,\mathrm{s}}(0) = 0$ and $\frac{\mathrm{d}u_{z,\mathrm{b}}(0)}{\mathrm{d}x} = 0$ the constants of integration result in $c_1 = c_2 = c_3 = 0$. This gives the particular solution:

$$u_z(x) = -\frac{F_0 L^3}{6\overline{EI_y}}\left(3\left(\frac{x}{L}\right)^2 - \left(\frac{x}{L}\right)^3\right) - \frac{F_0 L}{AG^{\mathrm{C}}}\left(\frac{x}{L}\right). \tag{9.432}$$

- Maximum value of deflection:

$$u_z(L) = -\frac{F_0 L^3}{3\overline{EI_y}} - \frac{F_0 L}{AG^{\mathrm{C}}}. \tag{9.433}$$

- Ratio of the partial deflections:

$$\frac{u_{z,\mathrm{b}}(L)}{u_{z,\mathrm{s}}(L)} = \frac{2}{3}\times\frac{L^2 G^{\mathrm{C}}}{E^{\mathrm{F}}\Delta h^{\mathrm{C}}\Delta h^{\mathrm{F}}} = 0.529 = 52.9\%. \tag{9.434}$$

5.6.8 Deflection of a Sandwich Beam (Cantilever) with Shear Force and Bending Moment Using the Partial Deflection Method

- Internal reactions:

$$Q_z(x) = -F_0, \tag{9.435}$$

$$M_y(x) = \tfrac{3}{2}F_0 L - F_0 x. \tag{9.436}$$

- General solution:

$$u_{z,\mathrm{b}}(x) = \frac{1}{EI_y}\left(\frac{F_0 x^3}{6}+\frac{c_1 x^2}{2}+c_2 x + c_3\right), \tag{9.437}$$

$$u_{z,\mathrm{c}}(x) = -\frac{F_0 x}{AG^{\mathrm{C}}} + c_4. \tag{9.438}$$

Fig. 9.37 Internal reactions for a sandwich beam according to Fig. 5.25, i.e. cantilever with end shear force F_0 at $x = L$ in negative z-direction: **a** shear force and **b** bending moment diagram

(a)

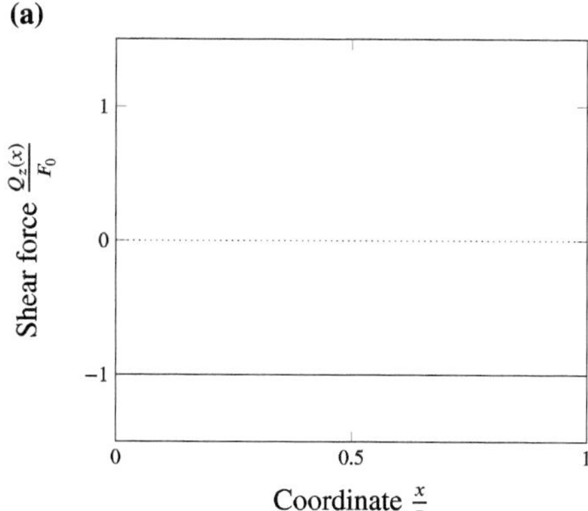

(b)

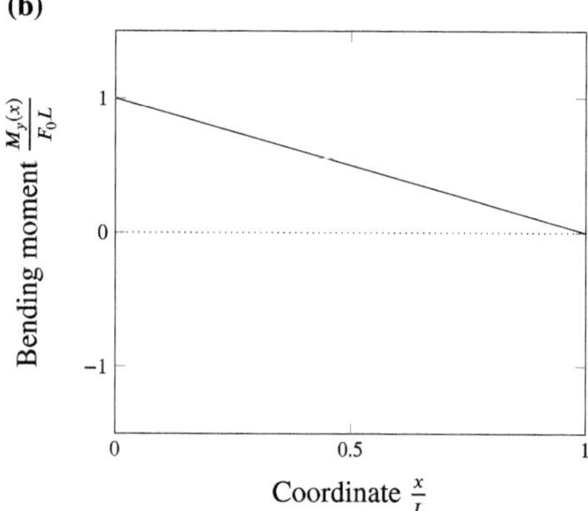

Using the boundary conditions $u_{z,\mathrm{b}}(0) = u_{z,\mathrm{s}}(0) = 0$, $\frac{\mathrm{d}u_{z,\mathrm{b}}(0)}{\mathrm{d}x} = 0$ and $M_y(L) = \frac{F_0 L}{2}$ the constants of integration result in $c_1 = -\frac{3F_0 L}{2}$ and $c_2 = c_3 = c_4 = 0$. This gives the particular solution:

$$u_z(x) = \frac{1}{EI_y}\left(\frac{F_0 x^2}{6} - \frac{3F_0 L x^2}{4}\right) - \frac{F_0 x}{AG^{\mathrm{C}}}.\tag{9.439}$$

- Maximum value of deflection:

$$u_z(L) = -\frac{7F_0 L^3}{12EI_y} - \frac{F_0 L}{AG^{\mathrm{C}}}.\tag{9.440}$$

- Ratio of the partial deflections:

$$\frac{u_{z,\mathrm{s}}(L)}{u_{z,\mathrm{b}}(L)} = \frac{6}{7} \times \frac{E^{\mathrm{F}}\Delta h^{\mathrm{F}}\Delta h_{\mathrm{c}}}{L^2 G^{\mathrm{C}}} = 1.12 = 112\%.\tag{9.441}$$

5.6.9 Deflection of a Sandwich Beam with Linearly Distributed Load Using the Partial Deflection Method

- Distributed load:

$$M_z(x) = -q_0 \left(\frac{x}{L} \right). \tag{9.442}$$

- Internal reactions:

$$Q_z(x) = \tfrac{1}{2} q_0 \frac{x^2}{L} - \frac{1}{6} q_0 L , \tag{9.443}$$

$$M_y(x) = \tfrac{1}{6} q_0 \frac{x^3}{L} - \frac{1}{6} q_0 L x . \tag{9.444}$$

- General solution:

$$u_{z,\,\mathrm{b}}(x) = \frac{1}{EI_y} \left(-\frac{1}{120} q_0 \frac{x^5}{L} + \frac{1}{36} q_0 L x^3 - \frac{7}{360} q_0 L^3 x \right) , \tag{9.445}$$

$$u_{z,\,\mathrm{c}}(x) = \frac{1}{AG^{\mathrm{C}}} \left(\frac{1}{6} q_0 \frac{x^3}{L} - \frac{1}{6} q_0 L x \right) , \tag{9.446}$$

where the boundary conditions $u_{z,\,\mathrm{b}}(0) = u_{z,\,\mathrm{s}}(0) = u_{z,\,\mathrm{b}}(L) = 0$, and the constants of integration $c_1 = -\frac{7 q_0 L^3}{360}$ and $c_2 = c_3 = 0$ were used.
- Ratio of the partial deflection at $x = \frac{2L}{3}$:

$$\frac{u_{z,\,\mathrm{s}}(2/3L)}{u_{z,\,\mathrm{b}}(2/3L)} = \frac{180}{17} \times \frac{\overline{EI_y}}{AG^{\mathrm{C}} L^2} = 6.9 . \tag{9.447}$$

9.5   Chapter 6

6.2.1 Failure Analysis of a Sandwich Beam with Distributed Load

The internal reactions and the corresponding maximum values can be taken from Fig. 9.38.

Checking whether one of the simplifying theories is applicable:

Fig. 9.38 Internal reactions for a sandwich beam with distributed load (configuration: hinged on both sides with constant distributed load q_0 in negative z-direction): **a** shear force and **b** bending moment diagram

(a)

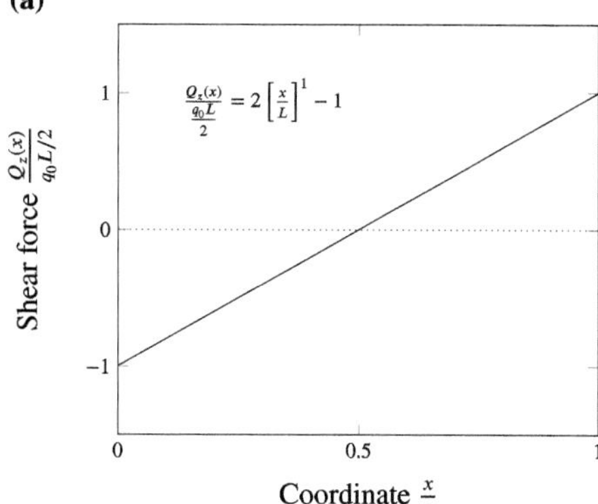

(b)

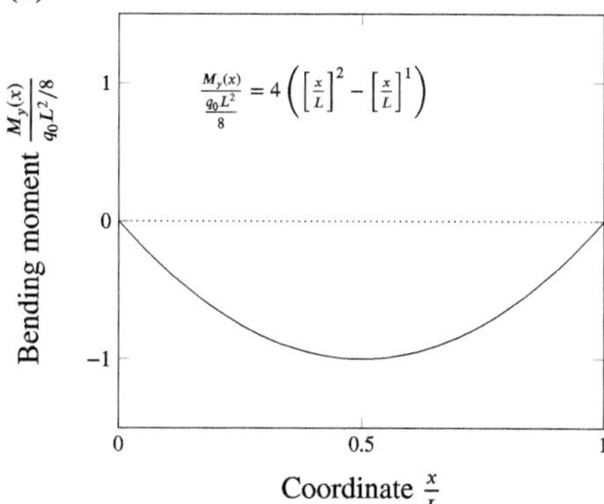

$$\frac{6E^{\mathrm{F}}\Delta h^{\mathrm{F}}(h_{\mathrm{c}})^2}{E^{\mathrm{C}}(\Delta h^{\mathrm{C}})^3} = 526.77 > 100\,, \tag{9.448}$$

$$\frac{\Delta h^{\mathrm{C}}}{\Delta h^{\mathrm{F}}} = 30.0 > 4.77\,. \tag{9.449}$$

Thus, the theory for soft cores and thin face sheets can be applied.

- Maximum normal stress in the face sheets:

$$\sigma_{x,\mathrm{F}} = 8.33\,\mathrm{MPa} < R^{\mathrm{F}}_{\mathrm{p}0.2}\,. \tag{9.450}$$

- Maximum shear stress in the core:

$$\tau_{zx,\mathrm{C}} = 0.081\,\mathrm{MPa} < \tau^{\mathrm{C}}_{\mathrm{aB}}\,. \tag{9.451}$$

- Shear stress in the adhesive layer:

$$\tau_{zx} = 0.081\,\mathrm{MPa} < \tau_{\mathrm{aB}}\,. \tag{9.452}$$

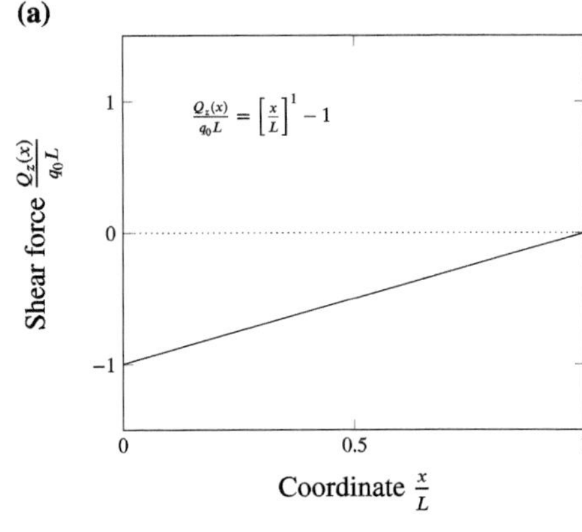

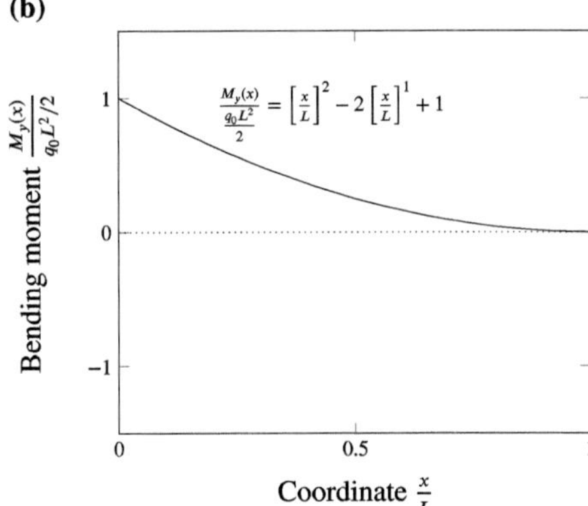

Fig. 9.39 Internal reactions for a sandwich beam with distributed load (configuration: cantilever beam with constant distributed load q_0 in negative z-direction): **a** shear force and **b** bending moment diagram

- Local wrinkling of the compression face sheet:

With $B_1 = 0.567$ it follows for the critical stress:

$$\sigma_{\mathrm{cr}} = 229.904\,\mathrm{MPa} > \sigma_{x,\mathrm{F}}\,. \tag{9.453}$$

Thus, there is no failure due to local wrinkling of the face sheet.

6.2.2 Failure Analysis of a Sandwich Beam (Cantilever) with Distributed Load

The internal reactions, i.e., shear force and bending moment, and the corresponding maximum values can be seen in Fig. 9.39.

Checking whether one of the simplifying theories is applicable:

$$\frac{6E^{\mathrm{F}}\Delta h^{\mathrm{F}}(h_{\mathrm{c}})^2}{E^{\mathrm{C}}(\Delta h^{\mathrm{C}})^3} = 526.77 > 100\,, \tag{9.454}$$

$$\frac{\Delta h^{\mathrm{C}}}{\Delta h^{\mathrm{F}}} = 30.0 > 4.77\,. \tag{9.455}$$

Thus, the theory for soft cores and thin face sheets can be applied.

- Maximum normal stress in the face sheets:

$$\sigma_{x,\mathrm{F}} = 33.22\,\mathrm{MPa} < R^{\mathrm{F}}_{\mathrm{p0.2}} \,. \tag{9.456}$$

- Maximum shear stress in the core:

$$\tau_{zx,\mathrm{C}} = 0.163\,\mathrm{MPa} < \tau^{\mathrm{C}}_{\mathrm{aB}} \,. \tag{9.457}$$

- Shear stress in the adhesive layer:

$$\tau_{zx} = 0.163\,\mathrm{MPa} < \tau_{\mathrm{aB}} \,. \tag{9.458}$$

- Local wrinkling of the compression face sheet:

With $B_1 = 0.567$ it follows for the critical stress:

$$\sigma_{\mathrm{cr}} = 229.904\,\mathrm{MPa} > \sigma_{x,\mathrm{F}} \,. \tag{9.459}$$

Thus, there is no failure due to local wrinkling of the face sheet.

6.2.3 Failure Analysis of a Sandwich Beam Under 4-Point Bending

The internal reactions, i.e., shear force and bending moment, and the corresponding maximum values can be seen in Fig. 9.40.

Checking whether one of the simplifying theories is applicable:

$$\frac{6E^{\mathrm{F}}\Delta h^{\mathrm{F}}(h_{\mathrm{c}})^2}{E^{\mathrm{C}}(\Delta h^{\mathrm{C}})^3} = 71.04 < 100 \,, \tag{9.460}$$

$$\frac{\Delta h^{\mathrm{C}}}{\Delta h^{\mathrm{F}}} = 1.0 < 4.77. \tag{9.461}$$

Thus, the exact theory must be applied.

- Maximum normal stress in the face sheets:

$$\sigma_{x,\mathrm{F}} = 1.709\,\mathrm{MPa} < R^{\mathrm{F}}_{\mathrm{p0.2}} \,. \tag{9.462}$$

- Maximum normal stress in the core:

$$\sigma_{x,\mathrm{C}} = 0.192\,\mathrm{MPa} < R^{\mathrm{C}}_{\mathrm{m}} \quad \left(< \left| \sigma^{\mathrm{C}}_{\mathrm{dB}} \right| \right) \,. \tag{9.463}$$

- Maximum shear stress in the core:

$$\tau_{zx,\mathrm{C}} = 0.119\,\mathrm{MPa} < \tau^{\mathrm{C}}_{\mathrm{aB}} \,. \tag{9.464}$$

- Shear stress in the adhesive layer:

$$\tau_{zx} = 0.114\,\text{MPa} < \tau_{aB}\,. \tag{9.465}$$

- Local wrinkling of the compression face sheet:

With $B_1 = 0.768$ it follows for the critical stress:

$$\sigma_{cr} = 27555.434\,\text{MPa} > \sigma_{x,F}\,. \tag{9.466}$$

Thus, there is no failure due to local wrinkling of the face sheet.

6.2.4 Failure Analysis of a Sandwich Beam Under 3-Point Bending with Point Load

The internal reactions, i.e., shear force and bending moment, and the corresponding maximum values can be seen in Fig. 9.41.

Checking whether one of the simplifying theories is applicable:

Fig. 9.40 Internal reactions for sandwich under 4-point bending: **a** shear force and **b** bending moment diagram

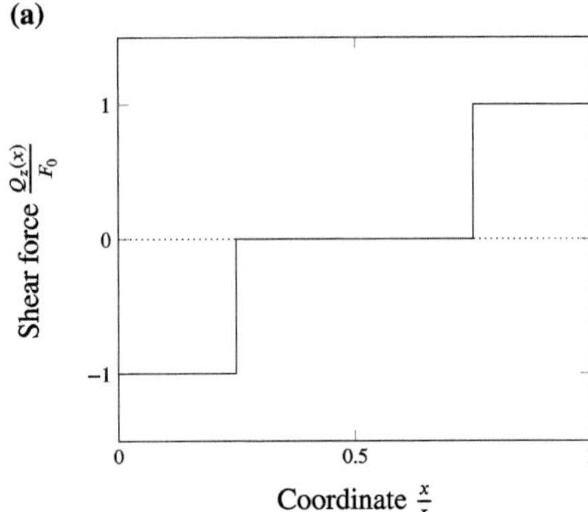

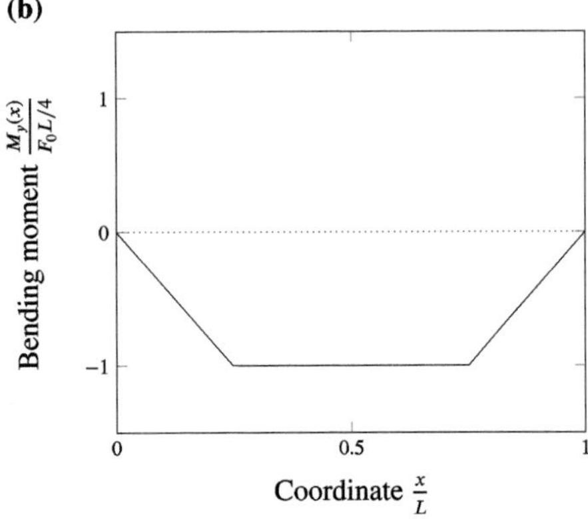

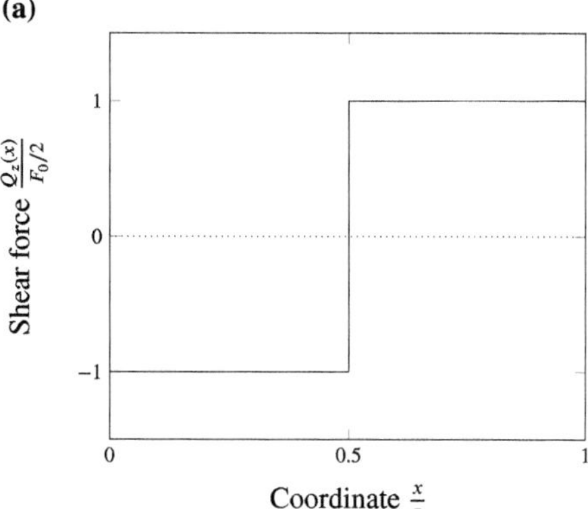

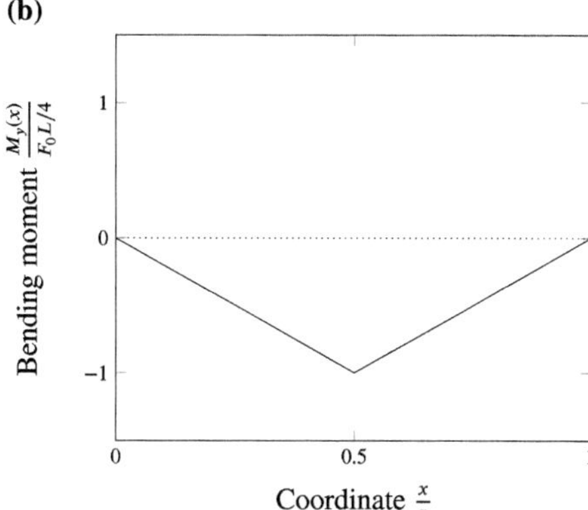

Fig. 9.41 Internal reactions for a sandwich under 3-point bending with point load F_0 in negative z-direction: **a** shear force and **b** bending moment diagram

$$\frac{6E^{\mathrm{F}}\,\Delta h^{\mathrm{F}}(h_{\mathrm{c}})^2}{E^{\mathrm{C}}(\Delta h^{\mathrm{C}})^3} = 115.625 > 100\,, \tag{9.467}$$

$$\frac{\Delta h^{\mathrm{C}}}{\Delta h^{\mathrm{F}}} = 4.0 < 4.77. \tag{9.468}$$

Thus, the simplifying theory for soft cores can be applied.

- Maximum normal stress in the face sheets:

$$\sigma_{x,\mathrm{F}} = 4.737\,\mathrm{MPa} < R^{\mathrm{F}}_{\mathrm{p}0.2}\,. \tag{9.469}$$

- Maximum shear stress in the core:

$$\tau_{zx,\mathrm{C}} = 0.0987\,\mathrm{MPa} < \tau^{\mathrm{C}}_{\mathrm{aB}}\,. \tag{9.470}$$

- Shear stress in the adhesive layer:

$$\tau_{zx} = 0.0987\,\mathrm{MPa} < \tau_{\mathrm{aB}}\,. \tag{9.471}$$

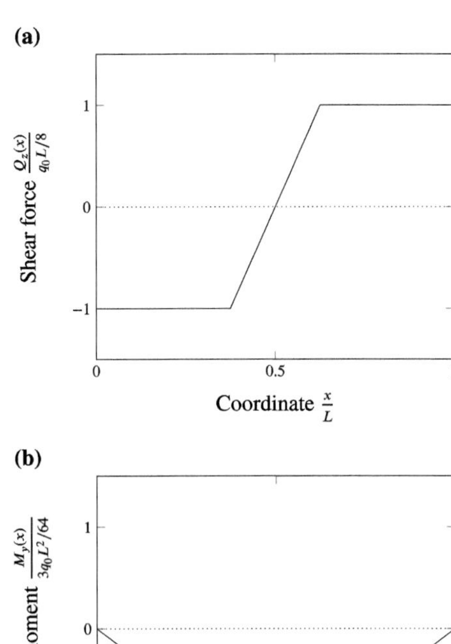

Fig. 9.42 Internal reactions for a sandwich beam under 3-point bending with negative distributed load q_0 in the range $3L/8 \leq x \leq 5L/8$: **a** shear force and **b** bending moment diagram

- Local wrinkling of the compression face sheet:

With $B_1 = 0.648$ it follows for the critical stress:

$$\sigma_{\mathrm{cr}} = 3564.461 \,\mathrm{MPa} > \sigma_{x,\mathrm{F}}\,. \tag{9.472}$$

Thus, there is no failure due to local wrinkling of the face sheet.

6.2.5 Failure Analysis of a Sandwich Beam Under 3-Point Bending with Distributed Load

The internal reactions, i.e., shear force and bending moment, and the corresponding maximum values can be seen in Fig. 9.42.

Checking whether one of the simplifying theories is applicable:

$$\frac{6E^{\mathrm{F}}\Delta h^{\mathrm{F}}(h_{\mathrm{c}})^2}{E^{\mathrm{C}}(\Delta h^{\mathrm{C}})^3} = 115.625 > 100\,, \tag{9.473}$$

$$\frac{\Delta h^{\mathrm{C}}}{\Delta h^{\mathrm{F}}} = 4.0 < 4.77\,. \tag{9.474}$$

Thus, the simplifying theory for soft cores can be applied.

- Maximum normal stress in the face sheets:

$$\sigma_{x,\mathrm{F}} = 4.145\,\mathrm{MPa} < R^{\mathrm{F}}_{\mathrm{p0.2}} \,. \tag{9.475}$$

- Maximum shear stress in the core:

$$\tau_{zx,\mathrm{C}} = 0.0987\,\mathrm{MPa} < \tau^{\mathrm{C}}_{\mathrm{aB}} \,. \tag{9.476}$$

- Shear stress in the adhesive layer:

$$\tau_{zx} = 0.0987\,\mathrm{MPa} < \tau_{\mathrm{aB}} \,. \tag{9.477}$$

- Local wrinkling of the compression face sheet:

With $B_1 = 0.648$ it follows for the critical stress:

$$\sigma_{\mathrm{cr}} = 3564.461\,\mathrm{MPa} > \sigma_{x,\mathrm{F}} \,. \tag{9.478}$$

Thus, there is no failure due to local wrinkling of the face sheet.

6.2.6 Failure Analysis of a Sandwich Beam Under 5-Point Bending with Point Loads

The free body diagram of the configuration and a system under consideration of symmetry is shown in Fig. 9.43.

The model according to Fig. 9.43b can be used to determine the internal reactions. The corresponding curves are shown in Fig. 9.44.

Checking whether one of the simplifying theories is applicable:

$$\frac{6E^{\mathrm{F}}\Delta h^{\mathrm{F}}(h_{\mathrm{c}})^2}{E^{\mathrm{C}}(\Delta h^{\mathrm{C}})^3} = 115.625 > 100 \,, \tag{9.479}$$

$$\frac{\Delta h^{\mathrm{C}}}{\Delta h^{\mathrm{F}}} = 4.0 < 4.77. \tag{9.480}$$

Thus, the simplifying theory for soft cores can be applied.

- Maximum normal stress in the face sheets:

$$\sigma_{x,\mathrm{F}} = 1.579\,\mathrm{MPa} < R^{\mathrm{F}}_{\mathrm{p0.2}} \,. \tag{9.481}$$

- Maximum shear stress in the core:

$$\tau_{zx,\mathrm{C}} = 0.0658\,\mathrm{MPa} < \tau^{\mathrm{C}}_{\mathrm{aB}} \,. \tag{9.482}$$

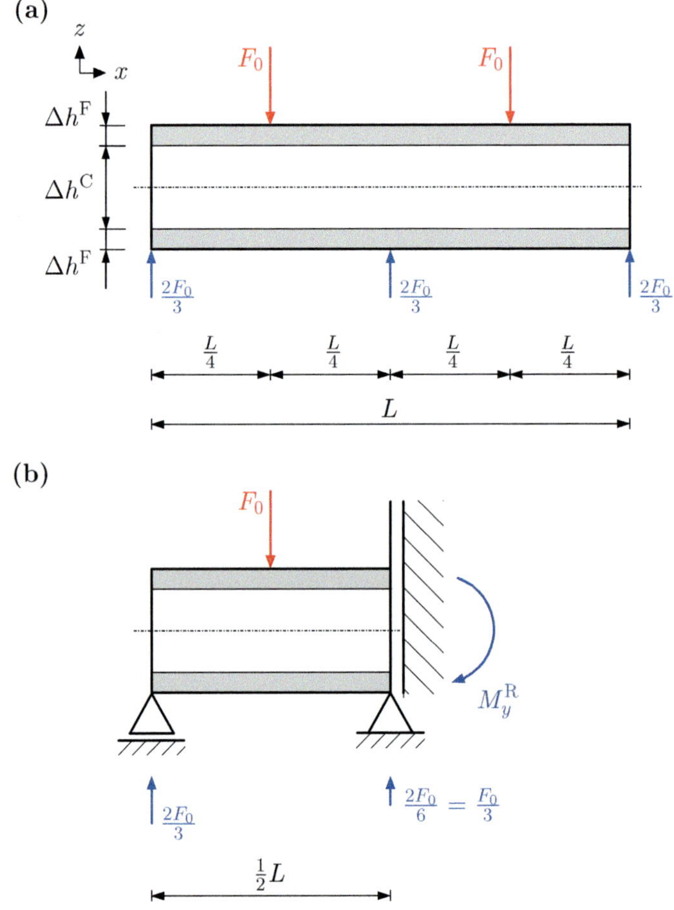

Fig. 9.43 Sandwich beam under 5-point bending with concentrated loads: **a** free body diagram; **b** free body diagram considering symmetry

- Shear stress in the adhesive layer:

$$\tau_{zx} = 0.0658\,\mathrm{MPa} < \tau_{\mathrm{aB}} \ . \tag{9.483}$$

- Local wrinkling of the compression face sheet:

With $B_1 = 0.648$ it follows for the critical stress:

$$\sigma_{\mathrm{cr}} = 3564.461\,\mathrm{MPa} > \sigma_{x,\mathrm{F}} \ . \tag{9.484}$$

Thus, there is no failure due to local wrinkling of the face sheet.

Fig. 9.44 Internal reactions for a sandwich beam under 5-point bending with concentrated loads: **a** shear force and **b** bending moment diagram

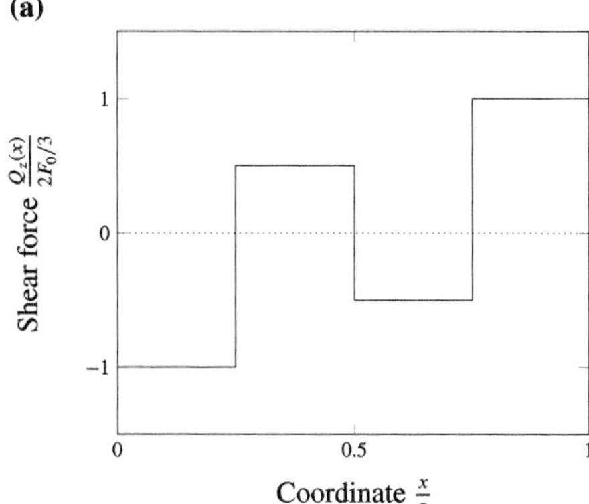

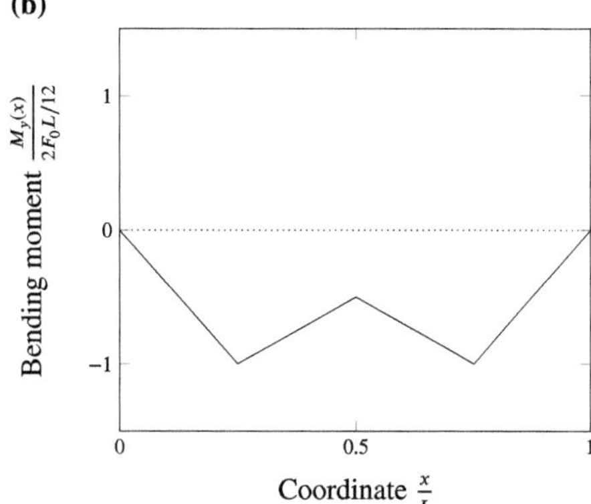

6.2.7 Failure Analysis of a Sandwich Beam Under 5-Point Bending with Distributed Load

The internal reactions, i.e., shear force and bending moment, are shown in Fig. 9.45.

Checking whether one of the simplifying theories is applicable:

$$\frac{6E^{\mathrm{F}}\Delta h^{\mathrm{F}}(h_{\mathrm{c}})^2}{E^{\mathrm{C}}(\Delta h^{\mathrm{C}})^3} = 115.625 > 100 , \tag{9.485}$$

$$\frac{\Delta h^{\mathrm{C}}}{\Delta h^{\mathrm{F}}} = 4.0 < 4.77 . \tag{9.486}$$

Thus, the simplifying theory for soft cores can be applied.

- Maximum normal stress in the face sheets:

$$\sigma_{x,\mathrm{F}} = 0.177\,\mathrm{MPa} < R^{\mathrm{F}}_{\mathrm{p0.2}} . \tag{9.487}$$

- Maximum shear stress in the core:

$$\tau_{zx,\mathrm{C}} = 0.0658\,\mathrm{MPa} < \tau^{\mathrm{C}}_{\mathrm{aB}} . \tag{9.488}$$

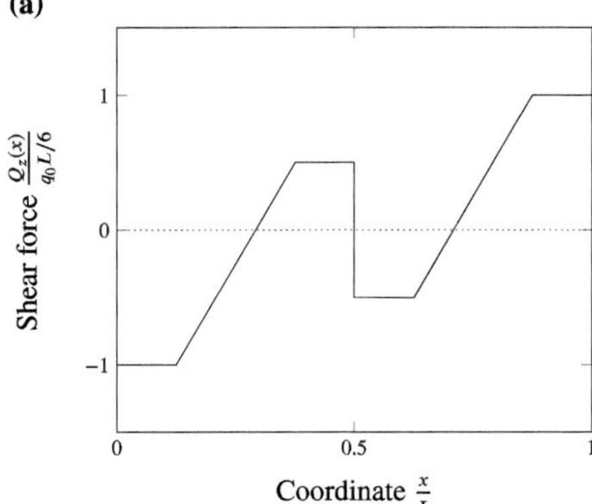

Fig. 9.45 Internal reactions for a sandwich beam under 5-point bending with distributed load: **a** shear force and **b** bending moment diagram

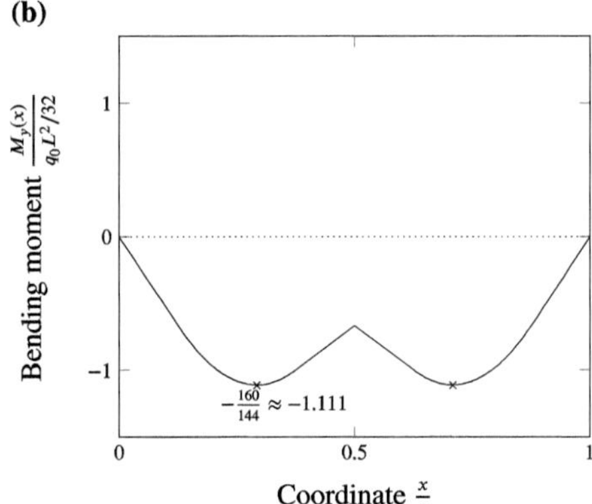

- Shear stress in the adhesive layer:

$$\tau_{zx} = 0.0658\,\text{MPa} < \tau_{\text{aB}} .$$
(9.489)

- Local wrinkling of the compression face sheet:

With $B_1 = 0.648$ it follows for the critical stress:

$$\sigma_{\text{cr}} = 3564.461\,\text{MPa} > \sigma_{x,\text{F}} .$$
(9.490)

Thus, there is no failure due to local wrinkling of the face sheet.

6.2.8 Global Instability Failure of a Sandwich Beam Clamped on Both Sides Under Compressive Loading

The configuration for determining the internal reactions is shown in Fig. 9.46.

This results in the following internal reactions:

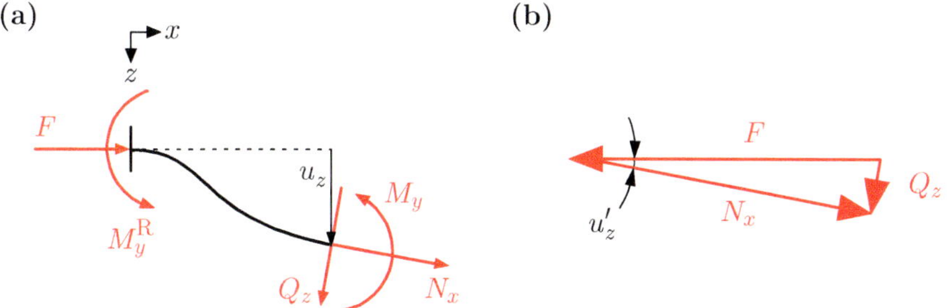

Fig. 9.46 Sandwich beam clamped on both sides under compressive load: **a** cutting free at location x; **b** force triangle (without considering the current sign)

$$M_y(x) = F u_z(x) - M_y^{\mathrm{R}}\,, \tag{9.491}$$

$$Q_z(x) = F\frac{\mathrm{d}u_z(x)}{\mathrm{d}x}\,. \tag{9.492}$$

The describing differential equation results in:

$$\frac{\mathrm{d}^2 u_z(x)}{\mathrm{d}x^2} + \lambda^2 u_z(x) = \lambda^2 \frac{M_y^{\mathrm{R}}}{F} \quad \text{with} \quad \lambda^2 = \frac{F}{EI_y\left(1 - \frac{1}{AG^{\mathrm{K}}}\right)}\,. \tag{9.493}$$

From the general solution[2], i.e. $u_z(x) = c_1\cos(\lambda x) + c_2\sin(\lambda x) + \frac{M_y^{\mathrm{R}}}{F}$, results with the corresponding boundary conditions, i.e. $u_z(0) = 0$ and $\frac{\mathrm{d}u_z(0)}{\mathrm{d}x} = 0$, the integration constants to $c_1 = -\frac{M_y^{\mathrm{R}}}{F}$ and $c_2 = 0$. Finally, the additional boundary condition $u_z(L) = 0$ results in the condition for determining the buckling force: $\cos(\lambda x) = 1$. This results in the buckling force for the boundary conditions under consideration:

$$F_{\mathrm{cr}} = \frac{\dfrac{4\pi^2 EI_y}{L^2}}{1 + \dfrac{1}{AG^{\mathrm{C}} \times \dfrac{4\pi^2 EI_y}{L^2}}} = \frac{F_{\mathrm{cr}}^{\mathrm{E}}}{1 + \dfrac{F_{\mathrm{K}}^{\mathrm{E}}}{AG^{\mathrm{C}}}}\,. \tag{9.494}$$

6.2.9 Instability Failure of a Sandwich Beam Hinged on Both Ends Under Compressive Loading

- Verification of conditions for soft core and thin face sheets:

---

[2] This can be determined with a computer algebra system (e.g. Maxima).

$$\frac{6E^{\mathrm{F}}\Delta h^{\mathrm{F}}(h_{\mathrm{c}})^2}{E^{\mathrm{C}}(\Delta h^{\mathrm{C}})^3} = 122.38 \geq 100 \,, \tag{9.495}$$

$$\frac{\Delta h^{\mathrm{C}}}{\Delta h^{\mathrm{F}}} = 20.0 \geq 4.77 \,. \tag{9.496}$$

- Global buckling:

$$F_{\mathrm{cr}} = \frac{\dfrac{\pi^2 \overline{EI_y}}{L^2}}{1 + \dfrac{1}{AG^{\mathrm{C}} \times \dfrac{\pi^2 \overline{EI_y}}{L^2}}} = \frac{F_{\mathrm{cr}}^{\mathrm{E}}}{1 + \dfrac{F_{\mathrm{cr}}^{\mathrm{E}}}{AG^{\mathrm{C}}}} = 614081.55 \text{ N} \,, \tag{9.497}$$

$$\sigma_{\mathrm{cr}} = \frac{F_{\mathrm{cr}}}{b(\Delta h^{\mathrm{C}} + 2\Delta h^{\mathrm{F}})} = 27.91 \text{ MPa} \,. \tag{9.498}$$

- Local wrinkling, antisymmetric:

$$k = 0.359 \,, \tag{9.499}$$
$$\Theta = 2.362 \,, \tag{9.500}$$
$$B_1 = 0.492 \,, \tag{9.501}$$
$$\sigma_{\mathrm{cr}} = 706.236 \text{ MPa} \,. \tag{9.502}$$

- Local wrinkling, symmetric:

$$k = 0.359 \,, \tag{9.503}$$
$$\Theta = 4.754 \,, \tag{9.504}$$
$$B_1 = 0.651 \,, \tag{9.505}$$
$$\sigma_{\mathrm{cr}} = 935.132 \text{ MPa} \,. \tag{9.506}$$

- Course of the critical buckling stress (Fig. 9.47)

Buckling stress limit for $L \to 0$:

$$\lim_{L \to 0} \sigma_{\mathrm{cr}} = \frac{1}{A} \times \frac{\dfrac{1}{L^2}}{\dfrac{1}{L^2}} \times \frac{\pi^2 \overline{EI_y}}{L^2 + \dfrac{\pi^2 \overline{EI_y}}{AG^{\mathrm{C}}}} \to G^{\mathrm{C}} \,. \tag{9.507}$$

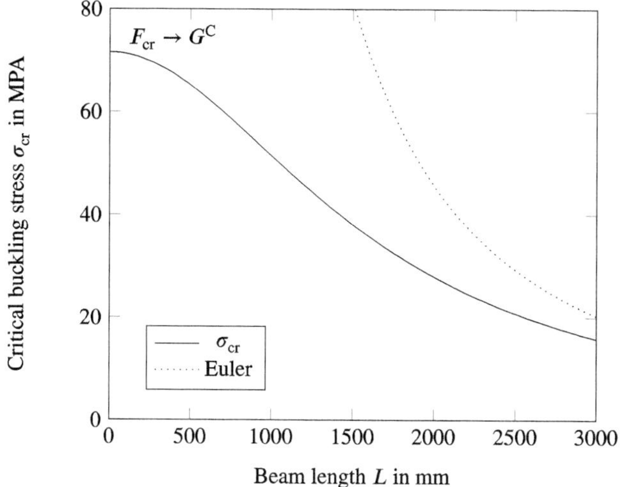

Fig. 9.47 Course of the critical buckling stress as a function of the beam length

9.6   Chapter 7

7.2.1 Optimization of a Sandwich Beam Under Compressive Loading

- Conversion of densities to consistent units
  Since the stiffnesses are given in MPa $= \mathrm{N/mm^2}$, the densities must be converted to the consistent unit $\mathrm{N/mm^3}$. The result is the following:

$$\varrho^{F} = 2691 \, \mathrm{kg/m^3} = 2691 \times 9.81 \times 10^{-9} \, \mathrm{N/mm^3} \,, \tag{9.508}$$

$$\varrho^{C} = 240 \, \mathrm{kg/m^3} = 240 \times 9.81 \times 10^{-9} \, \mathrm{N/mm^3} \,. \tag{9.509}$$

Case (a): $F = 2670 \, \mathrm{N}$:
- Calculation of point E

$$E \left(\Delta h_{E}^{C,n}; \Delta h_{E}^{F,n} \right) = \left(0.0191215; 2.923250 \times 10^{-5} \right) . \tag{9.510}$$

- Calculation of point G

$$G \left(\Delta h_{G}^{C,n}; \Delta h_{G}^{F,n} \right) = \left(0.00820895; 1.711632 \times 10^{-4} \right) . \tag{9.511}$$

- Optimal design
  Since $\Delta h_G^{C,n} \leq \Delta h_E^{C,n}$ holds, point G results as the optimal geometry with:

$$\Delta h_G^{C} = \Delta h_G^{C,n} \times L = 20.85\,\text{mm}\,, \tag{9.512}$$

$$\Delta h_G^{F} = \Delta h_G^{F,n} \times L = 0.43\,\text{mm}\,. \tag{9.513}$$

Under certain circumstances, however, one still has to respect the minimum thicknesses of sheet metal.

Case (b): $10 \times F = 26{,}700\,\text{N}$:

- Calculation of point E

$$E\left(\Delta h_E^{C,n};\ \Delta h_E^{F,n}\right) = \left(0.0242730;\ 2.923250 \times 10^{-4}\right). \tag{9.514}$$

- Calculation of point G

$$G\left(\Delta h_G^{C,n};\ \Delta h_G^{F,n}\right) = \left(0.0225247;\ 3.589804 \times 10^{-4}\right). \tag{9.515}$$

- Optimal design

Since $\Delta h_G^{C,n} \leq \Delta h_E^{C,n}$ holds, point G results as the optimal geometry with:

$$\Delta h_G^{C} = \Delta h_G^{C,n} \times L = 57.21\,\text{mm}\,, \tag{9.516}$$

$$\Delta h_G^{F} = \Delta h_G^{F,n} \times L = 0.91\,\text{mm}\,. \tag{9.517}$$

Under certain circumstances, however, one still has to respect the minimum thicknesses of sheet metal.

The position of the two minima is shown in Fig. 9.48. it is possible to conclude that the position of the minimum shifts further to the right with increasing external load.

7.2.2 Optimization of a Sandwich Beam Under Bending Load Due to a Single Force

- Second pole of the function g_3 in the $\Delta h^{C,n}$-$\Delta h^{F,n}$ coordinate system

$$\Delta h^{C,n} = \frac{F}{4bLG^C r_1} = 0.0832179\,. \tag{9.518}$$

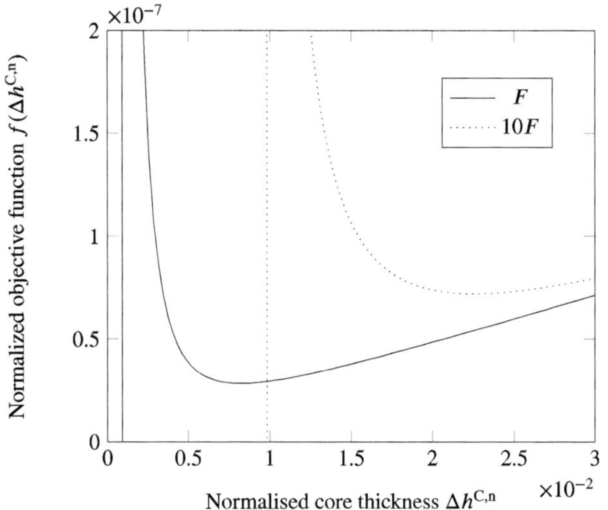

Fig. 9.48 Normalized objective function depending on the normalized core thickness with the position of the minima for different loads

- Intersection points in the $\Delta h^{C,n}$-$\Delta h^{F,n}$ coordinate system
  Point E (g_1-g_2)

$$E = \left(\Delta h_E^{C,n} \,\middle|\, \Delta h_E^{F,n}\right) = \left(\frac{F_0}{2bL\tau_p} \,\middle|\, \frac{\tau_p}{2\sigma_{cr}}\right) \tag{9.519}$$

$$= (0.0124827|\, 0.00116961) . \tag{9.520}$$

Point A (g_1-g_3)

$$A = \left(\Delta h_A^{C,n} \,\middle|\, \Delta h_A^{D,n}\right) = \left(\frac{4}{Lr_1}\left(\frac{2L\sigma_{cr}}{48E^D} + \frac{F_0}{16bG^K}\right) \,\middle|\, \frac{F_0}{4bL\sigma_{cr}\Delta h^{K,n}}\right) \tag{9.521}$$

$$= \left(0.130717|\, 1.116902 \times 10^{-4}\right) . \tag{9.522}$$

- Minima of the objective function f along the constraints
  Point B (minimum of f along g_1)

$$\frac{\partial f(\Delta h^{C,n})}{\partial \Delta h^{C,n}} \stackrel{!}{=} 0 \;\Rightarrow\; \Delta h_B^{C,n} = \sqrt{\frac{1}{2} \times \frac{\varrho^F}{\varrho^C} \times \frac{F_0}{bL\sigma_{cr}}} . \tag{9.523}$$

$$B = \left(\Delta h_B^{C,n} \,\middle|\, \Delta h_B^{F,n}\right) = \left(\sqrt{\frac{1}{2} \times \frac{\varrho^F}{\varrho^C} \times \frac{F_0}{bL\sigma_{cr}}} \,\middle|\, \sqrt{\frac{1}{8} \times \frac{\varrho^C}{\varrho^F} \times \frac{F_0}{bL\sigma_{cr}}}\right) \tag{9.524}$$

$$= (0.0180942 \,|\, 8.0687762 \times 10^{-4})\,. \tag{9.525}$$

Point C (minimum of f along g_3)

$$\frac{\partial f(\Delta h^{C,n})}{\partial \Delta h^{C,n}} \overset{!}{=} 0 \;\; \Rightarrow \;\; \Delta h_C^{C,n} \;\; \text{(Newton's iteration)}\,. \tag{9.526}$$

$$C = \left(\Delta h_C^{C,n} \,\middle|\, \Delta h_C^{F,n}\right) \tag{9.527}$$

$$= (0.0967536 \,|\, 5.295226 \times 10^{-4})\,. \tag{9.528}$$

- Optimum point
  It is case 1 and therefore point C is the optimal point.
  Optimal dimensions: $\Delta h^C = 245.754$ mm, $\Delta h^F = 1.345$ mm. With these geometric dimensions, the condition for thin face sheets and a soft core is also met:

$$\frac{6E^F \Delta h^F (h_c)^2}{E^C (\Delta h^C)^3} = 328.4 \geq 100\,, \tag{9.529}$$

$$\frac{\Delta h^C}{\Delta h^F} = 182.7 \geq 4.77\,. \tag{9.530}$$

- Graphical representation in the $\Delta h^{C,n}$-$\Delta h^{F,n}$ coordinate system (see Fig. 9.49)

7.2.3 Optimization of a Sandwich Beam Under Bending Load Due to a Distributed Load

- Objective function and constraints

$$f\left(\Delta h^{F,n}, \Delta h^{C,n}\right) = \varrho^C \Delta h^{C,n} + 2\varrho^F \Delta h^{F,n}\,. \tag{9.531}$$

$$g_1(\Delta h^{C,n}, \Delta h^{F,n}) = \frac{q_0}{8\Delta h^{C,n}\Delta h^{F,n}b} < \sigma_{cr}\,, \tag{9.532}$$

$$g_2(\Delta h^{C,n}, \Delta h^{F,n}) = \frac{q_0}{2\Delta h^{C,n}b} < \tau_p\,, \tag{9.533}$$

$$g_3(\Delta h^{C,n}, \Delta h^{F,n}) = \frac{10q_0}{384E^F b\Delta h^{F,n}(\Delta h^{C,n})^2} + \frac{q_0}{8bG^C\Delta h^{C,n}} < r_1\,. \tag{9.534}$$

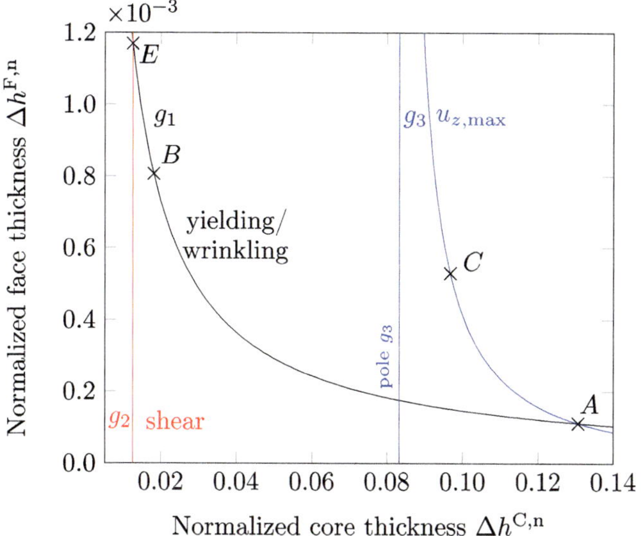

Fig. 9.49 Normalized face thickness as a function of normalized core thickness based on functions g_1, g_2 and g_3 for a bending beam with center load

Representation of the constraints in the $\Delta h^{C,n}$-$\Delta h^{F,n}$ coordinate system:

$$g_1: \qquad \Delta h_{g_1}^{F,n} > \frac{q_0}{8b\sigma_{cr}\Delta h^{C,n}}, \tag{9.535}$$

$$g_2: \qquad \Delta h_{g_2}^{C,n} > \frac{q_0}{2b\tau_p}, \tag{9.536}$$

$$g_3: \qquad \Delta h_{g_3}^{F,n} > \frac{\dfrac{10q_0}{384E^F b(\Delta h^{C,n})^2}}{r_1 - \dfrac{q_0}{8bG^C\Delta h^{C,n}}}. \tag{9.537}$$

- Intersection points in the $\Delta h^{C,n}$-$\Delta h^{F,n}$ coordinate system
  Point E (g_1–g_2)

$$E = \left(\Delta h_E^{C,n} \;\middle|\; \Delta h_E^{F,n}\right) = \left(\frac{q_0}{2b\tau_p} \;\middle|\; \frac{\tau_p}{4\sigma_{cr}}\right) \tag{9.538}$$

$$= (0.0124827 \,|\, 5.848035 \times 10^{-4}). \tag{9.539}$$

Point A (g_1–g_3)

$$A = \left(\Delta h_A^{C,n} \;\middle|\; \Delta h_A^{F,n} \right) = \left(\frac{1}{r_1 b} \left(\frac{80 b \sigma_{cr}}{384 E^F} + \frac{q_0}{8 G^C} \right) \;\middle|\; \frac{q_0}{8 b \sigma_{cr} \Delta h^{C,n}} \right) \tag{9.540}$$

$$= \left(0.100983 \;\middle|\; 7.228848 \times 10^{-5} \right) . \tag{9.541}$$

- Minima of the objective function f along the constraints
  Point B (minimum of f along g_1)

$$\frac{\partial f(\Delta h^{C,n})}{\partial \Delta h^{C,n}} \stackrel{!}{=} 0 \;\;\Rightarrow\;\; \Delta h_B^{C,n} = \sqrt{\frac{2 \varrho^F q_0}{8 b \sigma_{cr} \varrho^C}} . \tag{9.542}$$

$$B = \left(\Delta h_B^{C,n} \;\middle|\; \Delta h_B^{F,n} \right) = \left(\sqrt{\frac{2 \varrho^F q_0}{8 b \sigma_{cr} \varrho^C}} \;\middle|\; \frac{q_0}{8 b \sigma_{cr} \Delta h_B^{C,n}} \right) \tag{9.543}$$

$$= \left(0.0127946 \;\middle|\; 5.705486 \times 10^{-4} \right) . \tag{9.544}$$

Point C (minimum of f along g_3)

$$\frac{\partial f(\Delta h^{C,n})}{\partial \Delta h^{C,n}} \stackrel{!}{=} 0 \;\;\Rightarrow\;\; \Delta h_C^{C,n} \;\; \text{(Newton's iteration)} . \tag{9.545}$$

$$C = \left(\Delta h_C^{C,n} \;\middle|\; \Delta h_C^{F,n} \right) \tag{9.546}$$

$$= \left(0.0563599 \;\middle|\; 5.213426 \times 10^{-4} \right) . \tag{9.547}$$

- Optimum point

$$\Delta h_B^{C,n} > \Delta h_E^{C,n} \;\; \text{but} \;\; \Delta h_B^{C,n} < \Delta h_A^{C,n} \;\;\Rightarrow\;\; \text{not a valid point} , \tag{9.548}$$

$$\Delta h_D^{C,n} < \Delta h_C^{C,n} < \Delta h_A^{C,n} \;\;\Rightarrow\;\; \text{optimum point} . \tag{9.549}$$

Optimal dimensions: $\Delta h^C = 143.154$ mm, $\Delta h^F = 1.324$ mm. With these geometric dimensions, the condition for thin face sheets and a soft core is also met:

$$\frac{6 E^F \Delta h^F (h_c)^2}{E^C (\Delta h^C)^3} = 555.0 \geq 100 , \tag{9.550}$$

$$\frac{\Delta h^C}{\Delta h^F} = 108.1 \geq 4.77 . \tag{9.551}$$

- Graphical representation in the $\Delta h^{C,n}$-$\Delta h^{F,n}$ coordinate system (see Fig. 9.50)

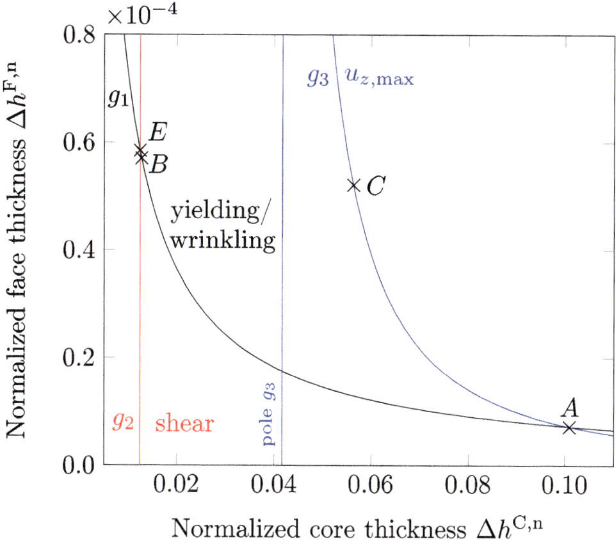

Fig. 9.50 Normalized face thickness as a function of normalized core thickness based on functions g_1, g_2 and g_3 for a bending beam with constant distributed load

7.2.4 Optimization of a Homogeneous Beam Under Bending Load Due to a Single Force

- Shear force and bending moment distribution
  The absolute values of the maxima can be taken from Fig. 9.41 as:

$$\left|Q_{z,\max}\right| = \frac{F_0}{2} \quad \text{and} \quad \left|M_{y,\max}\right| = \frac{F_0 L}{4}. \tag{9.552}$$

- Formulation of the mass and constraints

$$\sigma_{x,\max} = \frac{M_{y,\max}}{I_y} \times \frac{h}{2} < R_{\text{p0.2}}, \tag{9.553}$$

$$\tau_{xz,\max} = \frac{Q_{z,\max} h^2}{8 I_y} < \tau_{\text{p}}, \tag{9.554}$$

$$u_{z,\max} = \frac{F_0 L^3}{48 E I_y} < r_1 L, \tag{9.555}$$

$$h \leq 20b, \tag{9.556}$$

$$m = \varrho V. \tag{9.557}$$

- Normalized representation of mass and constraints
  The following formulation results from the normalization $b^{\text{n}} = b/L$ and $h^{\text{n}} =$

h/L:

$$g_1(h^n, b^n) = \frac{3F_0}{2b^n(h^n)^3 L^2} < R_{p0.2} , \tag{9.558}$$

$$g_2(h^n, b^n) = \frac{3F_0}{4b^n h^n L^2} < \tau_p , \tag{9.559}$$

$$g_3(h^n, b^n) = \frac{F_0}{4E b^n (h^n)^3 L} < r_1 L , \tag{9.560}$$

$$g_4(h^n, b^n) = h^n \leq 20 b^n , \tag{9.561}$$

$$f(h^n, b^n) = \frac{m}{L^3} = \varrho b^n h^n . \tag{9.562}$$

Representation in the h^n-b^n coordinate system, i.e. rearranged for b^n:

$$g_1 : \quad b_{g_1}^n > \frac{3F_0}{2L^2 R_{p0.2}(h^n)^2} , \tag{9.563}$$

$$g_2 : \quad b_{g_2}^n > \frac{3F_0}{4L^2 \tau_p h^n} , \tag{9.564}$$

$$g_3 : \quad b_{g_3}^n > \frac{F_0}{4E r_1 L^2 (h^n)^3} , \tag{9.565}$$

$$g_4 : \quad b_{g_4}^n \geq \frac{h^n}{20} . \tag{9.566}$$

Figure 9.51 shows the four constraints g_1, g_2, g_3 and g_4 in the h^n-b^n plane.

- Intersections

Point A (g_1–g_3)

$$A = \left(h_A^n \mid b_A^n\right) = \left(\frac{R_{p0.2}}{6E r_1} \;\middle|\; \frac{54 F_0 E^2 r_1^2}{L^2 (R_{p0.2})^3}\right) . \tag{9.567}$$

Point B (g_1–g_4)

$$B = \left(h_B^n \mid b_B^n\right) = \left(\left(\frac{30 F_0}{L^2 R_{p0.2}}\right)^{\frac{1}{3}} \;\middle|\; \frac{1}{20}\left(\frac{30 F_0}{L^2 R_{p0.2}}\right)^{\frac{1}{3}}\right) . \tag{9.568}$$

Point C (g_3–g_4)

$$C = \left(h_C^n \mid b_C^n\right) = \left(\left(\frac{5 F_0}{E r_1 L^2}\right)^{\frac{1}{4}} \;\middle|\; \frac{1}{2}\left(\frac{5 F_0}{E r_1 L^2}\right)^{\frac{1}{4}}\right) . \tag{9.569}$$

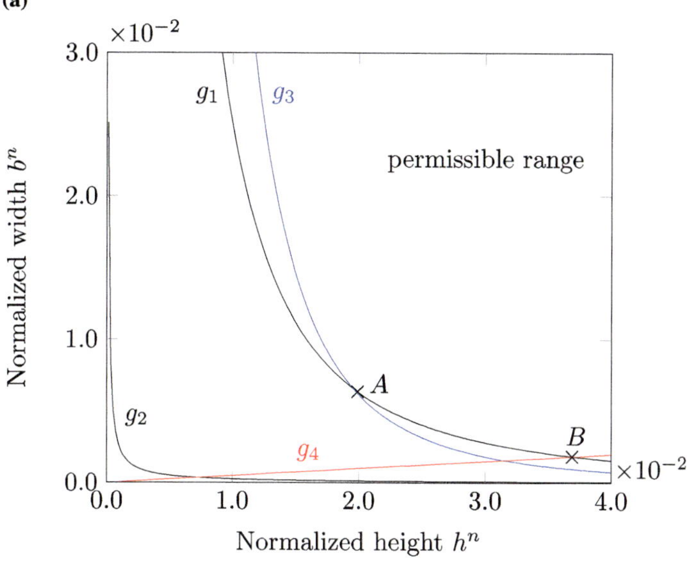

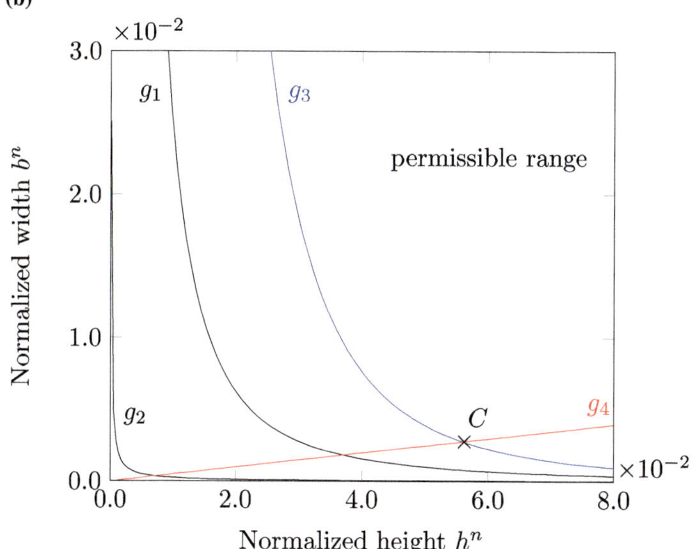

Fig. 9.51 Determination of the optimum point in the h^n-b^n coordinate system: **a** point B as the intersection of g_1 and g_4 ($r_1 = 0.03$). **b** Point C as the intersection of g_3 and g_4 ($r_1 = 0.003$)

Note: The objective function f has no local minimum along g_1 and g_3. The larger h^n, the smaller f (strictly decreasing functions). According to Fig. 9.51, $r_1 = 0.03$ results in point B and $r_1 = 0.003$ in point C as the optimal point.

References

1. Javanbakht, Z., Öchsner, A.: Advanced Finite Element Simulation with MSC Marc: Application of User Subroutines. Springer, Cham (2017)
2. Merkel, M., Öchsner, A.: Eindimensionale Finite Elemente: Ein Einstieg in die Methode. Springer Vieweg, Berlin (2014)
3. Öchsner, A., Merkel, M.: One-Dimensional Finite Elements: An Introduction to the FE Method. Springer, Berlin (2013)
4. Öchsner, A.: Elasto-Plasticity of Frame Structure Elements: Modeling and Simulation of Rods and Beams. Springer-Verlag, Berlin (2014)
5. Öchsner, A.: Computational Statics and Dynamics: An Introduction Based on the Finite Element Method. Springer, Singapore (2016)
6. Öchsner, A.: A Project-Based Introduction to Computational Statics. Springer, Cham (2018)

Appendix A

A.1 Mechanics and Mathematics

A.1.1 Second-Order Moment of Area

The second-order moments of area are generally defined as follows:

$$I_y = \int\limits_A z^2 \mathrm{d}A \,, \tag{A.1}$$

$$I_z = \int\limits_A y^2 \mathrm{d}A \,. \tag{A.2}$$

The formulas given in Table A.1 can be used for simple geometric cross-sections.

A.1.2 Derivation of the Shear Stress Distribution for the Beam

As a starting point for deriving the shear stress distribution for a beam, for example, a cantilever with shear force loading F_0 can be considered, see Fig. A.1. Furthermore, in the following we only consider a rectangular cross-section with side dimensions $b \times h$.

An infinitesimal beam element of this configuration is shown in Fig. A.2. The internal reactions shown are drawn according to the common sign definition as indicated in Fig. 2.8. Since there is no distributed load, i.e. $q_z = 0$, the vertical force equilibrium gives $Q_z(x) \approx Q_z(x + \mathrm{d}x)$.

Table A.1 Second-order moments of area around the y- and z-axes

Cross-section	I_y	I_z
	$\dfrac{\pi D^4}{64} = \dfrac{\pi R^4}{4}$	$\dfrac{\pi D^4}{64} = \dfrac{\pi R^4}{4}$
	$\dfrac{\pi b a^3}{4}$	$\dfrac{\pi a b^3}{4}$
	$\dfrac{a^4}{12}$	$\dfrac{a^4}{12}$
	$\dfrac{b h^3}{12}$	$\dfrac{h b^3}{12}$

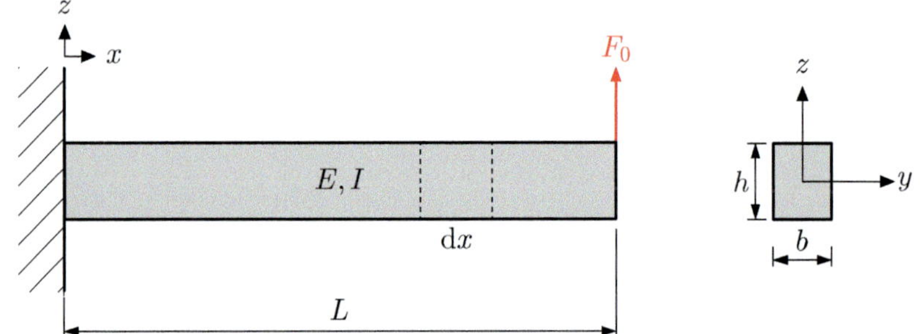

Fig. A.1 General configuration of a cantilever beam with shear force

The next step is to replace the inner reactions, i.e. the shear force and the bending moment, by the corresponding normal and shear stresses. To do this, a small element of height $h/2 - z'$ is cut out in the vertical direction from the infinitesimal (in the horizontal direction) beam element, see Fig. A.3.

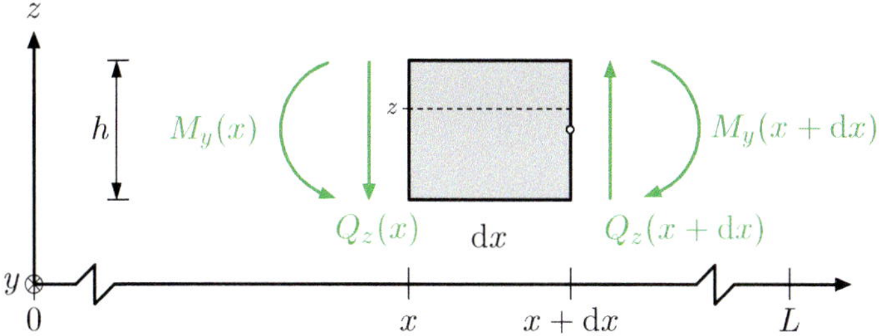

Fig. A.2 Infinitesimal beam element dx in the x-z plane with internal reactions

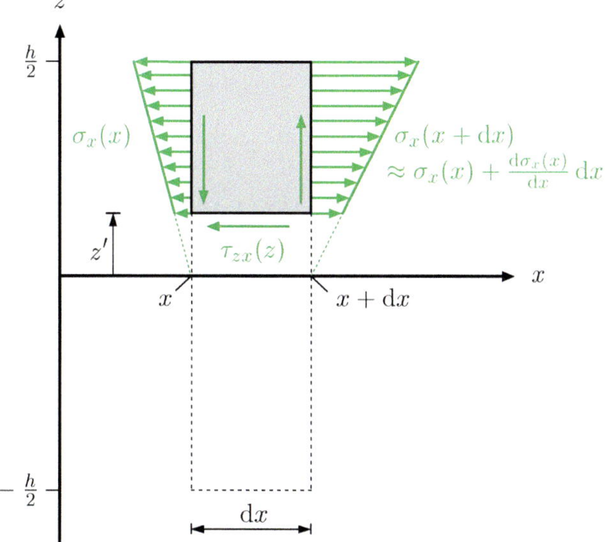

Fig. A.3 Infinitesimal beam element of dimensions d$x \times (h/2 - z')$. The overall configuration is shown in Fig. A.1

The horizontal forces equilibrium gives for this beam element:

$$-\int \sigma_x(x)\underbrace{b(z)\mathrm{d}z}_{\mathrm{d}A} + \int \sigma_x(x+\mathrm{d}x)\underbrace{b(z)\mathrm{d}z}_{\mathrm{d}A} - \tau_{zx}(z)b(z)\mathrm{d}x = 0\,, \qquad (A.3)$$

or simplified after to a Taylor series expansion of the stress at $(x + \mathrm{d}x)$:

$$\int \frac{\mathrm{d}\sigma_x(x)}{\mathrm{d}x}\mathrm{d}x\,\mathrm{d}A - \tau_{zx}(z)b(z)\mathrm{d}x = 0\,, \qquad (A.4)$$

or

$$\int \frac{\mathrm{d}\sigma_x(x)}{\mathrm{d}x}\,\mathrm{d}A - \tau_{zx}(z)b(z) = 0\,. \tag{A.5}$$

Rearranged for

$$\tau_{zx}(z) = \frac{1}{b(z)}\int \frac{\mathrm{d}\sigma_x(x)}{\mathrm{d}x}\,\mathrm{d}A\,, \tag{A.6}$$

gives with

$$\frac{\mathrm{d}\sigma_x(x)}{\mathrm{d}x} = \frac{\mathrm{d}}{\mathrm{d}x}\left(\frac{M_y(x)}{I_y}\times z\right) = \frac{z}{I_y}\times\frac{\mathrm{d}M_y(x)}{\mathrm{d}x} = \frac{Q_z(x)}{I_y}\times z \tag{A.7}$$

the shear stress distribution

$$\tau_{zx}(z) = \frac{1}{b(z)}\int \frac{Q_z(x)}{I_y}\times z'\mathrm{d}A = \frac{Q_z(x)}{I_y b(z)}\int z'\mathrm{d}A = \frac{Q_z(x)\mathcal{H}_y(z)}{I_y b(z)}\,, \tag{A.8}$$

where $\mathcal{H}_y(z)$ is the first-order axial moment of area for the part of the cross-section according to Fig. A.3. Assuming a rectangular cross-section with dimensions $b\times h$, this moment results in:

$$\mathcal{H}_y(z) = \int z'\mathrm{d}A = b\int_z^{h/2} z'\mathrm{d}z' = \frac{b}{2}\left(\frac{h^2}{4}-z^2\right) \tag{A.9}$$

$$= \frac{bh^2}{8}\left[1-\left(\frac{z}{h/2}\right)^2\right]\,. \tag{A.10}$$

This gives finally the shear stress distribution for a beam with a rectangular cross-section ($b\times h$) and the condition $\tau_{zx} = \tau_{xz}$ to:

$$\tau_{xz}(z) = \frac{Q_z(x)h^2}{8I_y}\left[1-\left(\frac{z}{h/2}\right)^2\right] \tag{A.11}$$

$$= \frac{3Q_z(x)}{2A}\left[1-\left(\frac{z}{h/2}\right)^2\right]\,. \tag{A.12}$$

The maximum of the shear stress distribution results for $z = 0$:

$$\tau_{xz,\mathrm{max}} = \frac{Q_z(x)h^2}{8I_y} = \frac{3Q_z(x)}{2bh} = \frac{3Q_z(x)}{2A}\,. \tag{A.13}$$

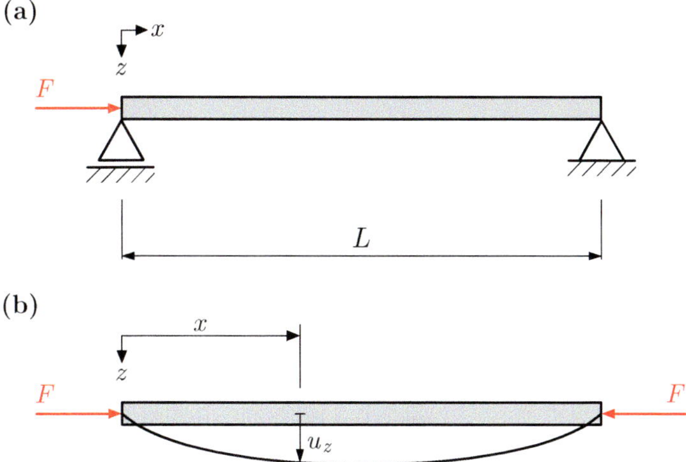

Fig. A.4 Beam hinged on both sides under a compressive load: **(a)** initial configuration and **(b)** deformation

Fig. A.5 Cutting free at location x

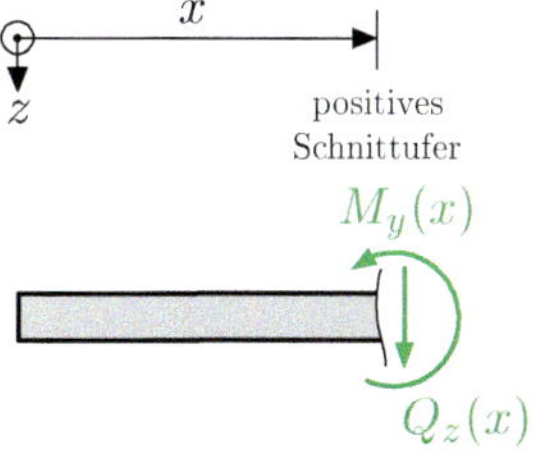

A.1.3   Derivation of the Euler Buckling Force for Homogeneous and Isotropic Euler-Bernoulli Beams

To derive the buckling formula according to Euler, consider a beam hinged on both sides under a compressive force F, see Fig. A.4 .

The equilibrium is now established for the first time on the deformed component[1] , see Fig. A.4b .

The equilibrium of moments at the location of the positive face (general position x) yields (see Fig. A.5):

$$\sum \overset{\frown}{M_y} = 0 \quad \Leftrightarrow \quad +F \times u_z - M_y(x) = 0 \, . \tag{A.14}$$

$$\Rightarrow \quad M_y(x) = +F \times u_z \, . \tag{A.15}$$

---

[1] For the derivation of the classical differential equations of Euler-Bernoulli beams under small deformations, the equilibrium is commonly established on the non-deformed component, see Sect. 2.2.1.

The deformation of an Euler-Bernoulli beam is generally described by a differential equation. For beams with constant bending stiffness EI_y, three classical formulations, based on the distributed load, based on the shear force distribution, or based on the bending moment distribution, can be given, see Eqs. (2.9)–(2.11). Using the formulation with the bending moment, we get in our case:

$$EI_y \frac{\mathrm{d}^2 u_z(x)}{\mathrm{d}x^2} = -F \times u_z(x) \,, \tag{A.16}$$

or transformed:

$$EI_y \frac{\mathrm{d}^2 u_z(x)}{\mathrm{d}x^2} + F \times u_z(x) = 0 \,. \tag{A.17}$$

Using the abbreviation $\lambda^2 = \frac{F}{EI_y}$ the following representation results:

$$\frac{\mathrm{d}^2 u_z(x)}{\mathrm{d}x^2} + \lambda^2 u_z(x) = 0 \,. \tag{A.18}$$

The general solution for such a differential equation is:

$$u_z(x) = c_1 \times \cos(\lambda x) + c_2 \times \sin(\lambda x) \,. \tag{A.19}$$

Using the boundary conditions $u_z(0) = 0$ and $u_z(L) = 0$, the unknown constants c_1 and c_2 can be approached.

From the first boundary condition, i.e. $u_z(0) = 0$, one gets:

$$0 = c_1 \times \cos(0) + c_2 \times \sin(0) \quad \Leftrightarrow \quad c_1 = 0 \,. \tag{A.20}$$

From the second boundary condition, i.e. $u_z(L) = 0$, one gets:

$$0 = c_2 \times \sin(\lambda \cdot L) \,. \tag{A.21}$$

If the product is to be zero, one of the two factors, i.e. c_2 or $\sin(\lambda \times L)$, must be zero. $c_2 = 0$ is a trivial solution (with $c_1 = c_2 = 0$ according to Eq. (A.19): $u_z = 0$, thus no deformation). Therefore, $\sin(\lambda \times L) = 0$ must be looked at more closely.

The condition $\sin(\lambda \times L) = 0$ means that $\lambda \times L = k \times \pi$ with $k = 0, 1, 2, \cdots$ (see Fig. A.6).

The condition $\lambda \times L = 0$ would mean $F = 0$ (see definition of λ). This results in the reasonable condition:

$$\lambda \times L = \pi \quad \Leftrightarrow \quad \lambda^2 = \frac{\pi^2}{L^2} \,. \tag{A.22}$$

And finally:

$$F_{\mathrm{cr}} = \frac{\pi^2 EI_{\min}}{L^2} \,. \tag{A.23}$$

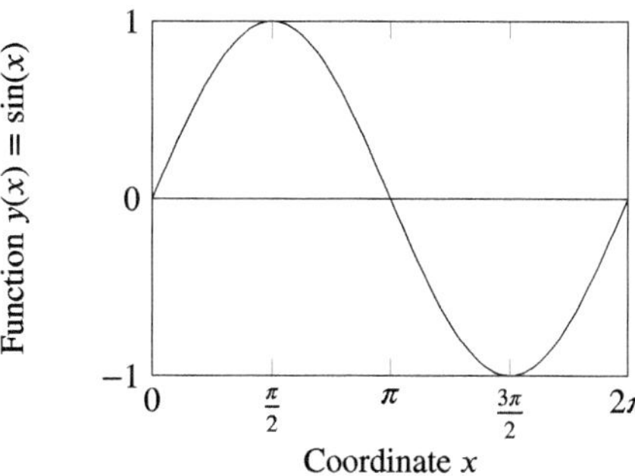

Fig. A.6 Representation of the trigonometric function $\sin(x)$

Table A.2 Characterization of the classical cases according to Euler

Case	Boundary condition	Buckling length
1	Free – fixed	$L_{cr} = 2L$
2	Hinged – hinged	$L_{cr} = L$
3	Hinged – fixed	$L_{cr} \approx 0.7L$
4	Fixed – fixed	$L_{cr} = \frac{1}{2}L$

This results in the buckling shape:

$$u_z(x) = c_2 \times \sin\left(\frac{\pi \cdot x}{L}\right), \tag{A.24}$$

where the constant c_2 remains undetermined. It should also be noted that the smaller value of I_y and I_z is to be taken for I_{min}.

Equation (A.23) can also be generalized according to Euler for other support cases by introducing the so-called buckling length L_{cr}. This results in the generalized buckling force according to Euler:

$$F_{cr} = \frac{\pi^2 E I_{min}}{L_{cr}^2}. \tag{A.25}$$

The different formulations of the Euler buckling force are summarized in Table A.2. The buckling stress σ_{cr} results from the buckling force using:

$$\sigma_{cr} = \frac{F_{cr}}{A} = \frac{\pi^2 E I_{min}}{A L_{cr}^2}. \tag{A.26}$$

If one defines the so-called slenderness ratio λ by means of the geometric quantities, i.e.

$$\lambda = \frac{L_{cr}}{\sqrt{\frac{I_{min}}{A}}} = L_{cr}\sqrt{\frac{A}{I_{min}}} \quad \Leftrightarrow \quad \lambda^2 = \frac{L_{cr}^2 A}{I_{min}}, \tag{A.27}$$

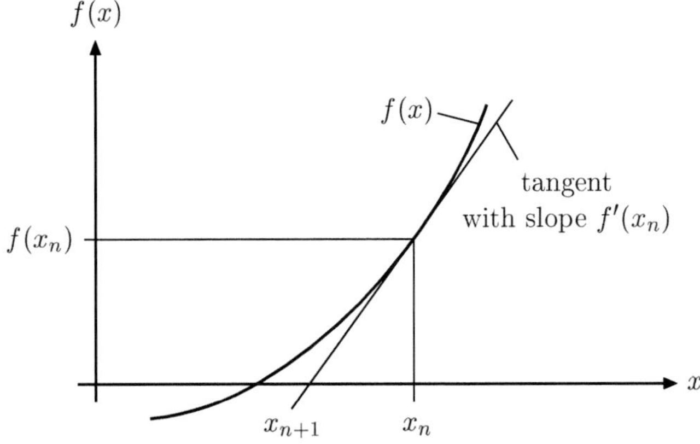

Fig. A.7 Schematic representation of Newton's iteration for the numerical determination of the root of a function

the buckling stress results in:

$$\sigma_{\mathrm{cr}} = \frac{\pi^2 E}{\lambda^2}.$$

(A.28)

A.1.4 Newton's Iteration

Newton's method is used for the numerical determination of the roots of functions, see [3]. To derive the iteration scheme, a function f at a point x_0 is expanded into a first-order Taylor series:

$$f(x) \approx f(x_0) + f'(x_0) \times (x - x_0) + \cdots .$$

(A.29)

Assuming the root at the location x, i.e. $f(x) = 0$, the following iteration rule results (see also Fig. A.7):

$$x_{n+1} = x_n - \frac{f(x_n)}{f'(x_n)}.$$

(A.30)

The iteration is performed until the difference between two consecutive abscissa values is less than a given tolerance: $tol = x_{n+1} - x_n$. It should also be noted that the convergence depends on the start value of the iteration. In order to select a meaningful start value, it may be necessary for the function to be represented graphically first.

A.1.5 Numerical Integration of Functions with Variables

When calculating the specific energy absorption SEA in Chap. 4, it is possible that difficult integrations have to be carried out. These integrals may contain general variables – such as in the integration in the limits from $x = 0$ to $x = L$ – and these can complicate the calculation of the integral. By using computer algebra systems (e.g. Maxima), some of these integrals can be analytically evaluated. However, no solution is found for other integrals with variables and one has to assign concrete numerical values to the variables in order to be able to carry out a *numerical* integration.

Alternatively, a simple numerical scheme should be presented here in order to be able to approximate integrals with variables numerically. For example, consider the following integral:

$$\int_0^L \underbrace{\left(-\left[\frac{x}{L}\right]^2 + \left[\frac{x}{L}\right]^1 \right)^{\frac{3}{2}}}_{y(x)} \, dx \, . \tag{A.31}$$

This integral can no longer be calculated *analytically* with the computer algebra system Maxima. Let us first consider the integrand, i.e. the function $y(x)$, which is to be integrated, see Fig. A.8a. It is a continuous function and all function values are greater than zero. The integral is represented by the area below the function graph (see the gray area in Fig. A.8a).

For the approximate calculation of the integral, the area under the function graph $y(x)$ is approximated by a series of n rectangles of constant width Δx (see Fig. A.8b), i.e.

$$\Delta x = \frac{L}{n}. \tag{A.32}$$

The abscissa values of the center points of each rectangle result for $1 \leq i \leq n$ in:

$$x_i = (i - 1) \times \Delta x + \frac{\Delta x}{2}. \tag{A.33}$$

Thus, the area A_i of a single rectangle based on the ordinate value $y(x_i) = y_i$ and the interval width is:

$$A_i = \Delta x \times y(x_i) = \Delta x \times y_i \tag{A.34}$$

Finally, the entire approximation of the area, and thus the integral, results from the summation of the partial areas:

$$\underbrace{\int_0^L y(x)dx}_{A} \approx \sum_{i=1}^n A_i = \sum_{i=1}^n \Delta x \times y(x_i) \, . \tag{A.35}$$

(a)

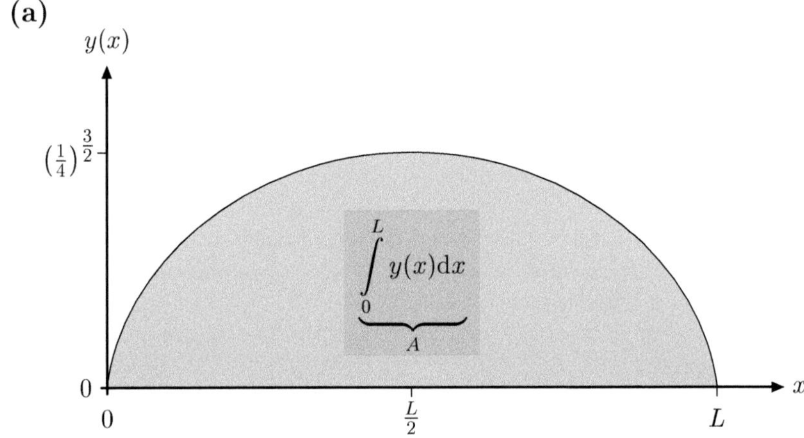

(b)

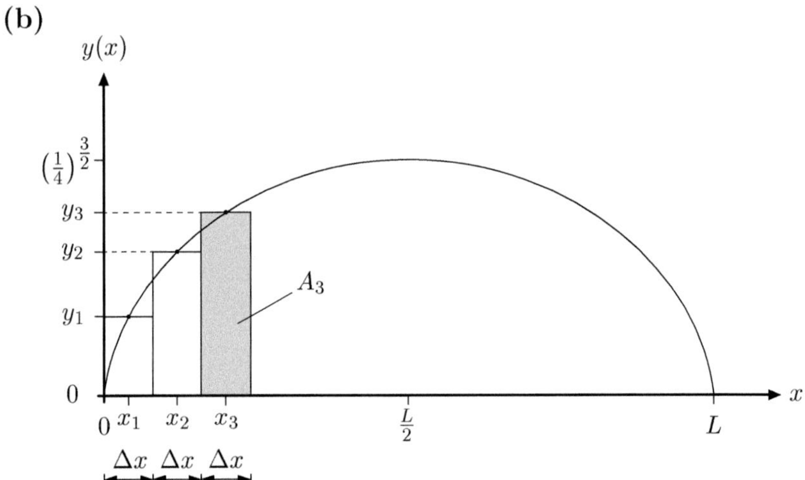

Fig. A.8 Approximation of the integration of a function (with variable), see Eq. (A.31)

The integration scheme according to Eqs. (A.32)–(A.35) can easily be implemented in a computer algebra system (see Fig. A.9) or realized using a classical programming language (see Sect. A.1.5).

The results of the numerical approximation of the integral according to Eq. (A.31) are summarized in Table A.3. Since no analytical solution to this problem is available, the difference between two approximations was evaluated to check the convergence of the calculation. For the given parameters one recognizes a quite good convergence. Finally, the integration scheme can also be tested on integrals for which there is an analytical solution, see Tables A.4 and A.5. Here, too, one can see that the integrals are approximated with sufficient accuracy from a certain interval width (the exact solution results for the constant function).

Numerical Integration of a Function with Variables

```
(%i3)    y : (-(x/L)^2 + (x/L))^(3/2)$

         n_list : [1,2,4,8,16,32,64,128,256,512,1024,2056]$

         for n in n_list do(
             dx : L/n,
             Area : 0,
             for i:1 thru n do(
                 current_x : (i-1)*dx + (dx/2),
                 current_y : at(y, x=current_x),
                 current_A : dx*current_y,
                 Area : Area + current_A
             ),
             print("n =",n, "Area =", float(Area)),
             print(" ")
         )$
```

```
n = 1 Area = 0.125 L
n = 2 Area = 0.08118988160479111 L
n = 4 Area = 0.07481934508844082 L
n = 8 Area = 0.07382736735303455 L
n = 16 Area = 0.0736645275266041 L
n = 32 Area = 0.07363687899496056 L
n = 64 Area = 0.07363209336908189 L
n = 128 Area = 0.07363125646066039 L
n = 256 Area = 0.07363110932005427 L
n = 512 Area = 0.07363108338032945 L
n = 1024 Area = 0.07363107880109886 L
n = 2056 Area = 0.07363107799047096 L
```

Fig. A.9 Numerical approximation of the integration of a function (with variable) using the computer algebra system Maxima, [2]

Table A.3 Approximation of the integration of a function (with variable), see Eq. (A.31)

n			A			$|\Delta A| = |A_{i+1} - A_i|$
1			$0.125L$			–
2			$0.08118988160479111L$			$0.04381011839520889L$
4			$0.07481934508844082L$			$0.006370536516350292L$
8			$0.07382736735303455L$			$9.919777354062687 \times 10^{-4}L$
16			$0.0736645275266041L$			$1.628398264304498 \times 10^{-4}L$
32			$0.07363687899496056L$			$2.764853164353986 \times 10^{-5}L$
64			$0.07363209336908189L$			$4.785625878675481 \times 10^{-6}L$
128			$0.07363125646066039L$			$8.369084214948641 \times 10^{-7}L$
256			$0.07363110932005427L$			$1.471406061159808 \times 10^{-7}L$
512			$0.07363108338032945L$			$2.593972482645146 \times 10^{-8}L$
1024			$0.07363107880109886L$			$4.579230591938988 \times 10^{-9}L$
2056			$0.07363107799047096L$			$8.10627898140126 \times 10^{-10}L$

Table A.4 Investigation of the convergence of the integration scheme according to Fig. A.8 based on the function $y(x) = 0$

n	A
1	$5.0L$
2056	$5.0L$
exakt	$5L$

Table A.5 Analysis of the convergence of the integration scheme according to Fig. A.8 based on the function $y(x) = \left(-\left[\frac{x}{L}\right]^2 + \left[\frac{x}{L}\right]^1 \right)^{1/2}$

n	A
1	$0.5L$
2	$0.4330127018922193L$
4	$0.4074209160795005L$
8	$0.39799115257648883L$
16	$0.39458586641222887L$
32	$0.39336897599008151L$
64	$0.39293642533352797L$
128	$0.39278308400169931L$
256	$0.39272879668973071L$
512	$0.39270959031025951L$
1024	$0.392702797544798\,L$
2056	$0.39270038787497651L$
4112	$0.39269954351696031L$
exact	$5\frac{\pi L}{8} = 0.39269908169872411L$

A.2   Computer Programs

Factors for Local Instability Calculation

The following Python program B1.py uses Newton's method to calculate the factor B_1 for Eqs. (6.28), (6.39) and (6.43) to be able to evaluate the critical crease stress σ_{cr}. Furthermore, the normalized wavelength $\frac{\lambda}{\Delta h^C}$ is calculated. The program requires a Python 3 installation[2], which provides the additional libraries sympy and numpy. Using klist, the start and end value or the associated subdivision into steps for the material and geometry parameter k can be specified, see line 8. A list of the Poisson's ratios of the core ν^C is specified in vlist, see line 9. By commenting out two of the three equations in the lines 56–58, one can distinguish between the cases in the Sects. 6.1.3, 6.1.4 and 6.1.4. The basics of the Python programming language can be found in [1,4,5].

---

[2] Thus, the program is called using python3 B1.py.

B1.py

```python
import sympy as sp
import numpy as np
sp.init_printing()

###PARAMETERS###

precision=0.000001
klist=np.linspace(0.05,4.00,100)
vlist=[0.50]
x0=100

###EQUATIONS###
v,k,x=sp.symbols('v k x')

expr=(k**2*x**2)/12 + (1/k)*((2/x) * (((3-v)*sp.sinh(x) *
    sp.cosh(x)+(1+v)*x)/((1+v)*(3-v)**2 *(sp.sinh(x))**2 -((1+v)**3)*x**2)))

expr2=(k**2*x**2)/12 + (1/k)*((2/x) * ((sp.cosh(x)-1)/((1+v)*(3-v)*sp.sinh(x) +
    ((1+v)**2)*x)))

expr3=(k**2*x**2)/12 + (1/k)*((2/x) * ((sp.cosh(x)+1)/(3*sp.sinh(x)-x)))

###FUNCTION###


def sympy_newton(expr,klist,vlist,precision,x0,label):
    """Takes a sympy expression, here with parameters v, k, and variable x, and
    lists of parameters k and v, and for a given x0 start value and precision
    calculates the minimum or maximum of function closest to x0, for each k and v.
    The function returns the k, v, wavelength, and y-value of the function at the
    minimum output is saved in a textfile, the name of which is specified with the
    variable label."""
    v,k,x=sp.symbols('v k x')
    first_der=sp.diff(expr, x)
    second_der=sp.diff(first_der, x)
    function=sp.lambdify((x,v,k),expr,("numpy", "math", "mpmath", "sympy"))
    function_first_der=sp.lambdify((x,v,k),first_der,("numpy", "math", "mpmath",
        "sympy"))
    function_second_der=sp.lambdify((x,v,k),second_der,("numpy", "math", "mpmath",
        "sympy"))
    wavelength=float((2*sp.pi)/x0)
    with open(label+".txt", 'w') as outfile:
        print("{:<16}{:<16}{:<16}{:<20}".format("K","B1","2pi/theta","v"),
    file=outfile)
        for k in klist:
            for v in vlist:
                if abs(function_first_der(x0,v,k)) < precision:

    print('{:.6f}'.format(k),'{:.6f}'.format(function(x0,v,k)),'{:.6f}'.
    format(wavelength),'{:.2f}'.format(v), sep='\t', file=outfile)
                else:
                    while abs(function_first_der(x0,v,k)) > precision:
                        if function_second_der(x0,v,k) == 0.0:
                            print("DivByZero error for: ",x0,v,k, file=outfile)
                            break
                        x = x0 -
    function_first_der(x0,v,k)/function_second_der(x0,v,k)
                        x0 = x
                        wavelength=float((2*sp.pi)/x0)
                        if abs(function_first_der(x0,v,k)) < precision:
```

```python
48      print('{:.6f}'.format(k),'{:.6f}'.format(function(x0,v,k)),'{:.6f}'.
        format(wavelength),'{:.2f}'.format(v), sep='\t', file=outfile)
49
50
51
52
53
54  ###RUN###
55
56  sympy_newton(expr,klist,vlist,precision,x0,"case1")
57  #sympy_newton(expr2,klist,vlist,precision,x0,"case2")
58  #sympy_newton(expr3,klist,vlist,precision,x0,"case3")
```

Optimal Dimensioning of Sandwich Beams Under Compressive Loads

The following Python program Optm_Comp.py calculates the optimal geometry of a sandwich beam under axial compression using the numerical Newton's method. The output file new.txt contains the coordinates of the point $E\left(\Delta h_E^{C,n}, \Delta h_E^{F,n}\right)$ in the Variables h_n_E, t_n_E and $G\left(\Delta h_G^{C,n}, \Delta h_G^{F,n}\right)$ in the variables h_n_G, t_n_G. Furthermore, the optimal point is given in the normalized variables h_n, t_n or in absolute values as h, t. At the end, a check is made as to whether the assumption of thin face sheets and a soft core is fulfilled. This program also requires a Python 3 installation, which provides the additional libraries sympy and numpy.

Optm_Comp.py

```python
1   import sympy as sp
2   import numpy as np
3   sp.init_printing()
4
5   ###PARAMETERS###
6   filename="new"
7   start=0.03
8   precision=0.000000000001
9
10  b=305.0
11  L=2540.0
12  rho_s=2691*9.81*10**(-9)
13  rho_c=240*9.81*10**(-9)
14
15  sig_Y=247.0
16  E_s=68948.0
17  E_c=6.8948
18  G_c=3.4474
19  P=10*2670
20  sig_wr=0.5*(E_s*E_c*G_c)**(1/3)
21
22  if sig_wr < sig_Y:
23      sig_limit=sig_wr
24  else:
25      sig_limit=sig_Y
26
27  ###CALCULATION_COORDINATES_POINT_E###
```

```python
h_n_E=np.sqrt((P/(2.0*b*L*G_c))**2.0 +
    4.0*sig_limit/(np.pi**2.0*E_s)) + P/(2.0*b*L*G_c)

t_n_E=P/(2.0*b*L*sig_limit)


###EQUATION_FOR_NEWTON_ITER###
x = sp.symbols('x')
f=((4*rho_s/(np.pi**2*E_s))/(x*((x*b*L/P)-1/G_c)))+rho_c*x



###CALCULATION_COORDINATES_POINT_G###

def sympy_newton(expr,precision,x0,label):
    """Takes a sympy expression with variable x, and for a given
    x0 start value and precision calculates the minimum or
    maximum of function closest to x0."""
    x=sp.symbols('x')
    first_der=sp.diff(expr, x)
    second_der=sp.diff(first_der, x)
    function=sp.lambdify((x),expr,("numpy", "math", "mpmath",
    "sympy"))
    function_first_der=sp.lambdify((x),first_der,("numpy",
    "math", "mpmath", "sympy"))
    function_second_der=sp.lambdify((x),second_der,("numpy",
    "math", "mpmath", "sympy"))
    iter=0
    with open(label+".txt", 'w') as outfile:
        if abs(function_first_der(x0)) < precision:
            return x0
        else:
            while abs(function_first_der(x0)) > precision:
                if function_second_der(x0) == 0:
                    print("DivByZero error for: ",x0,
    file=outfile)
                    break
                x = x0 -
    function_first_der(x0)/function_second_der(x0)
                x0 = x
                iter+=1
                if abs(function_first_der(x0)) < precision:
                    return x0
                if iter > 50:
                    print("Bad starting value or precision, over
    50 iterations", file=outfile)
                    break


###RUN###
h_n_G=sympy_newton(f,precision,start,filename)
```

```python
 72    if h_n_G is not None: #(check is Newton iteration was
          successfully completed)
 73        f_t_n_G=(((2*P)/(np.pi**2*b*L*E_s))/(x*(x-P/(G_c*b*L))))
 74        l_t_n_G=sp.lambdify((x), f_t_n_G, ("numpy", "math",
          "mpmath", "sympy"))

 76        t_n_G=l_t_n_G(h_n_G)


 79        ###OUTPUT###

 81        with open(filename+".txt", 'a') as outfile:
 82            print("h_n_G = ",h_n_G, file=outfile)
 83            print("t_n_G = ",t_n_G, file=outfile)
 84            print("h_n_E = ",h_n_E, file=outfile)
 85            print("t_n_E = ",t_n_E, file=outfile)

 87            if h_n_G <= h_n_E:
 88                h_n=h_n_G
 89                t_n=t_n_G
 90            else:
 91                h_n=h_n_E
 92                t_n=t_n_E
 93            print("h_n = ",h_n, file=outfile)
 94            print("t_n = ",t_n, file=outfile)


 97            h=h_n*L
 98            t=t_n*L
 99            print("h = ",h, file=outfile)
100            print("t = ",t, file=outfile)

102        ###CHECK_SANDWICH_THIN_WEAK_CORE###

104        ch_wt1=(6*E_s*t*h**2)/(E_c*h**3)
105        #print(ch_wt1)
106        ch_wt2=h/t
107        #print(ch_wt2)

109        with open(filename+".txt", 'a') as outfile:
110            if (ch_wt1 >= 100) and (ch_wt2 >=4.77):
111                print("Sandwich with WEAK core and THIN faces",
          file=outfile)
112            else:
113                print("NO weak core and thin faces", file=outfile)
```

Optimal Dimensioning of Sandwich Beams Under Bending Loads

The following Python program Optm_Comp_Bend_1.py calculates the optimal geometry of a sandwich beam under bending loads using the numerical Newton's method. Furthermore, the case of a simply supported beam with a central point force (see Fig. 5.19) is considered. The output file test3.txt contains the coordinates of the point $E\left(\Delta h_E^{C,n}, \Delta h_E^{F,n}\right)$ in the Variables h_n_E, t_n_E, $A\left(\Delta h_A^{C,n}, \Delta h_A^{F,n}\right)$

in the variables h_n_A, t_n_A, $D\left(\Delta h_D^{C,n}, \Delta h_D^{F,n}\right)$ in the variables h_n_D, t_n_D, $B\left(\Delta h_B^{C,n}, \Delta h_B^{F,n}\right)$ in the variables h_n_B, t_n_B and $C\left(\Delta h_C^{C,n}, \Delta h_C^{F,n}\right)$ in the variables h_n_C, t_n_C. Furthermore, the second pole of the function g_3 is provided as variable g3 pole. This program also requires a Python 3 installation, which provides the additional sympy and numpy libraries.

Optm_Comp_Bend_1.py

```python
import sympy as sp
import numpy as np
sp.init_printing()

###PARAMETERS###
filename="test3"
start=0.085
precision=0.000000000001

b=305.0
L=2540.0
r_1=0.003

rho_s=2691*9.81*10**(-9)
rho_c=240*9.81*10**(-9)

sig_Y=247.0
E_s=68948.0
E_c=6.8948
G_c=3.4474
tau=E_c/50.0

F=2667.0
Q_max=F/2.0
M_max=(F*L)/4.0

P=10*2670
sig_wr=0.5*(E_s*E_c*G_c)**(1/3)

if sig_wr < sig_Y:
    sig_limit=sig_wr
else:
    sig_limit=sig_Y

###CALCULATION_COORDINATES_POINT_E###
h_n_E=Q_max/(b*L*tau)
t_n_E=(M_max*tau)/(Q_max*L*sig_limit)

with open(filename+".txt", 'a') as outfile:
    print("Point E: h_n_E , t_n_E= ",h_n_E,", ",t_n_E,
    file=outfile)


###CALCULATION_COORDINATES_POINT_A###
```

```python
h_n_A=(F/(M_max*r_1)*(2*L*sig_limit/(48*E_s)+M_max/(4*b*L*G_c)))
t_n_A=M_max/(b*L**2*sig_limit*h_n_A)
with open(filename+".txt", 'a') as outfile:
    print("Point A: h_n_A , t_n_A= ",h_n_A,", ",t_n_A,
    file=outfile)


###CALCULATION_COORDINATES_POINT_D###
h_n_D=Q_max/(b*L*tau)
t_n_D=2*b*L*F*tau**2/(48*E_s*Q_max**2*(r_1-F*tau/(4*G_c*Q_max)))

with open(filename+".txt", 'a') as outfile:
    print("Point D: h_n_D , t_n_D= ",h_n_D,", ",t_n_D,
    file=outfile)


###CALCULATION_COORDINATES_POINT_B###
h_n_B=np.sqrt(2*rho_s*M_max/(b*L**2*sig_limit*rho_c))
t_n_B=M_max/(b*L**2*sig_limit*h_n_B)

with open(filename+".txt", 'a') as outfile:
    print("Point B: h_n_B , t_n_B= ",h_n_B,", ",t_n_B,
    file=outfile)


###CALCULATION_COORDINATES_POINT_C###
###EQUATION_FOR_NEWTON_ITER###
x = sp.symbols('x')
f=(2*rho_s*2*F/(48*E_s*b*L*x**2))/(r_1-F/(4*G_c*b*L*x))+rho_c*x

def sympy_newton(expr,precision,x0,label):
    """Takes a sympy expression with variable x, and for a given
    x0 start value and precision calculates the minimum or
    maximum of function closest to x0. The output is saved in a
    textfile, the name of which is specified with the variable
    label."""
    x=sp.symbols('x')
    first_der=sp.diff(expr, x)
    second_der=sp.diff(first_der, x)
    function=sp.lambdify((x),expr,("numpy", "math", "mpmath",
    "sympy"))
    function_first_der=sp.lambdify((x),first_der,("numpy",
    "math", "mpmath", "sympy"))
    function_second_der=sp.lambdify((x),second_der,("numpy",
    "math", "mpmath", "sympy"))
    iteration=0
    with open(label+".txt", 'a') as outfile:
        if abs(function_first_der(x0)) < precision:
            return x0
        else:
            while abs(function_first_der(x0)) > precision:
                if function_second_der(x0) == 0:
                    print("ZeroDivisionError for: ",x0," at
    iteration ",iteration, file=outfile)
```

```
            break
        x = x0 -
function_first_der(x0)/function_second_der(x0)
            x0 = x
            iteration+=1
            if abs(function_first_der(x0)) < precision:
                return x0
            if iteration > 50:
                print("Bad starting value or precision, over
    50 iterations", file=outfile)
                break


###RUN###
h_n_C=sympy_newton(f,precision,start,filename)
t_n_C=(2*F/(48*E_s*b*L*h_n_C**2))/(r_1-F/(4*G_c*b*L*h_n_C))

with open(filename+".txt", 'a') as outfile:
    print("Point C: h_n_C , t_n_C= ",h_n_C,", ",t_n_C,
    file=outfile)


###CALCULATION_POLE_G3###

g3_pole=F/(4*G_c*b*L*r_1)
print(g3_pole)

with open(filename+".txt", 'a') as outfile:
    print("g3 pole= ",g3_pole, file=outfile)
```

Numerical Integration of Functions with Variables

The following Python program numerical.py approximately calculates an integral,
whereby the area under the function graph $y(x)$ is approximated by a series of n
rectangles of constant width Δx (see Sect. A.1.5 for details).

numerical.py

```
import sympy as sp
import numpy as np
sp.init_printing()

x=sp.symbols('x')
L=sp.symbols('L')
y = (-(x/L)**(2) + (x/L))**(3/2)
n_list = [1,2,4,8,16,32,64,128,256,512,1024,2056]
for n in n_list:
    dx = np.divide(L,n)
    area = 0
    for i in range(1, n+1):
        current_x = (i-1)*dx + (dx/2)
        current_y = sp.lambdify((x),y)
        current_A = dx*current_y(current_x)
        area = area + current_A
```

```
[edus1929:Python_Codes andreas$ python3 numerical.py
n = 1 Area = 0.125*L
n = 2 Area = 0.08118988160047911*L
n = 4 Area = 0.07481934508884408*L
n = 8 Area = 0.0738273673530345*L
n = 16 Area = 0.07366452752660041*L
n = 32 Area = 0.0736368789949605*L
n = 64 Area = 0.0736320933690819*L
n = 128 Area = 0.07363125646066604*L
n = 256 Area = 0.07363110932005543*L
n = 512 Area = 0.0736310833803295*L
n = 1024 Area = 0.07363107880101989*L
n = 2056 Area = 0.07363107799047709*L
edus1929:Python_Codes andreas$ []
```

Fig. A.10 Numerical calculation of an integral, see program numerical.py

```
17    print('n =',n, 'Area =', area)
```

The result of running the numerical.py program is shown in Fig. A.10.

References

1. Linge, S., Langtangen, H.P.: Programming for Computations - Python: A Gentle Introduction to Numerical Simulations with Python. Springer, Cham (2016)
2. Makvandi, R.: Private communication. Institute of Mechanics, Otto von Guericke University Magdeburg, Germany (2020)
3. Mitchell, A.R., Griffiths, D.F.: The Finite Difference Method in Partial Differential Equations. John Wiley & Sons, New York (1980)
4. Padmanabhan, T.R.: Programming with Python. Springer, Singapore (2016)
5. Zhang, Y.: An Introduction to Python and Computer Programming. Springer, Singapore (2015)

Index

© The Author(s), under exclusive license to Springer Fachmedien Wiesbaden GmbH, part
of Springer Nature 2025
A. Öchsner, *Lightweight Design,*
https://doi.org/10.1007/978-3-658-48162-9